From Berlin to New York

Geschichte und Philosophie der Medizin

History and Philosophy of Medicine

Herausgegeben von
Andreas Frewer

Band 5

Brigitte Lohff / Hinderk Conrads

From Berlin to New York

Life and Work of the almost forgotten German-Jewish biochemist Carl Neuberg (1877–1956)

With a bibliography of Carl Neuberg's publications by Michael Engel and Brigitte Lohff

Translated from the German by Anthony Mellor-Stapelberg

Franz Steiner Verlag 2007

Cover:
Top: After the inauguration of the KWI for Experimental Therapy, Dahlem, Berlin (from left to right): Carl Neuberg, Emperor Wilhelm II, August von Wassermann, Adolf von Harnack (1913)
Bottom: Carl Neuberg Medal. American Society of European Chemists and Pharmacists (1947)

Series Editor / Reihenherausgeber:
Professor Dr. Andreas Frewer, M.A.
Medizinische Hochschule Hannover
Abt. Geschichte, Ethik und Philosophie der Medizin
Carl-Neuberg-Straße 1
30625 Hannover

Bibliografische Information der Deutschen Nationalbibliothek
Die Deutsche Nationalbibliothek verzeichnet diese Publikation in der Deutschen Nationalbibliografie; detaillierte bibliografische Daten sind im Internet über <http://dnb.d-nb.de> abrufbar.

ISBN 978-3-515-9062-9

ISO 9706

 Gedruckt auf säurefreiem, alterungsbeständigem Papier.
Druck: Printservice Decker & Bokor, München
Printed in Germany

We can see as far as we do because we stand on the shoulders of giants. A giant among the giants who laid the foundations for today's biomedical research was Carl Neuberg

Ines Hochmuth Mandl

CONTENTS

1. INTRODUCTION

CARL NEUBERG – AN ALMOST FORGOTTEN BIOCHEMIST

Carl Neuberg is described by his colleague Ines Mandl as a "giant among the giants" who turned biochemical research into one of the focuses of innovative development in medicine during the 20th century. Her generation was still very much aware that it was "standing on the shoulders of giants."[1] Many of these outstanding biochemists were of German origin, and had begun their scientific careers in Germany before being forced to flee the country. Only other emigrants were able to understand what a dislocation it represented in the career of a research scientist when he came to realize that he had no other alternative, if he wished to survive, than to escape from Germany. Erwin Chargaff,[2] who had also fled to the United States, described this fracture in the lives and careers of German-Jewish scientists very aptly in his book *"Heraclitean Fire"*:

> "At that time [the end of the 1920s – Authors' note] the Kaiser Wilhelm Institutes in green and pleasant Dahlem lived through their greatest period: physics with von Laue and Einstein; biology with Correns and Hartmann; physical chemistry with Haber, Polyani und Freundlich; cell physiology with Warburg; and there were a few more small principalities, with Neuberg, Herzog and Hess [...]. Never again would I have the feeling of belonging to a worthy and reasonable community of scholars [...] the last rays of the setting sun of the civilized nineteenth century [...]."[3]

Later, Chargaff described his experience of exile as follows:

> "It would be a great mistake to believe that they were received with open arms in those days. It was not too difficult for the young ones, with little offended pride to swallow; but the more distinguished, the more famous a man was, the greater was the reluctance to welcome him. These poor luminaries had a hard time [...].[4] Nobody benefited from the wrecking of Euro-

1 The quotation on the dedication page of this book is taken from a speech given by Ines Mandl (1978) to mark the 100th anniversary of Carl Neuberg's birth. She takes up the metaphor of the "nanus in humero gigantis" – the dwarf who is only able to see so far because he is standing on the shoulders of the giant – which originally goes back to Bernard of Chartres (d. c.1130): "Dicebat Bernadus Carnotensis nos esse quasi nanos gigantum humeris insidentes [...]." Another great man who memorably and modestly applied the quotation to himself was Sir Isaac Newton. See also Steudel (1953).

2 Erwin Chargaff (1905–2002), biochemist and writer, was born in the Ukraine. After studying in Vienna and at Yale University, he was an assistant for chemistry at the University of Berlin. He emigrated to Paris in 1933 and to New York in 1935, where he became an Assistant Professor in 1938, and from 1952 to 1974 was a full Professor in David Ritterberg's Department of Biochemistry at Columbia University. In 1958 he was awarded the *Carl Neuberg Medal* (cf. Chapter 8.3.). *Chargaff's Rule*, which states that in the DNA of any species the amount of adenine is always equal to the amount of thymine and the amount of cytosine to that of guanine, is named for him.

3 Chargaff (1978), 51.

4 Ibid., 76.

> pean science in the years 1930–1950 [...]. One country may have become poorer, but no other got much richer as a result [...]. Even the natural sciences, to say nothing of law or medicine, flourish in the bosom of a particular language and civilization, and what was torn up from there never grew again [...]. People like Neuberg and Meyerhof didn't have an easy time."[5]

Let these remarks of Erwin Chargaff, at once melancholy and critical, form the framework for this publication, which seeks to restore to collective awareness the almost forgotten life of Carl Neuberg.

An encyclopedia might summarize the first 57 years of this scientist's life as follows: Carl Neuberg, one of the founding fathers of dynamic biochemistry,[6] was born in Hanover, Germany, on July 29, 1877, as the son of a family of Jewish descent. He studied chemistry at the Universities of Würzburg and Berlin, where he took his doctorate under the supervision of Emil Fischer in 1900 with a thesis entitled "An investigation of acroleïn and glyceraldehyde. On acidic esters of boric acid. On the cleansing of osazones and the determination of their directions of optical rotation."[7] He then became an assistant to Ernst Salkowski, under whose direction he researched and taught at the Pathological Institute of the celebrated Berlin teaching hospital, the Charité, for ten years. After that he was appointed Head of the Department of Chemistry in the Institute for Animal Physiology at the Royal Agricultural College *(Königliche Landwirtschaftliche Hochschule)* in Berlin. In 1913, he was appointed Head of the Department of Chemistry at the Kaiser Wilhelm Institute (KWI) for Experimental Therapy, of which August von Wassermann was the Director. Thanks to his discovery of substitutes for glycerol, a material of strategic importance during the First World War, he was able to raise substantial funding for his project to set up an Institute of Biochemistry under the umbrella of the Kaiser Wilhelm Society (*Kaiser-Wilhelm-Gesellschaft* – KWG). Due to the impact of the inflation that followed the First World War, however, it took until 1925 for such an institute to be established as a successor institute to the KWI for Experimental Therapy, with Carl Neuberg, who had previously been Deputy Director of the KWI for Experimental Therapy and Biochemistry alongside Wassermann, becoming Director of the KWI for Biochemistry – until 1934.

Thus Neuberg's career, up to the age of 57, was that of a distinguished German academic scientist of Jewish origin. He was a highly reputed researcher, the recipient of numerous awards, the director of an institute of the internationally renowned Kaiser Wilhelm Society in the Dahlem district of Berlin, and the editor of the biochemistry periodical *Biochemische Zeitschrift* – but only until 1934.

5 "Von der Zertrümmerung der europäischen Wissenschaft in den Jahren 1930 bis 1950 hat niemand profitiert [...]. Das eine Land mag ärmer geworden sein, das andere nicht viel reicher [...]. Sogar die Naturwissenschaften, nicht zu reden von Jus oder Medizin, leben im Schoße einer bestimmten Sprache und Zivilisation, was da herausgerissen wurde, ist nie wieder nachgewachsen [...]. Solche Männer wie Neuberg und Meyerhof hatten es nicht leicht." Chargaff (German version, 1980), 109.

6 Nachmansohn (1979), 311–312.

7 "Zur Kenntnis des Acroleïns und Glycerinaldehyds. Über saure Ester der Borsäure. Über die Reinigung der Osazone und zur Bestimmung ihrer optischen Drehungsrichtung." Neuberg (1900). (Publications by Carl Neuberg, alone or in collaboration, are listed in Chapter 11.1.).

Neuberg was the recipient of many international honors. He was never awarded the Nobel Prize, although there was scarcely a scientist so frequently recommended for the Medicine and Chemistry Prizes as he was. Thanks to his tireless editorial work, the *Biochemische Zeitschrift* attained to an outstanding position internationally within just a few years. Neuberg researched into matters in the border areas between chemistry, biology, medicine and agricultural science. In an evaluation of his own scientific achievements in a letter to Henri Corbiere in 1953, he particularly emphasized the following results of his research: the discovery of carboxylase, the explanation of the various stages of fermentation, the development of a method to precipitate intermediate products of metabolic processes, the production of substitute materials for glycerol, the identification of various enzymes, including in particular a number of phosphatases, and the drawing up of a concept to explain the transportation of sparingly soluble substances in organisms.[8]

But the political events of the 20th century, and in particular the inhuman and murderous anti-Semitic ideology and policies of the Nazi dictatorship, struck at the roots of Neuberg's existence – in both the personal and the scientific sphere. In addition, the recollection of his scientific achievements and of his importance in the early development of biochemistry was almost completely extinguished in the post-war period, during which uncomfortable memories of the past were brushed under the carpet.

However, what should have been the task of German biochemists – that of preserving the memory of one of the most important pioneers of their subject – was taken up by David Nachmansohn in his book *German-Jewish Pioneers in Science 1900–1933*,[9] which appeared in the United States in 1979. Nachmansohn, who was a pupil of Otto Meyerhof's but always valued and sought Carl Neuberg's advice, speaks with great admiration of Neuberg's scientific achievements in his 1956 obituary for him:

> "The giants in physics and chemistry of the 20th century have revolutionized and transformed the whole existence of mankind. One of these giants was the man we mourn and honor tonight, Carl Neuberg."[10]

Just at the same time as a German translation of Nachmansohn's book appeared, almost ten years later in 1988,[11] a *Dictionary of Important Chemists*[12] was pub-

8 Neuberg mentioned these discoveries in response to a written inquiry from Corbiere as to what he considered his scientific "masterpiece." Neuberg to Corbiere, November 20, 1953. Carl Neuberg Papers (recatalogued by rsc 2002) [abbreviated below as CNP], Box 4: Corbiere, Henri.

9 Nachmansohn (1979).

10 Nachmansohn (1956), 82.

11 Nachmansohn/Schmid (1988): Die große Ära der Wissenschaft in Deutschland 1900 bis 1933. Jüdische und nichtjüdische Pioniere in der Atomphysik, Chemie und Biochemie. It is remarkable that the German translation of this work was published by the Wissenschafts-Verlag Stuttgart and not, like the American original, by the Springer Verlag, despite the fact that Neuberg had been an important Springer author for more than 30 years. Cf. Chapter 4.

12 Pötsch (1988): Lexikon bedeutender Chemiker.

lished by Pötsch in the German Democratic Republic (East Germany). This work draws attention to Neuberg's research on the enzymatic breakdown of sugars and the new insights this afforded into various enzymes and synthesized phosphorylated intermediate metabolic products, and also to his work on the role of insoluble substances in metabolism.[13]

Before this, in 1983, Nordwig had drawn attention to Neuberg's forced resignation from his position as Director of the KWI for Biochemistry in a four-page article in the *MPG Spiegel*;[14] while a degree thesis submitted at the East German Humboldt University in 1989 was concerned with Neuberg's role as a university teacher in Berlin.[15]

It was only from the 1990s onwards that the name of Neuberg began to appear more frequently again in the wide-ranging publications on the history of the KWG and the Max Planck Society (*Max-Planck-Gesellschaft* – MPG) as that of a Director of the KWI for Biochemistry.[16] Michael Engel, who is concerned with Neuberg's biochemical publications, wrote a comprehensive article for *Neue Deutsche Biographie* in 1999, which represented Neuberg's reappearance in biographical encyclopedias. Hinderk Conrads' doctoral thesis *Carl Neuberg (1877–1956). Eine Biographie,* which in a completely revised and expanded version forms the basis for this publication, appeared in 2002 at Hanover Medical School (*Medizinische Hochschule Hannover* – MHH), the main street of which had been renamed "Carl-Neuberg-Strasse" in his honor on January 1, 1997. The book on Adolf Butenandt edited by Schieder and Trunk in 2004 contains an article by Michael Schüring entitled *The Predecessor: Carl Neuberg's Relationship with Adolf Butenandt.*[17] Since 2002 an index to Neuberg's comprehensive correspondence with scientists from all over the world, compiled by the American Philosophical Society (APS), has been available on the internet under the title *Carl Neuberg Papers*: Neuberg's daughter, Irene Forrest, had made this material available to the APS in 1980, but Conrads was the first person to draw upon it for his work.

Various people repeatedly called upon Carl Neuberg in old age to write an autobiography. This, as David Nachmansohn remarked, would have been "an invaluable contribution to the history of that period of science," since Neuberg had known many of the important personalities of his time personally,[18] and his career, in the way it had fallen victim to the political upheavals of the time, provided

13 Pötsch (1988), 320.

14 Nordwig (1983).

15 Katterle/Peschel (1989).

16 Cf. Vierhaus/Brocke (1990) and the series of publications on the history of the Kaiser Wilhelm Society during the Nazi period, *Geschichte der Kaiser-Wilhelm-Gesellschaft im Nationalsozialismus,* commissioned by the MPG Presidential Commission, which has been appearing since 2000 [www.mpiwg-berlin.mpg.de/KWG/index.htm, 03/02/2006].

17 *Der Vorgänger: Carl Neubergs Verhältnis zu Adolf Butenandt.* Schüring (2004), 346–368. He does not, however, mention Conrads's thesis, although this had been available to the MPI for the History of Science since 2002.

18 Nachmansohn (1979), 322. For the autobiographical part Neuberg would have had what he himself considered to be a very apt title for a chemist, *"Erstunkenes und Erlogenes"* (roughly translatable as "Stinking Lies"), cf. Kisch (1966), 328.

an exemplary illustration of this particularly difficult period. Nevertheless, Neuberg was accustomed to dismiss such suggestions with the equivocal remark: "What I have published is not of general interest, and what is interesting I can't publish."[19] In addition, his customarily critical and very forthright manner would doubtless have provoked some annoyance and resentment among the scientific community, had he applied it to describing the characters of famous colleagues or developments and incidents in the history of science. A further argument put forward by Neuberg as to why he did not publish anything about himself and the history of biochemistry can be found in a letter to Hans Joachim Deuticke, later Rector of the University of Göttingen:

> "As fate has so far been merciful enough to allow me to continue doing experimental work at my age, and I even still have ideas, I am postponing the thought of writing an autobiography until I reach the stage of advanced imbecility."[20]

Neuberg was able to continue with his scientific work until just over two weeks before his death. Thus as late as May 15, 1956, a lecture of Neuberg's with the simultaneous demonstration of biochemical experiments was held at the New York Medical College, although owing to his poor health he was not able to present the experiments himself, but only to speak the introduction and the concluding words. The rest of the lecture was read by his daughter Irene, while his assistant Amelie Grauer performed the experiments. Afterwards, Friedrich F. Nord[21] and Severo Ochoa[22] spoke about his scientific achievements. Nachmansohn reports that the over 300 listeners were deeply moved, as an awareness spread through the lecture theater that this would be the very last lecture by a great representative of the science of biochemistry.[23]

19 "Was ich publiziert habe, ist nicht allgemein interessant, und was interessant ist, kann ich nicht publizieren." American Society of European Chemists and Pharmacists (1947), 6.

20 "Da mir das Schicksal bisher die Gnade erwiesen hat, dass ich bei meinem Alter noch experimentell arbeiten kann, und mir sogar noch etwas einfaellt, schiebe ich den Gedanken einer Autobiographie bis zum Zeitpunkt der fortgeschrittenen Verbloedung auf. " Neuberg to Deuticke, November 15, 1951. CNP, Box 4: Deuticke, Hans J.

21 Frans Friedrich Nord (1889–1973), biochemist, born in Budapest, worked with Neuberg at the Institute for Experimental Medicine. In 1921 he moved to the Physiological Institute of the Faculty of Veterinary Medicine at Berlin University; in 1938 he emigrated to the United States and was appointed to a Chair of Organic Chemistry and Enzymology at Fordham University, New York. Nord was the founder and editor of the periodicals *Ergebnisse der Enzymforschung, Advances in Enzymology* and *Archives of Biochemistry and Biophysics.*

22 Severo Ochoa (1905–1993), biochemist, born in Spain, was an assistant to Otto Meyerhof from 1929 to 1936; after emigrating to the United States he was appointed to a research professorship at New York University. In 1959 he was awarded the Nobel Prize in Medicine jointly with Arthur Kornberg (b. 1918).

23 Nachmansohn (1979), 322. Neuberg himself wrote to his daughter Marianne: "The lecture – which representatives of all the scientific institutes in New York came to – is considered to have been a success. The American idiots, you see, don't know what experimental lectures are [...]." ("Der Vortrag – zu dem Vertreter aller wissenschaftlichen Institute New Yorks gekommen sind – wird von den Leuten fuer einen Erfolg gehalten. Die amerikanischen Idioten kennen eben keine Experimental-Vorlesungen [...].)" Neuberg to Lederer, May 17, 1956. CNP, Box 8: Lederer, Marianne.

Neuberg did not, however, receive any really appropriate appreciation, and in the historiography of his subject he fell almost into oblivion. Carl Neuberg died in New York City on May 30, 1956, at the age of 78. Thus the year 2006 marked the 50th anniversary of his death, and in the same year the biochemical community also celebrated the centenary of the first international biochemistry periodical, whose first issue was edited by Carl Neuberg, then 29 years old, in October 1906.

It is only with the support and assistance of numerous people and institutions that it has been possible to reconstruct Neuberg's life and career. Above all, thanks are due at this point to Carl Neuberg's granddaughter, Joan Armer of Menlo Park, San Francisco. Without the personal conversations with her and the written information she provided it would have been impossible for us to understand the family background.[24] Our thanks are also due to Ines Hochmuth Mandl of New York for her hospitality and extensive conversations. She helped us to follow the course of Carl Neuberg's life in the United States.

Our thanks are also due to the Friends of Hanover Medical School *(Freundesgesellschaft der Medizinischen Hochschule Hannover)* for financial support that facilitated work on Neuberg's extensive correspondence during a research visit to the archives of the American Philosophical Society in Philadelphia. The Friends of Hanover Medical School have also assisted this publication by contributing to the printing costs. Further kind financial assistance with the printing costs was received from the German Biochemistry Society *(Deutsche Gesellschaft für Biochemie)*, which Professor Gaestel of Hanover Medical School did much to obtain for us. We would also like to express our special thanks to Dr. Michael Engel of Berlin for the many valuable conversations and for information that was of great assistance in understanding Neuberg's biochemical research. Thanks are due too to Hanim Aral and Florian Bruns for their assistance in compiling the index; to the staff of the Institute for the History, Ethics and Philosophy of Medicine at the MHH, in particular Dr. Sigrid Stöckel and Dr. Irene Hirschberg, for the patience and calmness with which they bore our sometimes utter immersion in the world of Neuberg; and to the staff of the Archives of the American Philosophical Society and the Max Planck Society for their constant support. Professor Andreas Frewer with his tireless commitment has made it possible, for this book to appear as Volume 5 of the series *History and Philosophy of Medicine*, and we are particularly grateful to him for this. We also thank most sincerely Dr. Thomas Schaber and Angela Hoeld of the Franz Steiner Verlag for helping to ensure that this publication could appear so quickly and in such a carefully presented form. Our very special thanks are due to Anthony Mellor-Stapelberg for his excellent translation and to Tim Ripperger for his constant assistance with tedious research and his patience in designing the layout.

24 The conversations with Joan Armer (1998) and Ines Mandl (1996) were conducted by Hinderk Conrads in the course of his research for his thesis, which he wrote under the supervision of Brigitte Lohff. The fact that he was able to make contact with Joan Armer in 1997 and to conduct an intensive correspondence with her thereafter was thanks to a wildly adventurous search mission which took him all the way from Sarstedt/Hanover, to the United States.

2. YOUTH AND EDUCATION

2.1. HOME AND FAMILY

Carl Neuberg's family tree, which can be traced back to the 17th century, shows that his ancestors lived mainly in the area around Hildesheim, Pyrmont and Hanover in what is today the State of Lower Saxony in northern Germany.[25]

His great-grandfather, the merchant Abraham Schey, who was probably born in 1775, had been a so-called "Schutzjude" or "Protected Jew"[26] in Sarstedt/ Hanover from 1802 onwards.[27] Schey had been granted a concession to carry on a drapery business in his house.[28] Within a short time, his business was very successful. He had four employees, and had to pay the office of the royal *Landdrost* or bailiff[29] the highest protection levy of all Jews resident in Sarstedt. In 1815 he is mentioned as President of the Jewish community in the Borough of Sarstedt.[30] The records for the year 1816 show that Schey and his wife Henriette, née Israel, had four sons, whom they named Sandel, Marcus, David and Levy. It is in 1828 that the name *Neuberg*, a supplementary name commonly adopted by Jewish families at the time, first appears in the list of Jews liable to pay the levy.[31]

In 1831, the eldest son, Sandel Abraham Neuberg, took over his father's business and was given permission to continue running it on the previous terms.[32] His brothers Marcus and David were employed in the family business for a time.[33] In 1839 Sandel married Henriette Esberg from Dankelshausen,[34] whose father, Isaac Moses Esberg, was elected President of the synagogue[35] there in 1844. There were six children of the marriage. The couple initially lived in Sarstedt; the birthplaces of the children indicate that they probably moved to the nearby city of Hanover, the royal capital of the eponymous Kingdom, in 1849 or 1850. Here

25 The information on Neuberg's family history is taken from the records of the Lower Saxony Central State Archives *(Niedersächsisches Hauptstaatsarchiv)* [abbreviated below as LSCSA], the Hanover City Archives *(Stadtarchiv Hannover)* and a family tree kindly made available by Carl Neuberg's granddaughter Joan Armer.

26 This term had been used in Germany since the end of the Middle Ages to indicate a Jew placed under the personal protection of a ruler, in return for the payment of particular duties.

27 LSCSA, Hanover 74, Hildesheim, 1841.

28 Ibid., 1870.

29 At this time, the *Landdrost* was the ruling prince's representative responsible amongst other things for law enforcement in an administrative district.

30 LSCSA, Hanover 74, Hildesheim, 1869.

31 LSCSA, Hanover 74, Hildesheim, 1841. There is no record of why Abraham Schey chose the name Neuberg.

32 LSCSA, Hanover 74, Hildesheim, 1841.

33 Ibid., 1871.

34 A place to the north of Hannoversch Münden, now part of Scheden municipality.

35 LSCSA, Hanover 83 b [A].

Sandel died in 1865 at the age of 63, and was buried, as his wife was too, in the Jewish cemetery *An der Strangriede*.[36]

Their eldest son, Julius Neuberg, who was born in Sarstedt in 1844,[37] purchased a town house in Karlstrasse in Hanover in 1873,[38] and set himself up in business there,[39] trading in cloth and buckskin.[40] A relationship by marriage with the family of his wife Alma, née Niemann, of Pyrmont[41] (now Bad Pyrmont), had existed even before their marriage, since his uncle, Marcus Neuberg, had married a sister of his mother-in-law's in 1846.[42]

Nachmansohn mentions a relationship[43] between the Neuberg family and that of the Nobel Prize winner Otto Meyerhof.[44] Krebs too indicates that the families were closely connected.[45] With regard to the connection between Levy Neuberg and Fanny, the daughter of one Abraham Meyerhof, the following facts have been established: Levy, who later also called himself Louis, married Fanny as his second wife in Hildesheim in 1861.[46] Their son Oskar Neuberg[47] was to play an important role in the career of his cousin, Carl Neuberg.[48] However, from the documents exisiting in the archives it has not proved possible to determine whether, and if so how, Fanny's family was connected to the family of Otto Meyerhof.

36 LSCSA, Hanover 83 b, [B] and information given to T. Ripperger by telephone by Mr. Riedmüller of the Jewish Community in Lower Saxony in March 2006.

37 Copy of an entry in the Register of Births of the Jewish Community in Sarstedt, July 11, 1892. CNP, Box 10: Neuberg, Carl, Biographical Material.

38 Contract of purchase of April 25, 1873. Hanover City Archives, mortgage records.

39 Directories of the City of Hanover (1874–1892).

40 Buckskin was a fulled woollen fabric, cropped of nap, made from carded yarn with a twill or herringbone weave, used for gentlemen's suits in the 19th century.

41 LSCSA, Hanover 83 b [D].

42 Ibid.

43 Nachmansohn (1979), 312.

44 Otto F. Meyerhof (1884–1951), a German-Jewish biochemist born in Hanover, worked at the KWI for Biology in the Dahlem district of Berlin; from 1929 onwards he was Director of the KWI for Medical Research in Heidelberg and at the same time a Professor at Heidelberg University. In 1922 he was awarded the Nobel Prize in Medicine jointly with Archibald Vivian Hill (1886–1977). He fled to Paris in 1938 and to the United States in 1940. There a chair was established for him at the School of Medicine of the University of Pennsylvania in Philadelphia. Martin (1988), 130, Hess (2001) and Werner (1995).

45 Krebs (1972), 14–15.

46 LSCSA, Hanover 83 b [C].

47 Hanover City Archives, verbal communication.

48 Oskar Neuberg (b. 1866) studied chemistry in Heidelberg and took his doctorate in 1890 with theses on "A synthesis of dehydrothiotoluidine" and "On the effect of acid chlorides on phenol ether." Cf. also Chapter 5.5.

2.2. CHILDHOOD, SCHOOL AND UNIVERSITY

Carl Sandel Neuberg was born in Hanover on July 29, 1877, as the first child of Julius and Alma Neuberg.[49] His birthplace, the house in Karlstrasse,[50] was completely destroyed during the air raids of 1943. Carl had two sisters:[51] Ella Gretel Neuberg, who died only a few months after her birth, and Anna Neuberg, who very probably fell victim to the persecution of the Jews during the Nazi tyranny.[52]

Neuberg lived in Hanover up to the age of 15, and attended the classical grammar school Lyceum I of the Ratsgymnasium, which stood on what is now Georgsplatz.[53] Nachmansohn reports that Neuberg retained the basic knowledge of ancient languages that he acquired there throughout his life.[54] In a letter to Andor Fodor,[55] for example, when the latter was working at the Hebrew University in Jerusalem, Neuberg used Hebrew[56] script in some places. However, Fodor wrote in his reply: "I had to clean up the spelling a little bit though; it really was too way out in places."[57]

In 1892 Carl Neuberg moved with his family to the capital, Berlin, where his father opened up a wholesale business in cloth and buckskin in Krausenstrasse, in the Friedrichstadt district.[58] Carl then attended the Friedrich Werdersche Gymnasium (grammar school) in the Moabit district, where, as his friend, fellow pupil

49 Neuberg's birth certificate. CNP, Box 12: Neuberg, Carl, Personal.

50 Despite intense efforts, no picture of the house could be found. Karlstrasse, which ran parallel to the railroad tracks to Bremen to the north-west of the central station, no longer exists.

51 Ella Gretel (Feb. 9 – Sept. 19, 1880) and Anna Neuberg (born Apr. 6, 1882, date of death unknown). Hanover City Archives, information given verbally in February 1997.

52 Federal Archives/International Search Service (1986), Vol. 2, 1277.

53 Degener (1911), 1013–1014.

54 Nachmansohn (1979), 321.

55 Andor Fodor (1884–1968), a Jewish biochemist born in Hungary, who after studying chemistry worked with Emil Abderhalden in Halle. In 1923 he was awarded the Leonard Prize of the *Deutsche Kolloidgesellschaft*, and in 1925 established the Institute of Chemistry of the Hebrew University in Jerusalem. Voswinckel (2002), 429.

56 Fodor, who taught chemistry in Hebrew at the University in Jerusalem, first had to develop Hebrew terms for many technical expressions. Voswinckel (2002), 429.

57 "Ein geringer Aufputz der Orthographie war allerdings vonnöten, sie war stellenweise doch zu abenteuerlich." Fodor to Neuberg, May 15, 1944, CNP, Box 5: Fodor, Andor.

58 Directories of the City of Berlin (1896–1909).

and later fellow student Felix Ehrlich[59] reported, he gained his "Abitur" school leaving certificate with distinction in March 1896.[60]

When Carl left school, his father wanted and expected his son to learn the brewing trade.[61] However, Carl did not fulfil this wish, but started initially to study astronomy, soon changing – with his father's approval – to chemistry.[62] In addition, he took courses in physics, mathematics and philosophy, studying at the Universities of Berlin and Würzburg and at Charlottenburg Technical College until he obtained his doctorate in the year 1900.[63]

His thesis, dedicated to his parents, some individual findings from which had already been published in the *Berichte der Deutschen Chemischen Gesellschaft* (Proceedings of the German Chemistry Society), bore as its full title: "An investigation of acroleïn and glyceraldehyde. On acidic esters of boric acid. On the cleansing of osazones and the determination of their directions of optical rotation."[64] The two reviewers, Emil Fischer[65] and Hans Landolt,[66] awarded the work the commendation *Eruditionis specimen laudabile*, and also certified that at the viva-voce defense of his thesis, which took place on July 14, 1900, Neuberg had displayed very sound knowledge in all the fields examined.[67] Neuberg completed his studies with the grade *magna cum laude*.[68]

59 Felix Ehrlich (1877–1941) was a *privatdozent* (lecturer) with Neuberg at the Royal Agricultural College in Berlin, and then a Associate Professor of Biochemistry in Breslau from 1909 onwards and full Professor of Biochemistry there from 1920. Cf. Neuberg to Bartens, March 8, 1954: "Felix Ehrlich was one of my best friends; we took our school leaving examinations and our doctorates on the same days, and our *habilitation* [the qualification for appointment as a professor – Translator's note] in the same year." ("Felix Ehrlich war einer meiner besten Freunde, wir haben am selben Tage das Abiturium und Doktor Examen gemacht, und uns im gleichen Jahr habilitiert.") CNP, Box 18: Zeitschrift für Zuckerindustrie.

60 Ehrlich (1937).

61 Neuberg later concerned himself with matters relating to the alcoholic fermentation of yeast, earning great merit in this field.

62 Nachmansohn (1979), 312.

63 Neuberg's resumé in his thesis. Archives of the Humboldt Universitäty, Berlin [abbreviated below as AHU], Faculty of Philosophy, No. 357.

64 "Zur Kenntnis des Acroleïns und Glycerinaldehyds. Über saure Ester der Borsäure. Über die Reinigung der Osazone und zur Bestimmung ihrer optischen Drehungsrichtung." Neuberg (1900).

65 Emil Hermann Fischer (1852–1919), after having held professorships in Munich, Erlangen and Würzburg, was appointed to the Chair of Chemistry of Berlin University in 1892. Fischer was a co-founder and Senator of the Kaiser Wilhelm Society (KWG) and a member of numerous other bodies within it. He was awarded the Nobel Prize in Chemistry in 1902 for his work on purines and the synthesis of coffein and glucose. Cf. Kazemi (2002), 90–94.

66 Hans Heinrich Landolt (1831–1910) was Professor of Chemistry and first Rector of the Royal Agricultural College in Berlin, which was founded in 1881. Together with Richard Börnstein he edited the standard work *Physikalisch-Chemische Tabellen*, which was published for the first time by Springer in 1883. He gave his name to the *Landolt reaction* or *iodine clock*, which designates the time lag with which iodine is formed from iodic acid and sulphuric acid.

67 AHU. Faculty of Philosophy, No. 357, 221.

68 Ibid., 190.

3. THE BERLIN YEARS (1898–1913)

3.1. ASSISTANT AT THE CHARITÉ

"As for your eulogies of my efforts, I am afraid that Virchow would have critically discounted them, although he felt extremely friendly toward me."[69] These were the words with which Carl Neuberg sought to play down the paeans of praise contained in the eulogy pronounced by Walter Fürst, Secretary of the American Rudolf Virchow Medical Society, on the occasion of Neuberg's 70th birthday. Although Rudolf Virchow[70] was not directly one of Neuberg's teachers, the quotation points to how significant, and from the point of view of medical history how exciting, the context of those times had been that Neuberg himself was part of.

As early as 1898, two years before completing his studies by taking his doctorate, Carl Neuberg took up a position in Berlin as an unsalaried private assistant in the Department of Chemistry of the Pathological Institute at the illustrious Charité teaching hospital.[71] This department had been set up on the initiative of Virchow, who even at that time was convinced of the outstanding importance of physiological chemistry to medicine.[72] In 1856, shortly after having been appointed to the Chair of Pathology in Berlin, Virchow had brought Felix Hoppe-Seyler[73] from Greifswald to be Prosector[74] and Head of the Department of Chemistry at the Charité.[75] Through his research, his introduction of new methods of analysis and his establishment and editing of the physiological chemistry journal *Zeitschrift für Physiologische Chemie*, Hoppe-Seyler became one of the founding fathers of his discipline. His work laid the foundations for the development of physiological chemistry in Berlin.

When Hoppe-Seyler was appointed to a position in Heidelberg in 1861, he was succeeded as Head of the Department of Physiological Chemistry first by Wilhelm Kühne[76] and then, in 1868, by Oskar Liebreich.[77] Finally, in 1871,[78] the

69 Neuberg to Fürst, October 10, 1947. CNP, Box 5: Fürst, Walter.

70 Rudolf Virchow (1821–1902), the most important pathologist of the 19th century and founder of cellular pathology, was also engaged in social medicine and involved in associated political issues.

71 AHU, Personnel File, UK–N 29, Vol. 3, page 1.

72 Eulner (1970), 66–94.

73 Felix Hoppe-Seyler (1825–1895), see Völckel (2003).

74 The prosector is attached to a dissection department. Hoppe-Seyler, although appointed to the position of a prosector at the institute, would certainly not, as a biochemist, have been engaged in dissection. See also Völckel (2003).

75 Sauer et al. (1960) and Völckel (2003).

76 Wilhelm Kühne (1837–1900) received his physiological training under Emil Du Bois-Reymond, Carl Ludwig and Ernst Brücke. In 1867 he invented the term *enzyme*, which he

position fell to a pupil of Hoppe-Seyler's and Kühne's, Ernst L. Salkowski, who was an important representative[79] of his discipline. During the years that followed, his department became a popular place for scientists to train and work at. One of the major focuses of Salkowski's department was the training of medical students in the field of chemical investigations to gain a better understanding of the composition of bodily fluids. At the same time, the students were to be put in the position of being able as doctors to carry out simple analysis procedures right at the patient's bedside. His book *Practicum der physiologischen und pathologischen Chemie*, published in 1893, became widely used by medical students. Salkowski described the purpose of his compendium in the preface, pointing out at the same time that there was a need, in the field of medical training, both for institutional improvements and for additional manpower:

> "But all these considerations are of little interest as long as only a not very large proportion of physicians are able to do practical chemical work. Hopefully, the time is not too far off when this will change; although it will require the state to undertake substantial expenditure, not only with regard to the equipment of workplaces, but also with regard to ancillary staff for teaching, the number of which must be far larger in a laboratory intended for physicians than in one intended for chemists."[80]

Carl Neuberg began his professional career in this area on the borders between biology and medicine, as a private assistant in the Department of Physiological Chemistry. At the beginning of July 1899, he was then appointed Chemical Assistant with an annual salary of RM 1,050.[81, 82]

About fifty years later, on the occasion of a meeting of American chemists, Neuberg described the experience of working at Salkowski's institute in a way that vividly depicts the state of physiological chemistry at that time, and at the same time gives an impression of the extent to which Neuberg was entering uncharted scientific territory:

derived from the Greek expression *en zymē* ("in yeast"). In 1871 he was appointed to the Chair of Physiology in Heidelberg.

77 Oskar Liebreich (1839–1908) studied chemistry and medicine, and held the Chair of Pharmacology at the Friedrich Wilhelm University from 1872 onwards. He discovered protagon as an important phosphorized substance in the brain and the narcotic effect of chloral hydrate, and introduced lanoline as a base for ointments.

78 Sauer et al. (1960) and Völckel (2003), 137.

79 Salkowski/Leube (1882).

80 "Alle diese Erörterungen haben freilich so lange geringes Interesse, als doch nur ein nicht sehr grosser Teil der Mediciner praktisch chemisch arbeitet. Hoffentlich ist die Zeit, da sich dieses ändert, nicht allzufern; freilich wird es dazu erheblicher staatlicher Aufwendungen bedürfen, nicht allein in Bezug auf Ausrüstung von Arbeitsplätzen, sondern auch in Bezug auf Hülfskräfte für den Unterricht, deren Anzahl in einem für Mediziner bestimmten Laboratorium weit grösser bemessen werden muss, als in einem für Chemiker bestimmten." Salkowski (1912), X.

81 RM = reichsmarks.

82 Letter from the Minister of Ecclesiastical, Educational and Medical Affairs to the Management of the Charité, May 18, 1899. AHU, Charité, 933, sheet 18.

> "When I came to Berlin to Ernst Salkowski in 1898 as an assistant I asked the great old man, how I should prepare for my job. He told me: 'This is easy. You come from Emil Fischers Institute, so you must know everything on carbohydrates, purines and pyrimidines. Fats and lipoids are not very difficult, and we shall soon know more about proteins since Emil Fischer just started working on them. Simply read all volumes of 'Zeitschrift für Physiologische Chemie.' There were 29 volumes at this time, and I read them carefully."[83]

Neuberg's principal teaching task as an assistant to Salkowski was to direct medical students during their "chemistry practicals." This gave him an early impression of chemistry seen from a medical angle, and of the questions that needed to be answered from a physician's point of view.

Many of the medical students that Neuberg had introduced to the fundamentals of general and physiological chemistry during the time he was teaching at the Charité later worked under him for a time.[84] It was also during this time that he made friends with Paul Mayer,[85] a friendship which was one of the most constant in Neuberg's life.

When Neuberg was appointed *privatdozent* – lecturer – at the Friedrich Wilhelm University in Berlin, the practical instruction he offered medical students together with Salkowski formed part of his official activities as a member of the university faculty.[86] Neuberg benefited greatly from his many years of cooperation with Salkowski and from the latter's profound knowledge and experience in the analysis of bodily fluids.[87] Neuberg's solid training and the good reputation of his teacher were no doubt additional reasons why Albert Albu[88] chose the 27-year-old Neuberg to write the chapter on chemistry in his textbook *Physiologie und Pathologie des Mineralstoffwechsels.*[89]

Furthermore, this early period of Neuberg's creative life was notable for the amount of work he published during it. Katterle and Peschel report a total of 120 works of Neuberg's written during the ten years he was at the Charité.[90] During this period, Neuberg concerned himself with a number of different aspects of physiological and pathological chemistry. He developed reactions to demonstrate the presence both of carbohydrates, in particular various sugars and their derivatives, and also of alcohols, aldehydes and cetons, as well as methods to isolate and synthesize these substances. He carried out investigations of the optical properties

83 Undated speech for a conference in Atlantic City. CNP, Box 11: Neuberg, Carl, Birthday Congratulations. All quotations from Neuberg's letters have been corrected only in the case of obvious errors; otherwise, his spelling and punctuation have been largely left unchanged (Authors' note).

84 Ehrlich (1937), XI.

85 Paul Mayer (1872–1949). See Chapter 5.7.; cf. also Engel/Conrads (1998).

86 See Chapter 3.3.

87 Also interesting in this connection are the parallels in the titles of the manuals by Salkowski (1882) and Neuberg (1911).

88 Albert Albu (1867–1921) [aka Albrecht Albu], research into vegetarianism and the influence of sport on the organism. In 1889 *habilitation* in internal medicine at the University of Berlin; later, he established a private clinic for metabolic disorders.

89 Albu/Neuberg (1906).

90 Katterle/Peschel (1989), 13.

of fats, and developed new procedures to demonstrate cholesterol and certain amino acids. In addition, he worked on putrefaction processes and the catalytic effect of sunlight on organic substances. As the Chemical Laboratory was directly attached to the Pathological Institute, he also involved himself with medical issues, in connection with which he wrote some works concerned with the chemical processes taking place in malignant tumors.

But despite his indubitable scientific achievements, his extraordinarily numerous publications and the establishment of the *Biochemische Zeitschrift,*[91] the first publication to be exclusively concerned with biochemical issues, Neuberg remained in the position of an assistant to Salkowski for ten years. Shortly before asking to be released from the Charité in 1909, because he wished to take up a new position the Royal Agricultural College in Berlin,[92] Neuberg had been granted the entitlement to bear the title of Professor.[93] But at the beginning of the 20th century, the opportunities for a German of the Jewish faith to pursue a university career were still very limited.[94]

3.2. HEAD OF DEPARTMENT, INSTITUTE OF ANIMAL PHYSIOLOGY

When the *Königliche Landwirtschaftliche Hochschule*, the Royal Agricultural College, was established in Berlin in 1881, an Institute for Animal Physiology was also set up. Nathan Zuntz[95] was appointed its Director; until receiving the call to Berlin he had taught at the Institute for Anatomy of the University of Bonn. The research carried out in Zuntz's department focused on the analysis of metabolic processes in animal organisms; in particular, the relationship between nutritional input and the animal's strength was investigated. Respiration experiments were carried out on live animals, and various methods were developed to measure the speed of metabolism. Members of the department were also concerned with a variety of animal diseases, and carried out observations of fish and pond life.[96] There were also investigations into poultry breeding and rearing. The wide range of issues covered by the Institute's research, and also the training of students in agricultural[97] subjects, led to an ever increasing demand for better premises. Thus very soon after the Institute had been founded, the need arose for a further

91 See Chapter 4.

92 Letter from the Management of the Charité to the Minister of Ecclesiastical, Educational and Medical Affairs, July 26, 1909. AHU, Charité, 933, sheet 140.

93 See Chapter 3.3.

94 Cf. Willstätter (1949), Glum (1964), Schwabe (1983), Ringer (1983), 127 and 218. See also Volkov (1997), 1–8, Deichmann (2002), Ch. 1, Kaplan (2003b).

95 Nathan Zuntz (1847–1920) studied medicine in Bonn. He worked at the same university as a physiologist after taking his doctorate and *habilitation*. He was appointed to the Royal Agricultural College in Berlin upon its establishment in 1881. Zuntz was also one of the pioneers of aviation medicine: See also Fn. 121.

96 Zuntz (1906), 160–162.

97 The Institute offered courses of study in the subjects "Agriculture," "Geodesy and Techniques of Cultivation" and "The Agrotechnical Trade."

building to be built, to house in particular the Institute for Animal Physiology. The new premises were finally ready to be moved into in 1908.[98]

In the following years, the growing volume of detailed physiological findings also led to further areas of work being opened up that were to be researched into using new methods. It was in this context that Carl Neuberg, as a chemist, was offered a position at the Royal Agricultural College's Institute for Animal Physiology. There is no record of exactly when he moved from the Pathological Institute of the Charité to the Royal Agricultural College. He was listed as a member of the college faculty from as early as April 1909,[99] even though negotiations dragged on until the end of May 1909. Neuberg had made it a condition that he should be given a 12-year contract and the position of Head of the Chemical Department.[100] He was still listed in the files of the Charité as being on the staff of the Department of Physiological Chemistry there until October 1909.[101] It was probably in the fall that he finally switched to the Institute for Animal Physiology of the Royal Agricultural College.[102]

In addition to teaching[103] the "chemistry exercises" course, Neuberg was to continue the research work that he had "conducted with such great success." A laboratory specially equipped for this purpose was made available to him.[104] Zuntz summarized the tasks of his new member of staff as follows:

> "The appointment of Professor Neuberg is intended to fulfill the need that has long been felt to take greater account of the organic structure of the body and of foodstuffs. In particular, in the investigations of metabolism the effects of the manifold complex compounds found in vegetable foodstuffs and what becomes of them in the body are to be studied."[105]

At the time of Neuberg's appointment, "Zuntz's Institute," as the department was universally known, had already moved into its premises in the College's new building. The Institute provided perfect conditions for experimental work. The researchers were satisfied; local residents, however, as is documented by their complaints, were less than thrilled by the Institute's new building with its hustle and bustle, the keeping of cattle, horses, sheep, dogs, rabbits and fish and the

98 Annual Report of the Royal Agricultural College, Berlin [abbreviated below as ARRACB] (1909), Volume XVII.

99 AHU, Personnel File, UK–N 29, Vol. 2, sheet 2.

100 AHU, UK–N 29, Vol. 2, sheets 5–9.

101 Letter from the Management of the Charité to the Minister of Ecclesiastical, Educational and Medical Affairs, July 26, 1909. AHU, Charité, 933, sheet 140.

102 ARRACB (1910). Vol. XVIII, 43.

103 For a list of the lectures given and courses taught by Neuberg, see the German edition of this book: Conrad/Lohff (2006), Chapter 10.2, 186–188.

104 ARRACB (1910). Vol. XVIII, 43.

105 "Durch die Anstellung von Prof. Neuberg soll dem seit langem gefühlten Bedürfnis einer intensiveren Berücksichtigung des organischen Aufbaus des Körpers und der Nahrungsmittel Rechnung getragen werden. Es sollen namentlich bei den Stoffwechseluntersuchungen die Wirkung der mannigfachen in der Pflanzennahrung vorkommenden komplizierten Verbindungen und das Schicksal derselben im Körper studiert werden." ARRACB (1910). Vol. XVIII, 43.

associated constant background noise.[106] Neuberg, to whom the new building afforded the most modern conditions for research, probably worked with his staff on the second floor of the building.[107] The capacity of the laboratories was soon fully utilized by Neuberg's staff and graduate and undergraduate students; but other scientists who were friends of his, such as Paul Mayer[108] and Arnold Durig, also made use of the premises.[109]

Neuberg later said of Durig, in a letter to H. Wastel written in 1946:

> "I am also very grateful to you for your lines concerning Professor Durig. I will write to him, although I have certain inhibitions about it, because I was in Vienna precisely at the time of the putsch. At that time, Durig was by no means averse to the 'new order,' and a short time later I received a letter from him in which he wrote that it was now time to cease corresponding with me. I do not by any means want to maintain that he was a Nazi, but he did not display any significant degree of courage."[110]

Among Neuberg's circle of friends at that time was also Emil Abderhalden,[111] who like Mayer and Durig worked at the Institute. Until very recently, Abderhalden was held in unreserved esteem both as a scientist and as the President of the German Academy of Sciences Leopoldina; in the light of recent studies,[112] however, he now needs to be viewed more circumspectly with regard not only to his research work,[113] but also to his attitudes:

> "The spectrum of his [Abderhalden's] reactions ranges from reserve, approbation and tactical delay to cooperation, over-obliging obedience and later mental suppression. No form of open criticism of the Nazi state, never mind rejection or resistance, is evident in his person."[114]

Neuberg too had himself noticed this ambivalence in his friend Abderhalden's behavior. In a letter to Ernst Wertheimer[115] in December 1945, Neuberg wrote:

106 ARRACB (1913). Vol. XXI, 67–68.

107 ARRACB (1911). Vol. XIX, 30–32.

108 Many publications arose from Mayer's work at the Institute, cf. ARRACB (1913). Vol. XXI, ARRACB (1914). Vol. XXII.

109 Arnold Durig (1872–1961), Professor of Physiology and Head of the Physiological Institute of the University of Vienna from 1918 to 1938.

110 "Sehr verbunden bin ich Ihnen auch fuer Ihre Zeilen Professor Durig betreffend. Ich werde ihm schreiben, obgleich ich gewisse Hemmungen habe, denn ich war gerade in den Tagen des Umsturzes in Wien. Durig war damals den 'Neuerungen' keineswegs abgeneigt und ich erhielt kurze Zeit später von ihm einen Brief, dass es an der Zeit sei, die Korrespondenz mit mir abzubrechen. Ich will keinesfalls behaupten, dass er ein Nazi war, aber bedeutenden Mut hat er nicht bewiesen." Neuberg to Wastel, April 22, 1946. CNP, Box 17: Wastel, H.

111 Emil Abderhalden (1877–1950), physiologist, biochemist, ethicist, eugenicist and "inventor" of the "protective enzyme" *("Abwehrferment")*, was Professor of Physiology and Physiological Chemistry at the University of Halle an der Saale from 1911 to 1950. In 1931 he became President of the German Academy of Sciences Leopoldina, a position he held until 1946. Cf. Deichmann/Müller-Hill (1998), Frewer (2000), Gerstengarbe/Seidler (2002).

112 Frewer (2000), Kaasch/Kaasch (2003).

113 Deichmann/Müller-Hill (1998), Kaasch (2000).

114 "Die Spannweite seiner Reaktionen reicht von Zurückhaltung, Übereinstimmung und Taktieren bis zur Kooperation, vorauseilendem Gehorsam und späterer Verdrängung. Eine Form offener Kritik oder gar Verweigerung und Widerstand gegen den NS-Staat sind in seiner Person nicht auszumachen." Frewer (2000), 275.

"What you write about Abderhalden I have heard in more or less the same way from someone else as well. It really is an irony of history that Abderhalden is today turning to his Jewish colleagues. When the first call of German researchers to back the *fuehrer* went out, Abderhalden was right out in front, and it has been alleged that he was actively involved in this appeal being made. He was often a dextrogyre and often a levogyre creature; today he would like to be a mesoform or to appear intramolecularly compensated."[116]

In the years before the First World War, however, no shadow yet fell on the collaboration between these three young scientist friends. The Annual Reports of the Royal Agricultural College's Institute for Animal Physiology document the wide diversity of Neuberg's scientific activities. During this time, he was intensively involved in investigating biochemical processes in fermentation. In addition, however, he was also concerned with the catalytic effect of various types of radiation; in particular, he investigated the effects of the sun's rays and summarized his findings in the 1913 monograph *Relationships between Life and Light.*[117] He assisted his colleagues with their experiments whenever chemical analyses, especially of animal urine samples, had to be carried out.[118] For his chemical analyses of mineral waters,[119] Neuberg was awarded the Cross of Merit, 3rd Class, of the Principality of Waldeck *(Fürstlich Waldeck'sches Verdienstkreuz III. Klasse)* in 1913.[120] But his scientific interests were not limited exclusively to biochemical problems. In 1910, together with Nathan Zuntz,[121] he took part in an excursion to Tenerife which was intended to provide insights into the effects of a high altitude climate on the human body.[122] But he was also involved in specifically clinical is-

115 Ernst Wertheimer (1893–1973), German-Jewish physiologist. While still a medical student, he performed his military service in a medical capacity. In 1921 Wertheimer became an assistant of Abderdalden's in Halle, where he was also Chief Assistant and Associate Professor (without tenure) until 1933. Although he was offered posts in Moscow and Canton, China, he went in 1934 to the Hadassah Hospital (Rothschild Hospital) of the University of Jerusalem, where he taught pathological physiology and later biochemistry until 1963.

116 "Was Sie mir ueber Abderhalden schreiben, hatte ich schon ungefaehr gleichlautend von anderer Seite gehoert. Es ist geradezu ein Witz des Weltgeschehens, dass sich Abderhalden heute an seine juedischen Kollegen wendet. Im ersten Aufruf der deutschen Forscher, sich hinter den Fuehrer zu stellen, stand Abderhalden an der Spitze und wie man behauptet hat, nahm er taetigen Anteil am Zustandekommen dieses Aufrufes. Er war oft ein dextrogyres und oft ein laevogyres Gebilde; heute moechte er gerne eine Mesoform sein oder intramolekular kompensiert erscheinen." Neuberg to Wertheimer, December 14, 1945. CNP, Box 17: Wertheimer, E.

117 *Beziehungen des Lebens zum Licht*, Neuberg (1913).

118 Cf. also the two-volume collection on animal urine, comprising 1,800 pages. Neuberg (1911).

119 These were concerned inter alia with the mineral waters of (Bad) Pyrmont, where Neuberg's mother Alma came from and where his sister Anna was born in 1882; cf. Neuberg (1913), see Chapter 11.1., Nos. 290/291. This celebrated spa town formed part of the Principality of Waldeck-Pyrmont.

120 AHU, UK–N 29, Vol. 1, sheet 11.

121 A commemorative plaque on the house Bleibtreustrasse 38/39 in Charlottenburg, Berlin, recalls Zuntz's work in the field of aviation medicine.

122 Excursion from March 15 to April 30, 1910. AHU, UK–N 29, Vol. 1, sheet 10.

sues. Together with Wilhelm Caspari,[123] Neuberg carried out a series of important investigations into the effect of chemical substances on malignant tumors during this time. Unlike Caspari, who from 1920 to 1933 was Director of the Department of Cancer Research at the State Institute for Experimental Therapy in Frankfurt am Main, Neuberg ceased to concern himself with these problems after moving to the Kaiser Wilhelm Society (KWG). Within a period of four years up to September 1913, when he quit the faculty of the Royal Agricultural College to move to the Institute for Experimental Therapy of the KWG, he published 65 works.[124]

As early as 1912, probably in response to a proposal of Neuberg's, the Department of Chemistry was designated in the Annual Reports of the Royal Agricultural College as the Department of Biochemistry.[125] This on the one hand illustrates the focus of Neuberg's research work and his understanding of physiological chemistry, but on the other hand also foreshadows the development of the following years, which he did all he could to advance, by which biochemistry became institutionalized as an independent discipline; he finally succeeded in achieving this goal when the first independent Institute for Biochemistry was established in 1925.

3.3. NEUBERG'S TEACHING CAREER AT THE CHARITÉ AND THE ROYAL AGRICULTURAL COLLEGE

Scarcely three years after taking his doctorate[126] at the Faculty of Philosophy of the Friedrich Wilhelm University in Berlin, Carl Neuberg undertook the next step in his academic career by seeking *habilitation* – the qualification conferring eligibility for a professorship. In June 1903 his teacher Emil Fischer stated in a letter of recommendation that he had not the slightest doubt about Neuberg's suitability to be a *privatdozent* or university lecturer.[127] To gain his *habilitation*, Neuberg, who at this time was still an assistant of Ernst Salkowski's, submitted four papers concerned with questions of the metabolism of carbohydrates and proteins.[128] In July 1903 he held his demonstration lecture, entitled "On the Formation of Sugar from Protein." Professors Hans Heinrich Landolt and Emil Fischer were appointed as assessors, although the latter fell ill, so that Professor Siegmund

123 Wilhelm Caspari (1872–1944), physician, physiologist and important cancer researcher. Caspari died in the Litzmannstadt/Łódź ghetto three years after his deportation. Voswinckel (2002), 233–234.

124 ARRACB (1909–1913), Vol. XVII–XXI.

125 ARRACB (1912), Vol. XX, 47.

126 Neuberg (1900).

127 Fischer to the Faculty of Philosophy, June 27, 1903. AHU, Faculty of Philosophy, No. 1226: *habilitation* documents, sheet 62.

128 Of the four handwritten titles of the *habilitation* papers, only three can be deciphered. These are: *"Über Chondroitinschwefelsäure"* ("On Chondroitin Sulfuric Acid"), *"Physiologischer Abbau von Jodalbumin"* ("The Physiological Breakdown of Iodine Albumin") and *"Synthese gepaarter Glucuronsäuren"* ("The Synthesis of Paired Glucuronic Acids"). AHU, Faculty of Philosophy, Nr. 1226: *habilitation* documents, sheet 64.

Gabriel had to deputize for him.[129] Neuberg received the *venia legendi,* the license to teach at a university, on July 31, 1903.[130] Three years later, in May 1906, he was granted the title of Professor "in recognition of his academic achievements";[131] however, this purely titular professorship carried with it neither a professorial chair nor any salary. But for Neuberg, who was only 28 years old, it represented an important affirmation and recognition of the work he had already done in the still young field of biochemistry. When Neuberg was appointed Head of Department at the Institute for Animal Physiology of the Royal Agricultural College in Berlin in 1909, he was at the same time granted an honorary tutorship[132] and a lectureship[133] in Animal Chemistry at the Friedrich Wilhelm University in Berlin.

Neuberg took up his teaching position at the university in the summer semester of 1904. He lectured principally in the fields of protein and carbohydrate chemistry and their physiological significance in the animal, and later also in the human, organism. Although the series of lectures he delivered bore a very diverse range of titles, in his teaching as in his own scientific research he was essentially concerned with the border area between chemistry and medicine.[134] Furthermore, as an assistant of Ernst Salkowski's Neuberg was also involved in "chemistry practicals" for medical students. After moving to the Royal Agricultural College, Neuberg only offered practical courses for students. According to the lecture lists, he delivered his last series of lectures in the winter semester of 1909/1910. At the Royal Agricultural College he offered "exercises in animal and plant chemistry" and taught special courses in organic chemistry and biochemistry. These practical exercises were discontinued at the Royal Agricultural College after Neuberg's move to the KWG, but he continued to offer practical courses in organic chemistry and biochemistry at his new institute, the KWI for Experimental Therapy. These courses of practical training, whose participants were integrated into the general working structures of the Institute, gave rise to a series of theses by young students of chemistry and medicine. Although to some extent this represented a way of saving on the Institute's staffing budget, Neuberg also emphasized that it was "[...] in the interests of his discipline – biochemistry – to cultivate relations with young students,"[135] and to introduce them to this still not very firmly estab-

129 Siegmund Gabriel (1851–1924), chemist. Fischer's permanent deputy from 1892 onwards.

130 Certificate of July 31, 1903. CNP, Box 10: Neuberg, Carl, Awards and Diplomas.

131 The Minister of Ecclesiastical, Educational and Medical Affairs to the Faculty of Philosophy of the Friedrich Wilhelm University, Berlin, May 25, 1906. AHU, Faculty of Philosophy, No. 1438, sheet 42.

132 AHU, Neuberg's Personnel File, UK–N 29, Vol. 1, sheet 2.

133 AHU, Faculty of Philosophy, No. 1439, sheet 188.

134 For a list of the lectures given and courses taught by Neuberg, see the German edition of this book: Conrad/Lohff (2006), Chapter 10.2, 186–188.

135 "[...] im Interesse der von ihm betriebenen Disciplin – der Biochemie – gelegen sei, die Beziehungen zur studierenden Jugend zu pflegen." Annual Report of the Department of Chemistry of the KWI for Experimental Therapy for the period July 1918 to March 1920. MPG Archives, 1459, sheets 154–157.

lished new subject.[136] Neuberg preferred to teach in a practical way, in order to communicate to young scientists how to go about their work in this field of research in the proper manner. He was a passionate experimenter, able to demonstrate impressively in practice the scope and the fundamentals of physiological and organic chemistry. He was, however, not a good speaker, and he was well aware of this fact.[137] After the Second World War he wrote to a colleague: "I have never been a good teacher, I was never a master of the art of lecturing as you are; only in the laboratory was I able to do my duty properly."[138] In all the tributes paid to Neuberg, there is hardly a mention of his teaching activities. One rare document in this respect is the speech given by Albert Gessler[139] at a meeting of the American Society of European Chemists in New York on the occasion of Neuberg's 70th birthday, in which he described the impact made by Neuberg as a young professor at the beginning of the twentieth century:

> "[...] I saw him through the eyes of a young student, while he looked across his lecture room with the eyes of experience and the great dignity of a professor. He was not a very old professor, to be sure, with his 28 years, but we certainly respected him deeply for his knowledge and his fiery enthusiasm that carried us along. In reality he was the youngest professor at the university, as I remember it, but he seemed to be constantly at work in his laboratory or at his desk. While he lectured about physiological chemistry, as it was called at that time, and protein chemistry, he seemed to represent to us something new that was coming over the horizon. Something larger than the physiological chemistry we were studying, something that he himself had begun to name Biochemistry."[140]

Even if such a description, in view of the occasion it was written for, may tend towards the extravagant, it is nevertheless clear that Neuberg, right from the beginning of his academic career, saw himself as representing a new direction of study that was concerned with the processes of life from a chemical point of view. He succeeded in conveying his enthusiasm for this subject to his students. Nachmansohn provides a further testimonial to Neuberg's qualities as a teacher:

> "Neuberg was a passionate teacher; he loved, and always found time, to discuss problems with young people. His extraordinary memory and his knowledge of literature made these discussions stimulating and an invaluable source of information. It was easy to get an appointment to see him, even in the Dahlem days, in spite of his many obligations. In one hour of discussion with Neuberg his students frequently learned more than they could have learned in days spent at the library."[141]

When it was a question of accepting and supervising graduate students working on their doctorates, all that ever counted for Neuberg was the academic

136 Cf. Deichmann (2002), Jaenicke (2002).

137 Information communicated personally by Ines Mandl, October 25, 1996.

138 "Nie bin ich ein guter Lehrer gewesen, habe nie die Vortragskunst besessen, über die Sie verfügen; allenfalls im Laboratorium habe ich meine Pflicht erfüllen können." Neuberg to Hofmann, March 2, 1952. CNP, Box 6: Hofmann, Eduard.

139 Albert Gessler was Vice President and Director of the Research Department of Interchemical Corporation, then based in New York, in the late 1940s and the early 1950s. Cf. also Hochberg (1993), 362, Deichmann (2002), 474.

140 American Society of European Chemists and Pharmacists (1947), 3.

141 Nachmansohn (1979), 323.

achievement and the aptitude of the student.[142] Thus the later Nobel Prize winner Fritz Lipmann,[143] who had already gained a doctorate in medicine, was able to take another one in chemistry under Neuberg in 1928. This is remarkable since Lipmann was working in Otto Meyerhof's laboratory; Meyerhof's work was to a certain extent in academic competition with Neuberg's, but his functions did not permit him to accept graduate students. Thus Lipmann found in Neuberg a *"Stiefdoktorvater,"* a "doctoral stepfather" or substitute supervisor, whom he always, as he himself declared, felt himself well treated by.[144]

Neuberg sometimes made very critical remarks about Meyerhof's research, especially after the war:

> "I never believed Meyerhoff's categorical statement with regard to oxybiosis. Peremptory as he was in many things, here too he overshot the target, and he was never able to endure being contradicted."[145]

Possessing as he did an encyclopedic knowledge of biochemistry, he was able to demonstrate that Embden[146] and Meyerhof had probably suppressed prior claims to authorship:

> "But Emden's systems, and Meyerhof's too, are based on a little bit of systematic thievery. The father of the whole business is the indeed richly plagiarized R. Nilsson, who discovered that phosphoglyceric acid occurred naturally and immediately related it to our synthetic compound, although this was racemic. In 1933 Embden and Meyerhof got hold of this key substance and did not even mention its discoverer, or were as dismissive of him as possible, in the same way as they also always concealed the fact that the essential point of the old fermentation scheme lay in our demonstration that the desmolytic decomposition of hexose into bodies of the 3-carbon family had been postulated by us as an essential feature,[147] whereby we repeatedly emphasized that the introduction of phosphoric acid had been omitted

142 Cf. Chapter 5.7.

143 Fritz Albert Lipmann (1899–1986) took a doctorate in medicine in 1924 and one in chemistry in 1928. In 1929 Lipmann went to the United States. Before switching to the Carlsberg Institute in Copenhagen in 1932, he was an assistant to the chemist Albert Fischer (1891–1956), and from 1949 onwards Professor of Biochemistry at Harvard Medical School. From 1957 to 1969 he taught at the Rockefeller University in New York. He received the Nobel Prize in Medicine jointly with Hans Adolf Krebs (1900–1981) in 1953. Cf. Kazemi (2002).

144 See Lipmann (1988), 14 and Krebs/Lipmann (1988), 120.

145 "Die kategorische Behauptung Meyerhofs betreffs der Oxybiose habe ich nie geglaubt. Peremptorisch wie in manchem, ist er auch hier ueber das Ziel hinausgeschossen und Widerspruch hat er nie vertragen." Neuberg to Windisch, February 4, 1954, in CNP, Box 18: Windisch, Fritz.

146 Gustav Embden (1874–1933), after studying medicine, became Director of the Chemical Laboratory of the Medical Clinic in Frankfurt am Main in 1907. After the establishment of the University of Frankfurt in 1914 he was appointed to the Chair of Vegetative Physiology. He discovered the important metabolic function of the liver, and described the natural metabolism of sugar in the organism.

147 The difficult structure of this sentence reflects that of the original German. [Translator's note]

for the sake of simplifying the model. All this happened in 1933, and the paralysis resulting from the Hitler business kept me from putting it right."[148]

148 "Aber den Systemen von Embden, wie auch Meyerhof liegt doch eine kleine systematische Gaunerei zugrunde. Der Vater der ganzen Herrlichkeit ist ja der weidlich begaunerte R. Nilsson, der das natuerliche Vorkommen der Phosphoglycerinsaeure entdeckt hat und sie alsbald mit unserer synthetischen Verbindung, die allerdings racemisch war, in Beziehung gesetzt. Im Jahre 1933 haben sich Embden und Meyerhof dieser Schluesselsubstanz bemaechtigt und den Entdecker gar nicht erwaehnt, oder so schlecht wie moeglich behandelt, wie sie auch immer unterschlagen haben, dass der wesentliche Punkt des alten Gaerungsschemas in unserem Nachweis lag, dass der desmolytische Zerfall der Hexose zu Koerpern der 3-Kohlenstoffreihe als wesentlicher Teil von uns postuliert war und zwar unter wiederholter Betonung, dass die Einfuegung der Phosphorsaeure wegen schematischer Vereinfachung unterblieben war. Alle diese Dinge geschahen im Jahre 1933 und die Laehmung durch die Hitlervorgaenge hat mich von der Rectification abgehalten." Neuberg to Bersin, March 7, 1955, CNP, Box 3: Bersin, Theodor.

4. THE "BIOCHEMISCHE ZEITSCHRIFT"

4.1. THE MAGAZINE'S FOUNDATION AND INTERNATIONAL PRESTIGE

In January 1904, when Albert Albu proposed to the publisher Ferdinand Springer[149] the publication of a new textbook,[150] he mentioned that the part of the work dealing with chemistry was to be written by a chemist called Carl Neuberg.[151] Springer therefore inquired of Langgaard[152] regarding Neuberg's qualifications, and received the following answer:

> "Dr. Neuberg is one of the best among our younger physiological chemists, and has produced a whole series of good papers. He is generally considered to be very able, is at any rate intelligent and very hard working, and is likely to have a future."[153]

Thanks to this positive evaluation and the high expectations that people had of the young research scientist, Neuberg was indeed engaged as co-author. Neuberg, who had just been appointed *privatdozent*, put a considerable amount of work into this publication, so that in the course of a year it expanded from the originally planned six printed sheets to no less than seventeen.[154] It may have been this first successful collaboration that prompted Ferdinand Springer, who was himself only 23 years old and had been "junior boss" of the publishing house for only a year, to invite Neuberg to a meeting on March 5, 1905, in order to explore with him the possibility of establishing a new physiological chemistry periodical. Neuberg was optimistic about the plan, and was commissioned by Springer to edit the *Biochemische Zeitschrift*. First of all, Neuberg had to enter into negotiations with a number of well-reputed scientific institutions in order to find suitable scientists to write articles and co-edit the periodical. This took the 27-year-old Neuberg on a tour of exploration taking in several German university cities, and he also visited numerous research institutions in Berlin.

149 The tale of how the *Biochemische Zeitschrift* was established has already been told at length in a chapter of Heinz Sarkowski's chronicle "Der Springer-Verlag"; but this makes no mention of the inglorious circumstances in which the publishing house parted from the periodical's editor in 1935, after 30 years of fruitful collaboration. See Sarkowski (1992), 166–170.

150 Albu/Neuberg (1906).

151 Albu to F. Springer, January 8, 1904. Archives of Springer Verlag, Heidelberg [abbreviated below as ASV], B. N. 32.

152 Alexander Langgaard (1847–1917) studied pharmacy, took his *habilitation* in 1888 and was granted a titular professorship at Berlin University.

153 "Herr Dr. Neuberg ist einer unserer besten jüngeren physiologischen Chemiker, der eine ganze Reihe guter Arbeiten geliefert hat. Er gilt ganz allgemein für sehr tüchtig, ist jedenfalls intelligent und sehr fleissig und dürfte wohl eine Zukunft haben." Langgaard to F. Springer, February 3, 1904. ASV, B. N. 32.

154 Springer to Albu, January 3, 1905. ASV, B. N. 32.

But at the end of 1905, serious differences that had arisen during the negotiation of the contract threatened to scupper the whole project. Neuberg was asked to accept a clause which, in the event of his quitting the editorial team or upon the expiration of his contract, would have prevented him from participating in the publication of any scientific journal that was in competition with the *Biochemische Zeitschrift* for a period of two years. But Neuberg found this condition unacceptable, and insisted on having the draft contract revised so that the only condition would be that such a publication might not bear the name "Biochemische Zeitschrift."[155] Neuberg initially declined an invitation from Springer to meet him in order to try and find a solution with the following words:

> "Besides, in my view there is no point in a personal meeting as long as you cannot bring yourself to accept my conditions, which I greatly regret I cannot materially depart from."[156]

At that time, Ferdinand Springer was more inclined to give up on the project or to look for another academic partner for the first periodical he had established himself than to accept Neuberg's demands. Finally, however, in December 1905, the partners reached agreement after all. Neuberg signed a contract to edit the magazine for a period of six years, and was given the right to work on other periodicals if the publisher should terminate his contract.[157] This episode in the story of the birth of the *Biochemische Zeitschrift* shows an admirable degree of self-assurance on the part of the young Neuberg. At the age of just 27, he felt himself capable of persuading the elite of the world of biochemistry that there was a need to launch a new publication for the young discipline, even though there was good reason to fear conflict with the publishers of already existing and well-established journals in the field, such as Hoppe-Seyler's *Zeitschrift für physiologische Chemie* and Hofmeister's *Beiträge zur chemischen Physiologie und Pathologie*. It is equally remarkable with what vehemence he was able to defend his interests against the Springer Verlag, which was already an academic publishing house of high renown, and at least in part to carry his point. It also testifies to Neuberg's great confidence in his own abilities that despite being himself only just at the beginning of his academic career, he had no doubts as to his ability to decide alone and on his own responsibility with regard to the acceptance or rejection of contributions submitted.[158]

The first number of the periodical finally appeared in October 1906; it contained 36 original papers, principally from Berlin institutes. According to the

155 Neuberg to F. Springer, December 6, 1905. ASV, B. B. 158.

156 "Außerdem erübrigt sich m. E. eine persönliche Zusammenkunft, solange Sie sich nicht entschließen können, meine Bedingungen anzunehmen, von denen ich zu meinem lebhaften Bedauern im wesentlichen nicht abgehen kann." Neuberg to F. Springer, December 8, 1905. ASV, B. B. 158.

157 Agreement of December 13, 1905, Section 6. ASV, B. B. 158.

158 Agreement of December 13, 1905, Section 2. ASV, B. B. 158.

title page, editorial responsibility was shared by the chemists Eduard Buchner[159] and Ernst Salkowski, the physicians Ehrlich[160] and Karl von Noorden[161] and the physiologist Nathan Zuntz. A total of 41 scientists from ten European countries plus Japan and the United States were listed as collaborating authors.[162]

The documents relating to the *Biochemische Zeitschrift* in the publishers' archives show that Neuberg always applied himself most intensively to the editorial and proofreading work for the periodical, even while he was abroad. There are, for example, numerous letters in which Neuberg informs the publishers of his addresses, so that they would be able to keep in touch with him during his travels and he would be able to fulfill his obligations as editor.[163]

How successful the journal was in the first few years after its foundation is shown amongst other things by the fact that in 1908 the academic periodical *Beiträge zur chemischen Physiologie und Pathologie; Zeitschrift für die gesamte Biochemie*, published by Vieweg in Brunswick, was absorbed by the *Biochemische Zeitschrift*. In advance of this takeover, its editor Franz Hofmeister[164] wrote to Carl Neuberg as follows:

> "In the interests of the subject and of the public, however, I am more inclined to allow the number of academic journals to be reduced on this occasion, and would therefore be inclined to have the Institute's work published in the *Biochemische Zeitschrift*, which has best

159 Eduard Buchner (1860–1917) studied chemistry and took his doctorate with a thesis on "The Influence of Oxygen on Fermentation" (1888) under the direction of Adolf Baeyer in Munich. In this study he disproved Pasteur's theory on the influence of oxygen in the fermentation process. He was a professor in Berlin and Warsaw before being appointed to the Chair of Biochemistry in Tübingen in 1911. Buchner died of a war wound in 1917.

160 Paul Ehrlich (1854–1915) worked with Emil Behring at Robert Koch's institute on a serum to provide protection against diphtheria. He established immunology as a field of research, and was the first director of the Royal Institute for Serum Research and Serum Testing in the Steglitz district of Berlin, which was renamed Royal Institute of Experimental Therapy after moving to Frankfurt in 1899, and Pasteur Institute in 1947. Ehrlich was a Senator of the KWG from 1911 to 1915, and won the Nobel Prize in Physiology and Medicine in 1908 together with Ilja Iljic Mecnikov.

161 Karl Harko von Noorden (1858–1944) was a specialist in metabolic disorders. From 1904 onwards he was Director of the Department of Internal Medicine at Frankfurt Municipal Hospital, and was appointed a professor at Vienna University in 1930, where he headed the Department for Metabolic Disorders until his death. He is considered to be the founder of the science of dietetics.

162 Cf. the title page of the first volume of the *Biochemische Zeitschrift*, 1906.

163 Neuberg to Springer, August 20, 1906. ASV, B. B.158.

164 Franz Hofmeister (1850–1922), the focus of whose work was in the field of peptide chemistry. After holding a professorship in Prague, he succeeded Hoppe-Seyler as Professor of Physiological Chemistry at the University of Strasbourg.

THE DEVELOPMENT OF THE "BIOCHEMISCHE ZEITSCHRIFT"

Year	Number of volumes	Number of articles	Neuberg with co-authors	Neuberg sole author
1906[165]	2	79	8	2
1907	5	156	16	4
1908	8	252	12	7
1909[166]	8	311	14	7
1910	7	299	20	11
1911	7	265	16	3
1912	10	302	14	7
1913[167]	10	365	12	8
1914[168]	10	378	35	8
1915	4	132	19	3
1916	7	172	4	2
1917	6	150	3	0
1918[169]	6	119	10	5
1919[170]	11	226	8	1
1920	11	221	7	1
1921[171]	13	321	9	0
1922	9	412	5	0
1923[172]	9	448	8	1
1924	11	567	13	2
1925	12	547	21	0
1926	13	596	14	0
1927	12	571	21	1
1928	12	551	13	0
1929	13	571	7	1
1930[173]	13	566	13	2
1931	14	636	11	0
1932	13	645	9	0
1933[174]	11	571	6	0
1934[175]	7	369	7	0
1935[176]	8	414	13	1
Total	282	11,212	368	77

165 Year of the foundation of the *Biochemische Zeitschrift*.
166 Neuberg at the Royal Agricultural College.
167 Neuberg moves to the KWI for Experimental Therapy.
168 Beginning of the First World War.
169 End of the First World War.
170 Neuberg made titular professor at the Royal Agricultural College in Berlin.
171 Neuberg Director of the KWI for Biochemistry.
172 Neuberg sole editor of the *Biochemische Zeitschrift*.
173 Jacoby joins the editorial staff of the *Biochemische Zeitschrift*.
174 Seizure of power by the National Socialists.
175 Grassmann joins the editorial board of the *Biochemische Zeitschrift*.
176 Neuberg's departure.

succeeded in realizing the idea that I was then seeking to pursue of creating closer links between chemistry and medicine."[177]

Ultimately, Vieweg sold its "continuity lists" with the names of 359 subscribers to Ferdinand Springer's publishing house,[178] after the latter had succeeded in bargaining the price down from 2,000 marks to 200 marks.[179]

The original plan to publish two to three issues a year[180] was impossible to keep to, thanks to the wealth of contributions that Neuberg received, worked over and assessed. As early as 1914, for example, the subscribers received no less than ten numbers, in which a total of more than 350 articles by scientists from eleven European and three non-European countries were published. The number of papers published in the periodical showed a continuous increase, interrupted only by the impact of the First World War.

The climax was reached in 1932 with 645 papers published in 13 issues. After the Nazis had come to power in Germany, the number of contributions declined again. A diagram of Slater's, in which he compares the number of pages published in the most important international biochemical journals, impressively demonstrates the extraordinary position that the *Biochemische Zeitschrift* held in the 1920s and 1930s:[181]

> "The *Biochemische Zeitschrift*, already a large journal in 1920, grew rapidly in the next 5 years and, after a slow further growth, declined precipitously in 1933, until 1939 it was smaller than Hoppe-Seyler."[182]

At the end of 1930, Neuberg took on the Head of the Institute of Chemistry at Moabit Hospital, Martin Jacoby,[183] as a member of the editorial team. With more than 500 articles appearing annually, the workload had understandably become too heavy for Neuberg, who in addition to his duties as Director of his Institute

177 "Im Interesse des Faches und des Publikums neige ich jedoch mehr dazu, bei dieser Gelegenheit eine Verringerung der Zahl der Fachjournale eintreten zu lassen, und wäre daher geneigt, die Arbeiten des Instituts in der Biochemischen Zeitschrift erscheinen zu lassen, die den mir seinerzeit vorschwebenden Gedanken, einen engeren Zusammenhang zwischen Chemie und Medizin herzustellen am besten verwirklicht hat." Hofmeister to Neuberg, April 18, 1908. ASV, B. B. 158. Hofmeister sold his periodical on condition that his name would appear on the title page of the *Biochemische Zeitschrift* as co-editor, with the note "Contributions on Chemical Physiology and Pathology." These conditions appear to have been accepted.

178 Sarkowski (1992), 170.

179 Helene Vieweg, the widow of Heinrich Vieweg, to Ferdinand Springer, May 7, 1908, and Springer to Vieweg, May 8, 1908. ASV, B. B. 158. In addition to the 200 marks, the Vieweg publishing house was promised a quarter of a page in the periodical on which to advertise its products for a period of three years.

180 Agreement of December 13, 1905, § 1. ASV, B. B. 158.

181 Slater (1986), 2.

182 Ibid., 2–3.

183 Martin Jacoby (1872–1941), born in Berlin, studied pharmacology and chemistry and worked in Heidelberg until 1906, latterly in the position of an associate professor. In 1907 he was appointed to the Institute of Chemistry at Moabit Hospital, emigrated to Great Britain in 1938 and died in Manchester at the age of 69.

was also bringing up two daughters as a single parent,[184] to be able to cope with it by himself any more. A remark by Charlotte Haber shows that Hela Neuberg supported her husband with great commitment in the editing of the journal right up until her death: "She proofread and corrected the articles that appeared in that periodical. Always ready to work for him whenever he called her."[185]

By the year 1930, Neuberg had edited and published a total of more than 8,500 scientific articles in 230 volumes. If the contributions which he worked on jointly with Jacoby in Volumes 231 (1931) to 264 (1933) and later with Jacoby's successor Wolfgang Grassmann[186] in Volumes 265 (1933) to 281 (1935) are included, the result is an impressive total of over 11,200 articles published in the *Biochemische Zeitschrift* under Neuberg's direction within 30 years.[187] But even this figure still does not take account of the contributions submitted but not published because their content did not fit into the periodical's concept, or which were not accepted because of agreements with the editors of other periodicals.[188]

In 1932, Neuberg himself, in a letter to his former pupil Kurt Jacobsohn, who was by then teaching at the University of Lisbon, commented as follows on the time taken up by his editorial work:

> "In any case, I was enormously pleased at the recognition given to my 26 years of working on the 'Bio.Z.'. It has certainly been hard labor, which has cost me 4 hours a day; that makes up 4.33 years of my life, or, if you calculate on the basis of a working day of only 12 hours, 8.7 years. Perhaps the time could have been spent more usefully in some other way."[189]

184 Hela Neuberg died of leukemia on March 24, 1929, at the age of 45. See also Chapter 5.3.

185 "Sie redigierte und korrigierte die Aufsätze, die in der genannten Zeitschrift erschienen. Immer bereit für ihn zu arbeiten, wenn er sie rief." Haber (1970), 134.

186 Wolfgang Grassmann (1898–1978), after studying and taking his doctorate and *habilitation* in chemistry at the University of Munich, also researched and taught there. In June 1934 he succeeded Max Bergman, who had emigrated in 1933, as Director of the KWI for Leather Research in Dresden. In 1947 he became Professor of Organic and Physical Chemistry in Regensburg, and from 1949 to 1964 Director of the MPI for Protein and Leather Research, which in 1972/73 was merged with the MPI for Biochemistry at Martinsried, near Munich.

187 Jacoby was co-editor from Vol. 231 (November 1930) to Vol. 264 (1933). During this time, 1,644 articles were edited. As from Vol. 265 (July 1933) Jacoby was replaced by Grassmann. From Vol. 280 onwards, Grassmann was sole editor, although documents in the ASV show that Neuberg continued to work on the magazine until the middle of Vol. 281. This represents a further 75 or so articles that he worked on.

188 One example of this is to be found in Kisch's papers: Neuberg had returned two articles that he had submitted for publication with expressions of great regret, since their content was purely physiological. Neuberg to Kisch, September 13, 1928. Staatsbibliothek zu Berlin, Preussischer Kulturbesitz (State Library of the Prussian Cultural Heritage, Berlin) [abbreviated below as SBPK], Manuscripts Department: Kisch papers, K2, sheet 1.

189 "Jedenfalls habe ich mich riesig über die Anerkennung gefreut, welche mein 26jähriges Wirken für die 'Bio.Z.' gefunden hat. Es ist ja gewiß eine Fronarbeit gewesen, die mich täglich 4 Stunden gekostet hat; das sind 4,33 Jahre meines Lebens, oder wenn Sie den Arbeitstag nur mit 12 Stunden ansetzen wollen, 8,7 Jahre. Vielleicht hätte man die Zeit nützlicher anwenden können." Neuberg to Jacobsohn, July 15, 1932, CNP, Box: Jacobsohn, K. P. The "recognition" referred to was Neuberg's election as an External Corresponding Member of the Academy of Sciences in Lisbon on June 16, 1932. Boletim da Academia das Ciências de Lisboa (1932), 324.

But it is not only the wealth of contributions and the large number of authors from all over the world that allow conclusions to be drawn as to the significance of the *Biochemische Zeitschrift* as one of the leading journals in its field. The high quality of the papers submitted is also indicated by the fact that work by no fewer than 18 winners of the Nobel Prize in Medicine and 11 of the Prize in Chemistry was published in Neuberg's journal.[190]

The names of other well-known scientists in the fields of medicine (Kimmelstiel,[191] Chiari,[192] Pick,[193] Liljestrand[194]) and chemistry (Michaelis,[195] Menten,[196] Harden,[197] Young,[198] Henderson[199] and Hasselbalch[200]) also give an idea of how important the *Biochemische Zeitschrift* was for these two disciplines, as well as for the development of biochemistry as a subject in its own right.

190 See Table on the following page.

191 Paul H. Kimmelstiel (1900–1970), a pathologist of Jewish origin, born in Hamburg, emigrated to the United States in 1933 and worked at the Harvard Institute of Pathology. Professor of Pathology in Milwaukee from 1958 to 1966. Together with Clifford Wilson he described diabetic glomerulosclerosis, also known as the *Kimmelstiel-Wilson disease*.

192 Hans Chiari (1851–1916), Viennese pathologist, assistant of Karl Freiherr von Rokitansky (1804–1878), took his *habilitation* in pathological anatomy in 1878. In 1882 he became *extraordinarius* in Prague and was a professor in Strasbourg from 1906 onwards. The *Arnold-Chiari syndrome* was described by him and the Heidelberg pathologist Julius Arnold (1835–1915).

193 Arnold Pick (1851–1924), psychiatrist and neurologist of Jewish descent, studied medicine in Vienna and then became Karl Alexander Westphal's assistant in Berlin. He worked on aphasia with Carl Wernicke. In 1886 he was appointed Professor and Director of the Psychiatric Clinic at the University of Prague.

194 Göran Liljestrand (1886–1968), Swedish pharmacologist, worked with the Nobel Prize winner Ulf von Euler (1905–1983) at the Karolinska Institute, and was secretary to the Nobel Committee for 40 years. The two scientists described the *Euler-Liljestrand mechanism* which is named for them.

195 Leonor Michaelis (1875–1949) studied medicine and became an assistant to Paul Ehrlich. From 1906 onwards he directed the Bacteriological Laboratory at the Charité. In 1922 he became Professor of Biochemistry at the University of Nagoya, Japan, in 1926 at the John Hopkins University and finally in 1929 at the Rockefeller Institute for Medical Research in New York. Michaelis did research in the field of the kinetics of enzymatic reactions.

196 Maud Menten (1879–1960), Canadian biochemist, was one of the first women to study medicine in Canada. In 1912 she did research with Michaelis in Berlin (*Michaelis-Menten equation)*, and from 1923 onwards was a pathologist at the University of Pittsburgh.

197 Arthur Harden (1865–1940), chemist, was awarded the Nobel Prize in Chemistry together with Euler-Chelpin after being proposed by Neuberg in 1929. Cf. also Chapter 8.2.

198 William John Young (1878–1942) was Harden's assistant at the Lister Institute in London. Together they researched into alcoholic fermentation (the *Harden-Young ester)*. Young went to the Institute of Tropical Medicine in Townsville, Queensland, Australia in 1912, and later became Professor of Biochemistry at the University of Melbourne.

199 Lawrence Joseph Henderson (1878–1942) studied medicine, and then went to Strasbourg for two years. In 1904 he was appointed Professor of Biological Chemistry at Harvard. The *Henderson-Hasselbalch equation* (buffer equation) describes the acid/base relationship in the organism.

200 Karl Albert Hasselbalch (1874–1962), Danish physician and chemist, was Director of Finsen's Lysinstitut laboratory in Copenhagen from 1905 to 1917.

NOBEL PRIZE WINNERS IN THE "BIOCHEMISCHE ZEITSCHRIFT"[201]

Nobel Prize	Subject	Name	e.g. Vol. (Year), page
1903	Chemistry	Arrhenius, Svante	11 (1908), 161
1907	Chemistry	Buchner, Eduard	15 (1908), 221
1908	Medicine	Ehrlich, Paul	co-editor from 1906
1913	Medicine	Richet, Charles	11 (1908), 273
1915	Chemistry	Willstätter, Richard	203 (1928), 308
1920	Medicine	Krogh, August	7 (1907), 24
1922	Medicine	Meyerhof, Otto Fritz	33 (1911), 291
1925	Chemistry	Zsigmondy, Richard A.	171 (1926), 198
1926	Chemistry	Svedberg, The (Theodor)	258 (1933), 1
1929	Medicine	Eijkman, Christiaan	11 (1908), 12
1929	Chemistry	Euler-Chelpin, Hans von	41 (1912), 215
1929	Chemistry	Harden, Arthur	32 (1911), 173
1929	Medicine	Hopkins, Frederick G.	159 (1925), 482
1930	Chemistry	Fischer, Hans	27 (1910), 223
1930	Medicine	Landsteiner, Karl	15 (1908), 33
1931	Medicine	Warburg, Otto H.	29 (1910), 414
1936	Medicine	Loewi, Otto	156 (1925), 295
1937	Chemistry	Karrer, Paul	209 (1929), 240
1937	Medicine	Szent-Györgyi von Nagyrapolt, Albert	110 (1920), 117
1943	Medicine	Dam, Henrik C. P.	152 (1924), 101
1943	Chemistry	Hevesy, George de	173 (1926), 175
1945	Medicine	Chain, Ernest Boris	247 (1932), 113
1945	Chemistry	Virtanen, Artturi Ilmari	193 (1928), 300
1947	Medicine	Cori, Carl F. u. Gerty T.	206 (1929), 39
1953	Medicine	Krebs, Hans Adolf	159 (1925), 311
1953	Medicine	Lipmann, Fritz Albert	191 (1927), 442
1955	Medicine	Theorell, Axel H. T.	175 (1926), 297
1958	Medicine	Tatum, Edward Lawrie	267 (1933), 360
1959	Medicine	Ochoa, Severo	227 (1930), 116
1970	Medicine	Euler, Ulf von	194 (1928), 197

201 *Biochemische Zeitschrift*, Vols. 1 to 279. See also Werner (1988).

In his introduction to Volume 100, Neuberg himself interpreted the periodical's success as being the consequence of a correct initiative taken at the right time:

> "The establishment of the *'Biochemische Zeitschrift'* in 1906 fell in a period when this science was just beginning to flourish in all civilized countries. This success is reflected in the sizable number of volumes devoted to the publication of new discoveries. And this development is by no means concluded, unless all the signs deceive us. Most of the branches of biological research have begun to be permeated by chemical aspects, and to be worked on using chemical procedures."[202]

The publisher Ferdinand Springer also had every reason to be thoroughly satisfied with the results achieved. The *Biochemische Zeitschrift* developed into one of the publishing house's best-selling periodicals. Its success was expressed in the figure of 800 subscribers, whereas the publisher's other journals were regularly bought by an average of only 275 people.[203] Springer knew well whom he owed this success to: "I am very well aware that the connection with you was the first rung on the ladder to success in my particular field of publishing activity."[204]

Neuberg received a further token of the publishing house's appreciation of his work after he had been forced to give up the editorship of the periodical in 1935. The circumstances under which he "left" the editorial team will be gone into in more detail later. But it may be pointed out at this point that as late as 1935 the publishers expressed their official appreciation of Neuberg's efforts to restore the international relations that had been disrupted after the First World War:

> "Even during the war, he [Neuberg, Authors' note] succeeded to an astonishingly high degree in maintaining the collaboration of neutral foreign countries on the magazine, and after the war in restoring the broken links very quickly, and creating new ones."[205]

The table of authors from abroad does indeed show that scientists from neutral countries continued to publish in the periodical throughout the entire duration of the war. The first post-war contributions from the victorious enemy powers were received by the editor in 1920 and 1921.

Carl Neuberg regarded his editing of the magazine as being an important part of his life's work, and particularly appreciated the fact that in his editorial activity he was able to act with complete independence – which was no doubt also very

202 "Die Begründung der 'Biochemischen Zeitschrift' im Jahre 1906 ist in eine Periode des Aufblühens dieser Wissenschaft in sämtlichen Kulturstaaten gefallen. Der Aufschwung spiegelt sich in der ansehnlichen Zahl von Bänden, die der Veröffentlichung gewonnener Erkenntnisse dienten. Diese Entwicklung ist, wenn nicht alle Zeichen trügen, keineswegs abgeschlossen. Es hat eine Durchdringung der meisten Zweige biologischer Forschung mit chemischen Betrachtungen und ihre Bearbeitung mit chemischen Verfahren eingesetzt." Biochemische Zeitschrift 100 (1919), 1.

203 Sarkowski (1992), 355, Note to Fig. 353.

204 "Ich bin mir durchaus bewusst, dass die Verbindung mit Ihnen die erste Sprosse auf der Leiter zum Erfolge auf meinem besonderen verlegerischen Betätigungsfelde gewesen ist." Springer to Neuberg, April 4, 1917. ASV, B. B. 158.

205 "Es ist ihm gelungen, auch während des Krieges die Mitarbeit des neutralen Auslandes in einem erstaunlich hohen Maße der Zeitschrift zu erhalten, nach dem Kriege abgerissene Bande sehr rasch wieder herzustellen und neue zu knüpfen." Biochemische Zeitschrift 280 (1935), Preface.

much in tune with his own nature. When later, in the United States, he was confronted with the concept and practice of teamwork and of the *peer review*, which was already customary in the editorial boards of academic journals at that time,[206] Neuberg complained in a letter to Andor Fodor that in his opinion far too much time passed between the submission of a contribution and its publication:

> "The Americans are totally unable to understand that at the 'Bio.Z.' I decided alone, autocratically and authoritarianly, and no later than immediately. The board method, which I consider to be an overboard method, is born of an immanent feeling of cowardice and a fear of taking any responsibility – the attributes of herd animals."[207]

The hard work that Neuberg invested in his editorial activity contributed substantially to his reputation and was also one of the things that academic societies and academies honored him for.[208] The appearance of Volume 250, for example, provided the occasion for the Academy of Sciences in Lisbon to elect Neuberg as an external Corresponding Member.[209] Neuberg's editorship of the periodical was also specially mentioned when he was elected an External Member of the Royal Danish Academy of Sciences. He was proposed for this honor in February 1929 by S. Sörensen, E. Biilman and V. Henriques. The citation mentioned the 205 numbers that had appeared up to then and laid stress on Neuberg's conscientious work, through which had performed meritorious service to the furtherance of biochemical research.[210]

206 The *"board method"* means that an article is first presented to a committee of editors or experts before it can be published. This standard procedure is nowadays known as *"peer review,"* and the acceptance of a scientific paper for publication is considered a seal of quality.

207 "Ganz unverständlich ist es den Amerikanern, daß ich bei der 'Bio.Z.' allein, selbstherrlich und autoritär spätestens sofort entschieden habe. Die Boardmethode, die ich für eine Abortmethode halte, entspringt dem immanenten Gefühle der Feigheit und der Angst vor jeglicher Verantwortung – den Attributen der Herdenviecher." Neuberg to Fodor, July 29, 1947. CNP, Box 5: Fodor, Andor.

208 In addition to the following examples, cf. the documents recording Neuberg's appointment as an Associate Professor of the University of Berlin in 1919 (AHU, Faculty of Philosophy, No. 1439, sheet 187) and his election as an Honorary Member of the Czechoslovak Academy of Agriculture on January 27, 1928. (The election took place on March 18, 1928. Archives of the Czechoslovak Academy of Agriculture in Prague, CAZ, inv. C. 27/N47, k. 109).

209 Cf. Paul Mayer to Kurt Jacobsohn, April 14, 1932, CNP, Box 7: Jacobsohn, Kurt. In this letter, Mayer makes the following suggestion: "Would this anniversary not be a suitable occasion to honor the research scientist and editor Neuberg [...] by awarding him an honorary Doctor of Science or electing him as a member of the Portuguese Academy of Sciences?" (Wäre dieses Jubiläum nicht ein geeigneter Anlaß, den Forscher und Redakteur Neuberg, [...], durch Promotion zum Dr. der Wissenschaften h.c. oder durch die Wahl zum Mitglied der portugiesischen Akademie der Wissenschaften auszuzeichnen?").

210 Archives of the Kongelige Danske Videnskabernes Selskab, Prot. No. 213 (1929): "I Aaret 1906 grundete Neuberg Biochemische Zeitschrift, af hvilket der indtill Dato er udkommet 205 Bind. Neuberg harr i hele dette Tidsrum vaeret Tidskrftet en omhyggelig og aarvaagen Redakteur og har ogsaa herigennem vaeret den biochemiske forskning til stor Nytte."

AUTHORS FROM ABROAD (1913–1922)

Year	Entente States	Central Powers etc.	Neutral States
1913	Belgium, France, Great Britain, USA, Italy, Japan, Russia (and Finland)	Austria-Hungary	Argentina, Denmark, Netherlands, Sweden, Switzerland
1914	Belgium, France, Great Britain, USA, Italy, Japan, Russia (and Finland), USA	Austria-Hungary	Brazil, Denmark, Netherlands, Sweden, Switzerland
1915	Italy, USA	Bulgaria, Austria-Hungary	Denmark, Netherlands, Spain, Sweden, Switzerland
1916		Bulgaria, Austria-Hungary	Denmark, Netherlands, Sweden, Switzerland
1917		Austria-Hungary	Denmark, Netherlands, Sweden
1918		Austria-Hungary	Denmark, Netherlands, Sweden
1919		Austria, Hungary, Czechoslovakia	Denmark, Netherlands, Sweden
1920	Italy (11/8/1920), Japan (11/22/1920)	Austria, Hungary, Czechoslovakia	Denmark, Netherlands, Sweden
1921	Great Britain (10/29/1921), Italy, Russia (4/23/1921), USA (7/28/1921)	Austria, Hungary, Estonia, Latvia, Poland, Czechoslovakia	Denmark, Netherlands, Spain, Sweden
1922	Italy, Japan, Romania, Russia, USA	Austria, Hungary, Latvia, Czechoslovakia	Denmark, Netherlands, Sweden, Switzerland

4.2. THE FORCED END OF NEUBERG'S EDITORSHIP

> "One day, a certain Dr. Salle,[211] the editor of a clinical journal published by Springer, appeared in my laboratory and rudely spoke pretty much the following words: 'On behalf of Dr. Ferdinand Springer I am to request you to cease your editorial work today, since you are no longer tolerable for the Springer publishing house. I suggest you approach Dr. Springer concerning any possible severance payment. You are not in a position to make any demands, so do not ask for compensation, but accept whatever Dr. Springer may offer you later, otherwise you will get nothing.' And with that, this unpleasant gentleman left my room."[212]

This is Carl Neuberg's own description of how he experienced the beginning of the forced end of his work of editing the *Biochemische Zeitschrift*. The circumstances that led to the editorship of the periodical changing hands are described differently by the various people concerned, in accordance with their different points of view. Even the vocabulary used to describe what happened makes this clear: Neuberg described himself as having been "thrown out,"[213] whereas the responsible people at the Springer Verlag referred to his "departure."[214] In the rest of this account, for the sake of form and to avoid making value judgments, the publishing house's choice of expression is used. It should also be pointed out, however, that almost all the documents that could be traced which relate to the circumstances of Neuberg's departure date from the post-war years, when Neuberg was attempting to obtain compensation for the injustice done to him. In the difficult task of assessing the events of that time from today's standpoint, these two circumstances must be borne in mind.

211 Victor Salle (1878–1942), born in Charkov, Russia, studied medicine in Russia and took his doctorate in Basle. Salle should not really have been able to practice medicine in Germany without having taken a German examination. He was, however, entered in the medical register without an official examination, thanks to an intervention by the Professor of Internal Medicine, Wilhelm His the younger (1863–1934). Through the mediation of the Swiss internist Rudolf Staehelin (1875–1943) he was appointed secretary of the *Zentralblatt für Innere Medizin* and from 1923 onwards of the *Klinische Wochenschrift*, both of which were published by Springer. As Sarkowski mentions in 1992, Salle went on to become Ferdinand Springer's private advisor on internal medicine. Cf. Sarkowski (1992), 178, 316, 332.

212 "Eines Tages erschien bei mir im Laboratorium ein Herr Dr. Salle, der Herausgeber eines im Verlag Springer erscheinenden klinischen Journals und sagte mir grob und ziemlich wörtlich: 'Im Auftrag des Herrn Dr. Ferdinand Springer habe ich Sie zu ersuchen, mit dem heutigen Tage Ihre Redakteurtätigkeit einzustellen, da Sie für den Verlag Springer nicht mehr tragbar sind. Ich stelle Ihnen anheim, wegen evtl. Abfindung sich mit Dr. Springer in Verbindung zu setzen. Zu fordern haben Sie nichts, verlangen Sie also keine Entschädigung, sondern nehmen Sie an, was Sie von Dr. Springer später eventuell herausschlagen können, sonst bekommen Sie gar nichts.' Damit verließ der unfreundliche Herr mein Zimmer." Neuberg to A. Hamburger, October 8, 1947. CNP, Box 6: Hamburger, Adolf. Hamburger was an attorney in New York, and first represented Neuberg in the compensation negotiations, which dragged on for five years.

213 "Herausschmiss": Neuberg to Hamburger, October 8, 1947. CNP, Box 6: Hamburger, Adolf.

214 "Ausscheiden": J. Springer to F. Springer, Memorandum of July 31, 1947. ASV, D. 38.7, Neuberg, Carl.

Once the Nazis had come to power, the Springer Verlag found itself under pressure in a variety of respects. On the one hand, the expulsion and emigration of numerous Jewish academics meant that the publishing house lost a large number of authors and editors of textbooks[215] and periodicals, while on the other hand it was itself exposed to suspicion on the part of the new regime of being a Jewish concern. Julius Springer, who was classed as a Jew under the National Socialist racial ideology, withdrew from the business in October 1935. His cousin Ferdinand Springer, who labored under the opprobrium of being a so-called "half-Jew," followed him seven years later in November 1942.[216]

Neuberg, whose authorization to teach at the University of Berlin and the Berlin Agricultural College had been withdrawn in 1934, and who was also faced with increasing difficulties in his position as Director of his Institute,[217] was now similarly subjected to growing pressure in his position as editor.

First of all, his co-editor Martin Jacoby, who like Neuberg was a German citizen of the Jewish faith, was replaced by Wolfgang Grassmann in September 1933, after the appearance of Volume 265. According to Neuberg, this occurred "in response to pressure from Springer, without any internal impulsion"[218] and "on the presumption that it would create more security for me."[219] Whether this change in the editorial team occurred under direct external pressure from party or state organizations, or whether it was effected as a matter of over-obliging obedience, must remain open at this point.

The text that appeared in the imprint of the post-war editions of the *Biochemische Zeitschrift* is rather odd in this respect. In 1957, the preface says:

> "Founded by C. Neuberg in 1908 and edited by C. Neuberg up to Vol. 246. Vols. 247-279 edited by C. Neuberg and W. Grassmann, Vols. 280-317 by W. Grassmann. Berlin, Springer."[220]

The incorrect date for the founding of the periodical was corrected to 1906 two numbers later; this may have been done after Neuberg had pointed out the error[221]

215 E.g. the Springer author Rudolf Höber with his *"Lehrbuch der Physiologie,"* the seventh edition of which appeared as late as 1933. The subsequent editor of Höber's standard work from 1936 onwards was the Göttingen physiologist Hermann Rein with his *"Einführung in die Physiologie."* Cf. also Lohff (2005), 216.

216 See Sarkowski (1992), 325–384.

217 See also Chapter 6.1.

218 "[...] auf Druck von Springer ohne jede innere Veranlassung": Neuberg to A. Hamburger, October 8, 1947. CNP, Box 6: Hamburger, Adolf.

219 "[...] unter der Voraussetzung, dass dadurch eine Sicherheit für mich geschaffen werden sollte." Neuberg to F. Springer, September 7, 1935. ASV, D. 38.7, Neuberg, Carl. This statement is also to be interpreted against the background of Neuberg's situation at the KWI for Biochemistry, which was becoming more and more precarious, and which he continued to direct merely in an acting capacity after his forced retirement.

220 "Begründet 1908 von C. Neuberg und bis Band 246 herausgegeben von C. Neuberg. Band 247–279 herausgegeben von C. Neuberg und W. Grassmann, Band 280–317 von W. Grassmann. Berlin Springer." *Biochemische Zeitschrift* 334 (1957), inside cover of No. 1.

221 Neuberg to Lynen, July 28, 1954. CNP, Box 8: Lynen, Fedor.

to the new co-editor Fedor Lynen.[222] But the date when Grassmann entered the editorial team continued to be placed too early right up until 1967, and is also repeated in the postscript to the final number of the journal.[223] It is unclear whether this misrepresentation arose merely in error, or was done deliberately. The possibility of its having merely been a mistake is rendered less likely by the fact that one of the errors appearing in the text of the imprint was later corrected. Even if it is assumed that it was simply a case of the digits of the volume numbers having been transposed,[224] the name "Jacoby" and not "Grassmann" should have appeared in the text. The fact that Volume 247 appeared in the year 1932, which was not yet politically critical, could indicate a deliberate attempt to mislead: it obscures the impression that the publishing house had bowed to the pressure of National Socialist racial ideology.

Neuberg and Grassmann do not appear to have found it all that easy to collaborate. There are two documents that indicate differences between the editors:

> "It is my impression that Grassmann is beginning to adopt attitudes that are not entirely objective with regard to the acceptance of articles. I don't know if he is doing this on his own initiative, or under influence."[225]

And in another place:

> "I received your manuscript and your letter at the same time. As far as the first is concerned, I am pleased to see your results, and will send the paper to my co-editor Dr. Grassmann immediately, in the hope that he does not make any difficulties about accepting it."[226]

222 Fe[o]dor Felix Konrad Lynen (1911–1979) was Professor of Biochemistry at the University of Munich from 1953 onwards, and was awarded the Nobel Prize in Medicine in 1964, jointly with the American Konrad Emil Bloch (1912–2000), for discoveries relating to the metabolism of cholesterol and other fatty acids.

223 *Biochemische Zeitschrift* 346 (1967), postscript of the publishers.

224 Neuberg had edited the periodical with Jacoby up to Vol. 264 (1933): the publishers gave Vol. 246 (1932) as the last edited by Neuberg alone.

225 "Meinem Eindrucke nach beginnt Grassmann sich bezüglich der Aufnahme von Arbeiten Gesichtspunkte zu eigen zu machen, die nicht ganz sachlich sind. Ich weiß nicht, ob er dies aus eigenem Antriebe oder unter Einflüssen tut." Neuberg to Kisch, February 10, 1934. SBPK, Manuscripts Department: Kisch papers, K3 (a–c), sheet 10.

226 "Gleichzeitig erhielt ich Ihre Manuskriptsendung und Ihren Brief. Was zunächst das erstere anlangt, so freue ich mich über Ihre Ergebnisse und werde die Abhandlung sofort meinem Mitredakteur, Herrn Dr. Grassmann senden in der Hoffnung, daß er bezüglich der Annahme keine Schwierigkeiten macht." Neuberg to Jacobsohn, February 12, 1934. CNP, Box 7: Jacobsohn, Kurt.

The recipients of both letters, Bruno Kisch[227] and Kurt Jacobsohn,[228] were mentioned on the title page of the *Biochemische Zeitschrift* as regular authors and contributors, and like Neuberg were of Jewish origin.[229] It would therefore not be far-fetched to assume that the difficulties that Neuberg mentions could have been concerned with the specific rejection by Grassmann of contributions from Jewish authors. In his appreciation of Neuberg's life's work on the occasion of his 70th birthday, Bruno Kisch himself remarked:

> "Two-hundred-seventy-eight volumes of this journal were edited by himself up to 1936 when he was compelled to give up his leading position in his journal to a Nazi."[230]

In the end, Carl Neuberg was excluded completely from editing the periodical, and its future fate was put entirely into the hands of Wolfgang Grassmann. It is not possible to give an exact date for this occurrence, since the original documents are not to be found either in the ASV or among Neuberg's papers.[231] As already mentioned, the documents that do exist originate almost exclusively from the post-war period, when Neuberg was involved in negotiations about compensation for the injustice that had been done him. From these documents it may be concluded that the Springer Verlag dismissed Neuberg in around August 1935, since the only letter to mention the matter dates from early September of that year.[232] It may have been that Neuberg's dismissal was the result of external pressure: "Neuberg's departure as editor was forced by the Nazi party."[233] And Neuberg wrote in a letter to his lawyer Hamburger:

> "I did of course conduct the subsequent negotiations with Dr. Ferdinand Springer in a civilized manner, and as I also mentioned in my letter of 9.7., I did not hold any grudge against him personally. He explained to me that he was dismissing me on the orders of the *Reichsschrifttumskammer* [the National Writers' Guild in the Nazi period – Translator's note]. In

227 Bruno Kisch (1890–1966), born in Prague as a scion of a famous Jewish family, studied medicine and botany and became Professor of Physical Chemistry and Physiology at the University of Cologne in 1925. He was a renowned cardiovascular researcher. In 1927 he founded the *Deutsche Gesellschaft für Kreislaufforschung* (German Association for Circulation Research) in Bad Nauheim. In 1934 his *venia legendi* was withdrawn, and he emigrated to the USA in 1938. It was only with the greatest difficulty, and after having to repeat all his medical examinations, that Kisch was again able to find a place for himself in clinical research. He did research in the field of cardiac irregularity and calcium metabolism. He described the "overdrive suppression phenomenon." Cf. Schaper/Schaper (1995).

228 Kurt Jacobsohn (1904–1991) took a doctorate in chemistry, was an ancillary assistant at the KWI for Biochemistry and took over the direction of the Department of Biochemistry at Bento da Rocha Cabral Hospital in Lisbon in 1929. He was Professor of Chemistry at the University of Lisbon from 1934 until his retirement.

229 Kisch and Jacobsohn are both mentioned in Neuberg's unpublished article "Jews in Biochemistry." CNP, Box 11: Neuberg, Carl, Manuscript "Jews in Biochemistry."

230 Kisch (1947), 100.

231 Neuberg officially left after the completion of Vol. 279 (1935). However, as the documents in the ASV show, he was still involved in the production of Vol. 280 (September/October 1935) and to some extent also of Vol. 281 (October/November 1935).

232 Neuberg to F. Springer, September 7, 1935. ASV, D. 38.7, Neuberg, Carl.

233 "Neubergs Ausscheiden als Herausgeber ist von der NSDAP erzwungen worden." Memo from J. Springer to F. Springer, July 31, 1947. ASV, D. 38.7, Neuberg, Carl.

my letter of 9.7., I requested a copy of the letter from the *Reichsschrifttumskammer* and an official notice of dismissal from the Springer publishing house, but I never received them – at least, they are not in my files."[234]

Understandably under the prevailing circumstances, in the negotiations following his dismissal Neuberg was only able to obtain part of what he was seeking to achieve. Neuberg had wanted Ferdinand Springer to agree to the following: 1) that he be paid a lump sum of RM 20,000 and the proportion of his editor's salary still outstanding for Volumes 280 and 281; 2) that he be sent two copies of each issue of the *Biochemische Zeitschrift* and one copy each of the periodicals *Pflügers Archiv*, *Naturwissenschaften* and *Berichte der gesamten Physiologie*; 3) that he should be able to obtain books published by Springer Verlag at the author's price; 4) that he be supplied free of charge with offprints of articles by himself and his staff; 5) conditionally, that his name be included in the list of contributors on the title page; and finally 6) financial compensation if the remark "Founded by Carl Neuberg" should be dropped from the journal's title page.[235] In the end, he received a severance payment of RM 17,500 and one free copy of the magazine that he had founded and nurtured.[236] The contractually agreed provision that the phrase "Founded by C. Neuberg" should continue to appear on the title page after his departure was observed up to Volume 295 in 1938. After that it disappeared, as did the list of German and foreign contributors, which was replaced by the phrase "with contributions from numerous experts in the field."[237]

The negotiations on compensation were carried on in a calm atmosphere appropriate to the 30-year-long collaboration. This is repeatedly emphasized in the documents, especially by the Springer side, in connection with the "restitution debate" of the post-war years:

"His [Neuberg's, Authors' note] relations with my publishing house and in particular with my cousin Dr. Ferdinand Springer were close and friendly. It was in this friendly manner that the

234 "Selbstverständlich habe ich in urbaner Form die späteren Verhandlungen mit Dr. Ferdinand Springer geführt und habe gegen ihn persönlich, wie ich auch in meinem Brief vom 7.IX. erwähn(t)e, kein Ressentiment gehegt. Er setzte mir auseinander, daß er mir im Auftrage der Reichsschrifttumskammer kündige. Die im Brief vom 7.IX. erbetene Kopie des Schreibens der Reichsschriftumskammer und die offizielle Kündigung des Verlages Springer, habe ich nie erhalten – jedenfalls ist sie nicht in meinen Akten." Neuberg to A. Hamburger, October 8, 1947. CNP, Box 6: Hamburger, Adolf.

235 Neuberg to F. Springer, September 7, 1935. ASV, D. 38.7, Neuberg, Carl.

236 From a legal opinion of the attorney Meilicke for the Springer Verlag of March 8, 1948. ASV, D. 38.7, Neuberg, Carl. Note: Otherwise than is stated here, Sarkowski writes that Neuberg also received one copy each of the other periodicals he had asked for. See Sarkowski (1992), 393.

237 "[...] unter Mitwirkung zahlreicher Fachgenossen." Cf. *Biochemische Zeitschrift*, title pages of Vols. 280 (September/October 1935) to 296 (March/May 1938).

> discussions regarding a severance payment for Professor Neuberg, which proved to be necessary in 1935, took place […]."[238]

Neuberg's loyal attitude towards his publishers was also expressed in his letter of September 1935 to the Springer publishing house that has already been mentioned:

> "Now that I finally have to leave the circle of your collaborators, I would like to take the opportunity to assure you once again that I will always take the sincerest interest in the affairs of your publishing house and in your personal wellbeing, and remember the always very cordial atmosphere of our thirty-year relationship without any feeling of bitterness."[239]

In correspondence with friends among his colleagues, on the other hand, Neuberg's tone with regard to the publishing house was not without a certain degree of bitterness. He expressed his disappointment at the conduct of the publishing house and of his successor Grassmann openly to Bruno Kisch:

> "I can answer your question quite clearly: of course nothing happened voluntarily from my side, but the matter was induced from higher places. All that is certain is that neither my colleague Gr. nor the publishing house did anything at all to avert the situation. Perhaps the not purely Aryan publishing house was in a position in which the wishes expressed were not even unpleasant to it, although it naturally averred the opposite."[240]

With Kurt Jacobsohn he even discussed what the proper way to react to the events might be:

> "The most important thing would be for all those concerned to have their names deleted from the list of contributors. The French have now informed me that they intend to do so, as they do not wish there to be any appearance of their allowing their names to be used to endorse sanctions against a colleague. There are also some Polish research scientists who share this

238 "Seine Verbindungen zu meinem Verlag und insbesondere meinem Vetter Dr. Ferdinand Springer waren eng und freundschaftlich. In diesem freundschaftlichen Sinne hat dann auch die im Jahre 1935 sich als notwendig erweisende Auseinandersetzung über die Abfindung des Professors Neuberg stattgefunden [...]." J. Springer to A. Hamburger, August 13, 1947. ASV, D. 38.7, Neuberg, Carl.

239 "Die Gelegenheit benütze ich, da ich nunmehr endgültig aus dem Kreise ihrer Mitarbeiter ausscheiden muß, ihnen nochmals die Versicherung zu geben, daß ich an den Interessen Ihres Verlages sowie an Ihrem persönlichen Wohlergehen stets den allerwärmsten Anteil nehmen werde und ohne ein Gefühl der Verbitterung immer freundschaftlichster Gesinnung unserer dreißig Jahre währenden Beziehung eingedenk bleiben." Neuberg to F. Springer, September 7, 1935. ASV, D. 38.7, Neuberg, Carl.

240 "Ihre Frage kann ich klar beantworten, freiwillig ist selbstverständlich von meiner Seite nichts geschehen, sondern die Angelegenheit ist von höherer Seite angeregt [worden]. Sicher ist nur, daß weder mein Kollege Gr. noch der Verlag das geringste getan haben, die Angelegenheit abzuwenden. Der nicht rein arische Verlag ist vielleicht in der Situation gewesen, daß ihm das Ansinnen nicht einmal unangenehm war, obgleich er natürlich das Gegenteil versichert." Neuberg to Kisch, September 30, 1935. SBPK, Manuscripts Department: Kisch papers, K3 (a–c), sheet 12.

> view. That seems to me honorable under the circumstances. I naturally ask you to treat this information as utterly confidential."[241]

If, in view of these remarks, one compares the list of contributors on the title page of Volume 279 (1935), the last to appear under Neuberg's regime, with those of the following volumes,[242] it is noticeable that the names of more and more scientists disappear in the course of time.[243] Whereas in Volume 279, 98 contributors from 20 states were listed, Volume 295 (December 1937) lists only 70 contributors from 17 states.

As compared with 1935, the names of 28 scientists no longer appeared on the periodical's title page two years later. With regard to the representatives of the countries mentioned in the letter quoted above, it is noticeable that almost none of the researchers working in France are still listed, whereas all their Polish colleagues remain on the list.

There is unfortunately no clear evidence as to why so many scientists were no longer listed. On the one hand, it could indeed have been a form of protest and thus a display of solidarity with Neuberg, as is suggested in the letter quoted above. But on the other hand, it is also conceivable that the new editor or the publishers removed some names of Jewish scientists on their own initiative. But this interpretation is less likely, since this would mean that the people responsible had acted selectively.[244] Neuberg himself assumed that the publishing house "will not simply delete the names of all non-Aryans from the title page, if only for business reasons."[245] Regrettably, no documents could be found in the ASV that might have illuminated these circumstances.

If, however, one accepts the supposition that the withdrawal of names from the list of contributors on the title page could have been a form of protest, the conclusion is unavoidable that by no means all scientists joined in this protest. Furthermore, among those scientists whose names continued to appear on the title pages, there were some who were linked to Neuberg by close personal or

241 "Das Wichtigste wäre, daß die interessierten Herren ihre Namen von der Liste der Mitarbeiter streichen lassen. Wie mir jetzt die Franzosen mitgeteilt haben, wollen sie dies tun, weil sie nicht den Anschein erwecken wollen, eine Sanktionsmaßregel gegen einen Kollegen mit ihrem Namen zu decken. Diesen Standpunkt teilen übrigens auch einige polnische Forscher. Mir scheint es nach Lage der Dinge honorig. Selbstverständlich bitte ich Sie, diese Mitteilung als vollkommen vertraulich zu betrachten." Neuberg to Jacobsohn, December 9, 1935. CNP, Box 7: Jacobsohn, Kurt. The letter is signed only "Carolus".

242 The issues concerned are Vols. 280 to 295. Vol. 295 of December 1937 was the last to contain such a list.

243 See Tables, pp. 45-47. An attempt has been made to discover biographical reasons why the scientists concerned might have ceased to contribute to the *Biochemische Zeitschrift*: Such reasons might be religious affiliation (13 of the researchers were of Jewish descent), or official or enforced retirement.

244 A number of Jewish scientists remained on the list of contributors, for example Otto Meyerhof, Otto Warburg, Israel Parnas and the Bohemian scientist Otto von Fürth (1867–1938), who had been the first to describe suprarenin in 1898.

245 "[...] schon aus geschäftlichen Gründen nicht etwa die Namen aller Nichtarier vom Titelblatt streichen werde." Neuberg to F. Springer, September 7, 1935. ASV, D. 38.7, Neuberg, Carl.

academic relationships, so that they might have been expected to display solidarity with Neuberg.[246]

CHANGES IN THE LIST OF CONTRIBUTORS [247]

Country	Contributors up to Vol. 279	Names removed by Vol. 295
Germany	35	10
USA	7	2
Switzerland	5	1
France	4	3
Great Britain	4	2
Denmark	4	2
Netherlands	4	2
Italy	3	1
Czechoslovakia	2	2
Finland	2	1
Palestine	1	1
Belgium	1	1
Russia	6	0
Austria	6	0
Sweden	5	0
Poland	4	0
Japan	2	0
Hungary	1	0
Portugal	1	0
Latvia	1	0
Total	98	28

246 This thought is provoked by the observation that scientists in Germany and abroad reacted in a variety of different ways to the removal, for political and ideological reasons, of a colleague who in the 30 years of his editorship had done such great service to the science they were all involved in, and thus no doubt also to the careers of one or the other of them.

247 Analysis based on the following figures: Vol. 279 (1935), the last volume edited by Neuberg: 98 scientists from 20 states. Vol. 295 (1937), the last volume with a list of contributors: 70 scientists from 17 states.

AUTHORS WITHDRAWING AFTER NEUBERG'S DEPARTURE – PART I [248]

Vol.	Name	Place	Possible reasons for withdrawal
279	Ehrlich, Felix	Breslau	Jewish descent, friend of Neuberg's
279	Fodor, Andor	Jerusalem	Jewish descent, later worked to bring about Neuberg's appointment in Jerusalem
279	Freundlich, Herbert M.	London	Jewish descent, forced resignation 1933, emigration to Great Britain 1933/34
279	Friedmann, Friedrich	Cambridge	Probably of Jewish descent
279	Höber, Rudolf	Philadelphia	Jewish descent, dismissed and emigrated to Great Britain 1933, to the United States 1940
279	Jacoby, Martin	Berlin	Jewish descent, emigrated to Great Britain 1939
279	Laqueur, Ernst	Amsterdam	Probably of Jewish descent, forbidden to teach after German invasion of the Netherlands
279	Loewy, Adolf	Davos	Jewish descent
279	Mayer, Paul	Karlsbad	Jewish descent, close friend of Neuberg's
279	Pringsheim, Ernst G.	Berlin	Jewish descent, emigrated to Great Britain 1939
279	Rona, Peter	Berlin	Jewish descent, dismissed 1933, emigrated to Budapest 1938

248 The volume given is that in which the names were listed for the last time. The last column gives biographical reasons, to the extent that these could be found, which may explain why the person was deleted or had himself deleted from the list. See also Part II of the table.

AUTHORS WITHDRAWING
AFTER NEUBERG'S DEPARTURE – PART II

Vol.	Name	Place	Possible reasons for withdrawal
279	Wohl, Alfred	Danzig	Jewish descent, Neuberg's teacher, premature emeritus status 1933
279	Wohlgemuth, Julius	Berlin	Jewish descent, dismissed 1934 and emigrated to Palestine
280	Snapper, Isodore	Amsterdam	
281	Kisch, Bruno	Cologne	Jewish descent, Neuberg's friend and physician
282	Aubel, E.	Paris	
282	Fromageot, Claude	Lyons	Pupil of Neuberg's, helped Neuberg to escape through France
282	Porcher, Ch.	Paris	
285	Stoklasa, J.	Prague	Proposed Neuberg for honorary membership
286	Bonanni, A.	Rome	Probably went into retirement
286	Flexner, Simon	New York	Given emeritus status in 1935
286	Henri, V.	Liège	
286	Kochmann, Martin	Halle	Jewish descent, suicide while in Nazi imprisonment 1936
286	Lemmermann, O.	Berlin	Retired 1934
286	Orla-Jensen, Sigurd	Copenhagen	
286	Pohl, Julius	Hamburg	Already retired since 1928
286	Wendt, Georg von	Helsinki	
289	Henriques, Valdemar	Copenhagen	

Carl Neuberg, who fled to the USA and was an American citizen from 1946 onwards, was in written contact with many of his colleagues after the Second World War, including those from Germany. This is likely to be how he found out that the Springer Verlag was intending to revive the *Biochemische Zeitschrift*, which had been so successful before the war. As early as the beginning of March 1947, Neuberg wrote in a letter to Adolf Butenandt:

> "I don't even want to raise the question of whether I might not be entitled to claim the editorship again; but I'd be interested to know if they are at least fulfilling the contractual obligation that as long as the journal exists, it must say on the title page: Founded by Carl Neuberg."[249]

In October of the same year, Neuberg informed the publishing house through his attorney Hamburger that he was demanding retrospective compensation for the injustice he had suffered.[250]

One reason for this step may well have been the fact that in the early years after his emigration Neuberg was having to live in very modest circumstances. Secondly, it will become clear that he was concerned less about money than with finding a way to regain access to the European academic literature he was missing so painfully in America. And furthermore, Neuberg took objection on principle to the lack of any will to make restitution that he experienced from the German side. However, the Springer Verlag rejected Neuberg's demands categorically:

> "Neuberg's claims are totally without foundation. There can be no question of royalties, since Neuberg was only the editor, paid a salary volume by volume, while the publishers remained sole proprietors of the periodical. [...] The publishers would be very pleased, in view of the many years of collaboration, to make one copy of each number of the revived *'Biochemische Zeitschrift'* available to Neuberg and to assist him from time to time with other literature that he may be interested in. But any obligation is to be rejected out of hand."[251]

249 "Ich will die Frage gar nicht aufwerfen, ob ich nicht ein Recht hätte, die Redaktion wieder zu beanspruchen, aber interessieren würde mich, ob wenigstens die kontraktliche Verpflichtung gehalten wird, daß solange das Journal besteht, auf dem Titelblatt stehen muß: Begründet von Carl Neuberg." Neuberg to Butenandt, March 7, 1947. CNP, Box 3: Butenandt, Adolf.

250 This appears from a letter from J. Springer to F. Springer dated July 26, 1947, which contains a quotation from Attorney Hamburger's letter to the publishing house: "I represent the interests of Professor Neuberg, who is claiming royalties as the founder and editor of the *'Biochemische Zeitschrift.'* [...] Prof. Carl Neuberg maintains that although he received a severance payment, his acknowledgement of having been adequately compensated is subject to challenge as having been made under the duress arising out of the circumstances of the time." ("Ich vertrete die Interessen von Herrn Prof. Neuberg, welcher Lizenzgebühren als Gründer und Herausgeber der 'Biochemischen Zeitschrift' geltend macht. [...] Prof. Carl Neuberg behauptet, dass er zwar eine Abfindungssumme erhalten hat, dass aber die Ausgleichsquittung als unter dem Zwang der Zeitumstände gegeben, anfechtbar ist.") ASV, D. 38.7, Neuberg, Carl.

251 "Die Ansprüche von Neuberg sind ohne jede Grundlage. Von Lizenzgebühren kann keine Rede sein, da Neuberg bandweise bezahlter Herausgeber, der Verlag alleiniger Besitzer der Zeitschrift war. [...] Der Verlag ist gern bereit, Neuberg mit Rücksicht auf die langjährige gemeinsame Arbeit laufend ein Exemplar der jetzt wiedererscheinenden ‚Biochemischen Zeitschrift' zur Verfügung zu stellen und ihn auch sonst mit für ihn in Betracht kommender Literatur gelegentlich zu unterstützen. Irgend eine Verpflichtung ist aber glatt abzulehnen." J. Springer to F. Springer, memo of July 31, 1947. ASV, D. 38.7, Neuberg, Carl.

At first, the management of the publishing house tried to persuade its former collaborator to withdraw his claim by reminding him that until then he had always been well disposed towards the House of Springer. The publishers tried to present things as if the negotiations on a severance payment that had taken place at the earlier time had not taken place under external duress, and explicitly pointed out that the phrase "Founded by C. Neuberg" did indeed appear on the title page of the recently revived periodical.[252] Neuberg, who is not likely to have felt the events of 1935 to have been particularly free and unconstrained,[253] was not able to share the publishers' view. In any case, he appears to have been concerned not so much about compensation legally due to him, but rather about moral restitution for the injustice he had suffered, and which had been inflicted on him despite his merits and his good relationship with Springer. An indication of this circumstance is to be found in a letter of Neuberg's to his attorney Hamburger:

> "The Springer publishing house, which I have always been in good standing with, owes quite a lot to me over and above this. As far as I know, the 'Albu and Neuberg' was the first medical book that Springer published, on my initiative, and I also hassled old Mr. Springer into the medical business in negotiations that took place still at Mon' Bijouplatz [sic: Authors' note]. I advised the Springer publishing house repeatedly when they were establishing new journals, e.g. *'Naturwissenschaften,'* or about accepting or rejecting books. So I am quite certain that the Springer Verlag, which as long as I had to do with it always acted in a gentlemanly and irreproachable manner, will, on mature consideration, feel the desire to make recompense, and implement it. Since I have still not learned to be a crass materialist, although I was not able to rescue a single penny, I would also accept appropriate and worthy compensation in the form of books."[254]

The publishing house, however, stuck to its formalistic point of view and had a four-page legal opinion drawn up, which came to the conclusion that from a legal

252 J. Springer to A. Hamburger, August 13, 1947. ASV, D. 38.7, Neuberg, Carl.

253 "But there cannot be even the slightest doubt that there can be no question of Professor Neuberg having surrendered his rights voluntarily. No Jewish scholar in the 'Third Reich' voluntarily renounced contractual rights that under normal circumstances would have ensured him a secure livelihood for life." ("Aber deshalb kann doch nicht der geringste Zweifel darüber bestehen, daß von einer freiwilligen Aufgabe seiner Rechte durch Herrn Professor Neuberg keine Rede sein kann. Kein jüdischer Gelehrter hat im ‚Dritten Reich' freiwillig auf Vertragsrechte verzichtet, die ihm unter normalen Umständen auf Lebensdauer ein sicheres Auskommen gesichert hätte[n].") Attorney Barstorf to Springer, October 18, 1948. ASV, D. 38.7, Neuberg, Carl.

254 "Der Verlag Springer, mit dem ich immer gut gestanden habe, verdankt mir über dies noch einiges. Soviel ich weiß, war der Albu-Neuberg das erste medicinische Buch, das Springer auf meine Veranlassung herausbrachte und ich habe namentlich auch den alten Herrn Springer in Verhandlungen, noch am Mon' Bijouplatz auf das medicinische Geschäft gehetzt. Ich habe den Verlag Springer wiederholt bei Gründung neuer Journale, z. B. der ‚Naturwissenschaften', und bei der Annahme oder Ablehnung von Büchern beraten. Daher glaube ich ganz gewiß, daß der Verlag Springer, der solange ich mit ihm zu tun hatte, vornehm und einwandfrei war, bei reiflicher Überlegung den Willen zur Wiedergutmachung fühlen und in die Tat umsetzen wird. Da ich, obwohl ich nicht einen Pfennig gerettet habe, immer noch nicht gelernt habe, ein krasser Materialist zu sein, würde ich eine angemessene und würdige Entschädigung auch in Büchern nehmen." Neuberg to A. Hamburger, October 8, 1947. CNP, Box 6: Hamburger, Adolf.

point of view it need have no fear of having to pay compensation. The points made in this opinion may be summarized as follows:

1. At the present time (1948) no composition settlement is possible, since no law of restitution exists that applies to Berlin.
2. According to a legal precedent, contracts that Jewish citizens seek to challenge are in fact challengeable only from November 9, 1938 (*"Reichskristallnacht"*) onwards.
3. Precise arithmetical calculations are set out to show that, taking into account the value of the free volumes of the *Biochemische Zeitschrift* that he would now receive until the probable time of his death and the severance payment of RM 17,500 that had been paid to him, translated into the present-day value of the currently valid currency, Neuberg had received his editor's fee for the length of the contract he had previously had with the publishing house (up to the end of 1942) without for his part having performed any services for the publishing house, since he had left the editorial team in 1935.[255]

The dispute was declared ended by a decision of the Office of Restitution, which decided in favor of the publishing house and rejected Neuberg's claims.[256]

As has already been mentioned, the relationship between Neuberg and the Springer Verlag during the years of their close collaboration might have been described as particularly good. When Carl Neuberg and Ferdinand Springer launched their academically ambitious and commercially risky project of establishing a new periodical, neither of them was yet 30 years old. It may be assumed that the response to the periodical and its level of sales even during the initial years after its foundation led to a more than purely business-like relationship between the two young men. The publisher's archives, for example, contain documents relating to family affairs of Neuberg's.[257] Springer printed 200 wedding announcements at the request of Neuberg's sister, Anna Salomon, for delivery on May 18, 1907.[258] And some of the passages already mentioned from the correspondence between Neuberg and Ferdinand Springer also indicate a good personal relationship.

Both sides profited from the joint project. In the *Biochemische Zeitschrift* the publishing house had one of its most successful journals, and was able to use it to build up and extend contacts with authors and institutions all over the world. Neuberg, through his editorial activity, enhanced both his academic reputation and his fame and influence. But this work must also have frequently given him additional ideas for his own research activities.[259] Furthermore, the task of editing the periodical also required a wide-ranging study of the literature, through which

255 Legal opinion of Attorney Meilicke, March 8, 1948. ASV, D. 38.7, Neuberg, Carl.

256 J. Springer to Meilicke, February 6, 1952 and June 6, 1952. ASV, E 72.1, Neuberg, Carl.

257 For example, the announcement of the engagement of Hela Lewinski and Carl Neuberg (ASV, N. 32.) or of the death of Neuberg's mother Alma (ASV, B. B. 158).

258 Anna Salomon to F. Springer, May 8, 1907. ASV, B. B. 158. Carl Neuberg married his wife Hela on May 21, 1907.

259 An example is the academic contact to his colleague Otto H. Warburg. Neuberg reported in a letter to the American biochemist Dean Burk, who studied under Warburg for a time after the Second World War: "As editor of the *Biochemische Zeitschrift* I published the greater part of Warburgs publications and discussed them often before they were printed." Neuberg to D. Burk, April 21, 1952. CNP, Box 3: Burk, Dean.

Neuberg acquired an ever deeper knowledge of his subject and at the same time kept himself up to date with the latest developments. Among Neuberg's outstanding characteristics were his remarkable memory, his profound knowledge of biochemistry and his encyclopedic knowledge of the academic literature.

These qualities were much appreciated by his colleagues, who were often able to make use of them.[260] Ines Mandl, for example, recalls:

> "He was always more than generous in sharing his ideas as well as his expertise and to the very end the scientific community of New York gratefully benefited from his encyclopedic knowledge of the literature. I remember many a phone call during the time I worked with him when a colleague would ask him to recall some obscure reference of long ago. Invariably he knew the answer and undoubtedly saved many a scientist valuable time and effort."[261]

The good and friendly relations that Neuberg had cultivated with the publishing house and with Ferdinand Springer in particular were impaired by the unpleasant and wounding dispute after the war. It was not without bitterness that Neuberg complained, in a letter to Ludwig Anschütz,[262] of the lack of any desire to make restitution that he experienced from the German side after the war, saying of the behavior of the Springer publishing house:

> "What might someone not as mild-mannered as I am say when Springer now declares that I laid down the editorship of the '*Biochemische Zeitschrift,*' which I myself had founded, voluntarily. And that is the publishing house which I helped to make great."[263]

When the question of compensation had finally been decided in favor of the publishers, his attitude to them became even more forthright:

> "Dr. Keller passed on to me your request for me to work with you on the Loewi festschrift. I am of course willing to do so in principle; but basically I refuse to publish anything at all in any of Springer's magazines."[264]

And in another letter Neuberg writes:

> "The fact that the old rogue Springer, who is again earning millions, is miserly with offprints is no surprise to me after everything I have experienced throughout the whole of my life. What do they say in Vienna? 'Der Teufel soll ihn holen.' ('Devil take him.')"[265]

260 Cf. also David Nachmansohn's account of an occurrence that demonstrated this quality of Neuberg's. Nachmansohn (1979), 323–325.

261 Mandl (1978), 42.

262 Ludwig Anschütz (1889–1954) was Professor of Organic Chemistry in Brünn (now Brno) from 1930–1945, and later in Würzburg.

263 "Was soll jemand, der nicht so friedlich ist wie ich, dazu sagen, wenn Springer heute erklärt, ich hätte freiwillig die Redaktion der von mir begründeten 'Biochemischen Zeitschrift' niedergelegt. Es ist der Verlag, den ich mit groß gemacht habe." Neuberg to L. Anschütz, February 1, 1949. CNP, Box 2: Anschütz, Ludwig.

264 "Dr. Keller übergab mir Ihre Aufforderung, an der Loewi-Festschrift mitzuarbeiten. Selbstverständlich bin ich im Prinzip dazu bereit. Doch grundsätzlich lehne ich es ab in irgendeiner Zeitschrift des Verlages Springer irgendetwas zu publizieren." Neuberg to H. Häusler, October 23, 1952. CNP, Box 6: Häusler, H. The subsequent correspondence shows that Neuberg did then contribute to the Loewi festschrift with an article by himself and his assistant Amelie Grauer.

In view of the persistence with which Neuberg attempted to prosecute his claim for damages, it is little wonder that the publishing house too no longer displayed such a friendly attitude as before. A memo from Julius Springer to his cousin Ferdinand shows that Neuberg, on one of his visits to Germany in the 1950s, had also visited the publishing house and once again repeated his demands emphatically, although he had previously promised not to do so. At this point Springer shows a great lack of respect, and he does not appear to have any sympathy at all for Neuberg and his demands:

> "He [Neuberg, Authors' note] is 75 years old, still mentally alert, but physically already somewhat decrepit. He is constantly chewing at something, talks at great length and never comes to an end. Despite our differences, he expressly asked me to remember him to you. [...] From an incautious remark that he made, I gathered that he had been saying bad things about us to Warburg."[266]

Such was the inglorious end of 30 years of very successful collaboration between the publisher and the editor of the *Biochemische Zeitschrift*. Certainly, Neuberg was by nature no easy person to negotiate with, and was thoroughly able be very forthright when pursuing his demands, which could easily arouse feelings of resentment in his interlocutor. Despite this, it would have been becoming to the publishing house to have been more obliging towards Neuberg in his difficult situation as an emigrant, at least with regard to the literature that he wanted to have. But no more than it had stood up for its Jewish editor in 1934, in order to keep him in his position, was the Springer concern able to bring itself after the war to accept its share of responsibility for the injustice that Neuberg had suffered. It may be that there were commercial interests that caused the firm to behave in this way.

The post-war volumes of the *Biochemische Zeitschrift* were initially edited by Franz Gottwald Fischer[267] and Konrad Lang.[268] It was the latter who also in-

265 "Daß der alte Gauner Springer, der wieder Millionen verdient, mit Sonderabdrucken knausert, wundert mich nach Erfahrungen, die ich ein ganzes Menschenleben lang gemacht habe, nicht. Wie sagt man in Wien: Der Teufel soll ihn holen." Neuberg to H. Häusler, November 29, 1954. CNP, Box 6: Häusler, H. The remark related to the delay in supplying offprints of the festschrift.

266 "Er ist 75 Jahre alt, geistig noch frisch, körperlich aber schon etwas greisenhaft. Er mümmelt dauernd, redet viel und breit und kann kein Ende finden. Er bat mich trotz unserer Differenzen ausdrücklich, auch Dich von ihm zu grüßen. [...] Daß er uns bei Warburg schlecht gemacht hat, konnte ich aus einer unvorsichtigen Bemerkung heraushören." J. Springer to F. Springer, September 23, 1951. ASV, E. 72.1, Neuberg, Carl.

267 Franz Gottwald Fischer (1902–1960) studied medicine, and then worked with the Nobel Chemistry Prize winner Hans Fischer (1881–1945). In 1928 he discovered the structure of phytol. From 1954 onwards he edited the series *"Theoretische Biochemie – Physikalisch-Chemische Grundlagen der Lebensvorgänge."*

268 Konrad Lang (1898–1985) studied medicine and chemistry in Freiburg, and after taking his doctorate went to Kiel. In 1936 he took his *habilitation* in Berlin, refused to join the Nazi *"Dozentenbund"* ("Faculty League") and was therefore given no further opportunity to work at the university. In 1945 he became Acting Professor of Physiological Chemistry, and from 1946 onwards full Professor, in Mainz.

formed Neuberg in May 1947 that preparations were being made to revive the journal, which he was to manage jointly with Fischer.[269]

Although he was at odds with the publishing house, Neuberg displayed generosity towards the new editors.[270] He corresponded with Fischer and Lang during the post-war years, and even gave them some editorial tips as to how best to boost the sales figures.[271] It was presumably the emotional ties to "his" periodical that persuaded him to act in this way:

> "I shall always be fond of the *Biochemische Zeitschrift*, which I edited for 30 years; it's like a favorite child – you don't go on bearing a grudge against it for having fallen under bad influence for a while."[272]

In 1967, the periodical was finally absorbed by the *European Journal of Biochemistry* as the official publication of the "Federation of European Biochemical Societies,"[273] which was now published in English as the customary international language; published initially by Springer, it now appears under the Blackwell Publishing imprint as *The FEBS Journal*.[274]

269 Fischer to Neuberg, May 13, 1947. CNP, Box 8: Lang, Karl. In this letter Fischer also informed Neuberg that the commitment to print the sub-title "Founded by C. Neuberg" had been honored. The new editors' attention had been drawn to this old agreement by their former tutor at the University of Munich, Heinrich Wieland (1877–1951).

270 "I am all the more pleased because I know that the editorship is in good hands." ("Meine Freude ist umso größer als ich die Redaktion in guten Händen weiß.") Neuberg to K. Lang, June 16, 1947. CNP, Box 8: Lang, Karl.

271 Neuberg recommended that the number of volumes to appear every year should be fixed in advance, as this would increase the chances of selling the magazine, particularly in England. Neuberg to K. Lang, June 16, 1947. CNP, Box 8: Lang, Karl.

272 "An der Biochemischen Zeitschrift, die ich 30 Jahre redigiert habe, werde ich immer hängen, wie an einem Lieblingskind, dem man nicht nachträgt, daß es zeitweise unter schlechtem Einfluß stand." Neuberg to K. Lang, June 16, 1947. CNP, Box 8: Lang, Karl.

273 European Journal of Biochemistry 1 (1967).

274 Information on *The FEBS Journal* [Federation of European Biochemical Societies] can be found on the internet at www.febsjournal.org/, 03/30/2006.

5. NEUBERG AND THE KAISER WILHELM SOCIETY

5.1. THE FOUNDING OF THE KAISER WILHELM SOCIETY

There is probably no other scientific organization in Germany that enjoys such high renown both at home and abroad – not only in the scientific community but among the general public as well – as the *Max-Planck-Gesellschaft zur Förderung der Wissenschaften* (Max Planck Society for the Advancement of Science – MPG). This association, which has existed for over sixty years now,[275] together with its numerous institutes scattered throughout the whole of Germany and the scientists who work there, has for decades been one of the most important pillars upon which the whole structure of scientific research in Germany rests. The high regard in which the MPG is held was also enjoyed to the same, if not to an even greater, extent by its predecessor organization founded in 1911, the *Kaiser-Wilhelm-Gesellschaft zur Förderung der Wissenschaften* (Kaiser Wilhelm Society for the Advancement of Science – KWG).[276] The history of the KWG, its institutes and the persons who were associated with it has been the subject of comprehensive historical research for a number of years now. As an introduction to this chapter, which deals with the close links between Carl Neuberg and the KWG and their importance in Neuberg's life, it is apposite to provide a short summary of the historical context and the history of the KWG.[277]

The emergence of the KWG as a nationally acclaimed scientific organization alongside the research establishments financed by university and state funds was closely linked to the upsurge in the German economy after the creation of the German Empire in 1871 and to the high international standing of German science in the second half of the 19th century. Much of the upper middle class was experiencing growing private affluence at this time, and its members were beginning to amass fortunes. Parallel to this, education and research were undergoing an upturn of unprecedented magnitude, giving rise to numerous discoveries and inventions that established Germany's reputation as a nation of science. This in its turn strengthened the social standing of education and scientific work, giving rise to an increasing willingness to provide financial support to educational and research establishments.[278]

275 See also Heinemann (1990), Brocke/Laitko (1996) and Beyler (2004).

276 For the pre-history and foundation of the KWG, cf. Brocke (1990a). See also Vierhaus/Brocke (1990), Brocke/Laitko (1996), Beyler (2004) and Schieder/Trunk (2004), as well as other volumes of the series on the history of the Kaiser Wilhelm Society during the Nazi period edited by Reinhard Rürup and Wolfgang Schieder for the Presidential Commission of the Max Planck Society since 2000 [www.mpiwg-berlin.mpg.de/KWG/index.htm, 03/02/2006].

277 Unless otherwise stated, the following is based on information from Vierhaus/Brocke (1990).

278 Ringer (1983), Schwabe (1983).

At the same time, the number of students in Germany was increasing incessantly, with dire consequences for a university system that might once have been exemplary, but was no longer in a position to meet the new requirements. At this point in time, the universities could keep up neither with the growing number of students nor with the needs of research, which was moving more and more into the area of applied science. In particular, the professors and lecturers who taught at the universities were so busy that they scarcely found time for research work alongside their teaching duties. Although the financial support provided for research projects by the Imperial German Government had been substantially increased at the end of the 19th century, the funds were nevertheless scarcely adequate to finance in particular experimental research with its enormous requirement for equipment. At the same time, capital expenditure by the state was increasing in order to take account of the increasing fragmentation into more and more narrowly specialized branches of science.

But as has already been stated, in wide areas of the business world and among the prosperous middle classes there was a growing willingness to set up private foundations to contribute to the advancement of science – sometimes in areas corresponding to the personal interests of the donors – through financial support. This type of funding led at the same time to a reduction in the state influence on research, and made it necessary for new forms of organization to be developed. In addition, however, nationalistic aspirations to secure a leading international position for the German Empire in the scientific arena probably exerted an influence that is not to be underestimated on the foundation of the KWG.

On the occasion of the bombastic festivities to celebrate the centenary of the foundation of Berlin University, held in the Great Hall of the Royal Library on October 11, 1910, Kaiser Wilhelm II proclaimed his "desire to establish under My Name and Protection a Society which will make it its task to establish and maintain research establishments."[279] To explain his plans, the Emperor drew attention to the gap that had opened up in Germany between the state of science at the universities and the need for more and more application oriented research, a gap that the establishment of a new independent research society was intended to close. He announced that he had already been promised funds amounting to between nine and ten million marks for this purpose – a stock of starting capital that, as Brocke points out, had been provided predominantly by Jewish citizens, and to a far lesser extent by the German nobility.

The constituent session of the KWG took place on January 11, 1911. The main governing body of the Society was the Senate, which had 20 members, of whom ten were elected by the General Meeting of members and ten appointed by the Emperor. The Senate was responsible for appointing directors of institutes and for electing and monitoring the activities of the President and the Executive Committee. In addition, the individual institutes also had supervisory bodies called "Boards of Trustees" *(Kuratorien)*. The State exercised its influence through ministerial representatives both in the Senate and in the various Boards of

279 Brocke (1990a), 27.

Trustees. The Berlin theology professor Adolf von Harnack[280] was elected first President of the KWG; it was he who had also composed the memorandum submitted to the Emperor in 1909 in which the need to establish the institution and an outline of its structure were explained. Anyone who wished to become an ordinary member of the Society had to pay an admission fee of at least 2,000 marks and an annual contribution of 1,100 marks,[281] which would be the equivalent of about five times these figures in today's euros.

As has already been mentioned in connection with the foundation of the Society, a glance at the sources of donations to the KWG and at the composition of its first Senate, a quarter of whom were of the Jewish faith, demonstrates the great interest of this group in the KWG and their heavy involvement in it. It should be pointed out, however, that in Prussia – as in other parts of Germany too – it was practically impossible for Jewish scientists to attain to full professorships, through which they would have become Prussian civil servants with all the benefit and pension entitlements that this implies. The structures of the KWG, on the other hand, opened up the way for Jewish scientists – at least until the 1930s – to pursue an academic career, as the appointment as Director of an Institute carried with it the title of an Honorary Professor.[282]

As early as October 23, 1912, the first two Institutes of the KWG – the Kaiser Wilhelm Institute (KWI) for Chemistry and the KWI for Physical Chemistry and Electrochemistry – were inaugurated by Kaiser Wilhelm II in the Dahlem district of Berlin. The KWI for Chemistry[283] was headed by Ernst Otto Beckmann[284] as Director, with Richard Martin Willstätter,[285] Lise Meitner[286] and Otto

280 Adolf von Harnack (1851–1930), from 1878 onwards Professor of Protestant Theology in Leipzig, Marburg and Berlin, co-founder of the KWG and its first President from 1911 to 1930.

281 Henning/Kazemi (2002), 21.

282 Willstätter (1949), Glum (1964).

283 On the history of the KWI for Chemistry see Henning/Kazemi (2002), 153–167.

284 Ernst Otto Beckmann (1853–1923) took his doctorate in chemistry in Leipzig and his *habilitation* in Brunswick. He developed the *Beckmann thermometer* with its differential scale; the intramolecular rearrangement of ketoximes also bears his name (*Beckmann rearrangement).*

285 Richard Martin Willstätter (1872–1942), studied chemistry; after his *habilitation* Professor in Munich and Zurich, 1912–1916 Head of the Department of Organic Chemistry at the KWI for Chemistry. In 1915 he became the first KWG scientist to be awarded the Nobel Prize (in Chemistry). In 1925 he resigned from his position as Director of the Institute of Chemistry at Munich University on account of the anti-Semitic attitude of his fellow faculty members, and fled to Switzerland in 1939. Cf. Kazemi (2002), 299–303.

286 Lise Meitner (1878–1968), German-Jewish physicist, worked at the Department of Radioactivity of the KWI for Chemistry from 1913–1915, and from 1917–1938 headed the Physical and Radioactivity Department of the KWI for Chemistry. In 1922 she was the first woman to achieve *habilitation* in physics. She fled to Sweden in 1938. She was awarded the Enrico Fermi Prize in 1966 for her work on nuclear fission with Otto Hahn and Fritz Strassmann (1902–1980), but not the Nobel Prize. See also Vogt (1996).

Hahn[287] as scientific assistants. The later Nobel Prize winner Fritz Haber[288] was appointed Director of the KWI for Physical Chemistry and Electrochemistry; in 1952, the Max Planck Institute for Chemistry was renamed in his honor.[289]

5.2. THE KAISER WILHELM INSTITUTE FOR EXPERIMENTAL THERAPY

On March 19, 1912, on the recommendation[290] of Paul Ehrlich and his pupil August von Wassermann,[291] the Senate of the Kaiser Wilhelm Society passed a resolution establishing the KWI for Experimental Therapy,[292] which was to concern itself with biomedical research projects in the border area between medicine and chemistry.[293] Unlike the two institutes that already existed – the KWI for Chemistry and the KWI for Physical Chemistry and Electrochemistry – the focus of this establishment was to be less on basic research and more on application oriented issues in the fields of infectious diseases and cancer. With this in mind, a Department of Chemistry and a Department of Biology were established under the umbrella of the one institute; they were to devote themselves to chemical and biological investigations respectively, but in close cooperation with each other.

The decision to establish the Institute had, however, been preceded by controversy as to whether it was necessary.[294] The President of the KWG, Adolf von Harnack, was initially against the proposal. In his view, the "purely practical

287 Otto Hahn (1879–1972) was Head of the Department of Radioactivity at the KWI for Chemistry from 1912–1946, and from 1917 also of the Chemical Radioactivity Department. 1914–1918 engaged in First World War work, including Fritz Haber's gas warfare group. 1928–1946 Director of the KWI for Chemistry; 1946–1948 President of the MPG in the British Occupation Zone, 1948–1960 President of the MPG. In 1944 Hahn was awarded the Nobel Prize in Chemistry alone, the work of Meitner and Strassmann not being acknowledged.

288 Fritz Haber (1868–1934), 1911–1933 Director of the KWI for Physical Chemistry and Electrochemistry. 1914–1918 development of numerous types of poison gas weapon, Head of the Central Unit for Chemical Issues in the Ministry of War. Nobel Prize in Chemistry 1918. In early 1919 Haber was wanted by the Allies as a war criminal. Co-founder in 1920 of the *Notgemeinschaft der deutschen Wissenschaft*, the "Emergency Association of German Science," set up to rescue research work in Germany from the catastrophic economic effects of the First World War and the subsequent great inflation, he resigned in 1933 due to the dismissal of Jewish staff, and emigrated to England. Cf. Stoltzenberg (1996), (1998), Kazemi (2002), 100–107 and Schmaltz (2005).

289 For the history of the KWI for Physical Chemistry and Electrochemistry and of the later Fritz Haber Institute of the MPG, see Henning/Kazemi (2002), 122–152.

290 Henning/Kazemi (2002), 189.

291 August von Wassermann (1866–1925), immunologist, a pupil of Robert Koch and Paul Ehrlich. He was concerned amongst other things with infectious diseases, especially syphilis (the Wassermann reaction) and tuberculosis and with research into cancer, and made important contributions to the development of modern immunology. From 1913–1925 he was Director of the KWI for Experimental Therapy. See Chapter 5.2.

292 For the history of the KWI for Experimental Therapy see also Henning/Kazemi (2002), 188–201.

293 Henning/Kazemi (1990), 915.

294 Wendel (1975), in particular Chapter 2c.

objectives" that the Institute was to pursue were irreconcilable with the basic principles of the KWG. Furthermore, he feared that there would be overlapping with the Royal Institute for Experimental Therapy in Frankfurt am Main,[295] of which Paul Ehrlich was the Director.

Despite the skepticism of the KWG's President, the establishment of the KWI for Experimental Therapy was approved, probably on the recommendation of the member of the Senate and moving spirit[296] behind the project, Paul Ehrlich himself. It may be assumed that the decision was also influenced by the particular interest of the Emperor and the military top brass in such a research establishment, since they were hoping for progress in the field of combating infectious diseases. Another far from negligible factor in the decision-making process was the interest of some financially powerful elements in industry who hoped the outcome would be profitable for their businesses.[297]

The basic funding of this third KWI rested on two pillars: The Prussian State Government provided the site alongside the KWI for Chemistry, which was already under construction, and paid the Director's salary of 15,000 marks. The KWG itself guaranteed to fund the actual building – 500,000 marks – and the running costs of the institute – 79,000 marks.[298] In addition, the industrialist Dr. August Oetker,[299] who in view of the nature of his business had an economic interest in biochemical research, furnished an endowment of 100,000 marks, the interest on which was budgeted at around 4,000 marks a year. This money was to be used to finance young scientists on the one hand, and new research projects on the other – on condition that these projects were carried out exclusively at the KWI for Experimental Therapy.[300]

The Institute, which was designed by Privy Councillor of Works Ernst E. von Ihne[301] together with Max Guth,[302] was completed after a building time of only seven months and ceremoniously opened by Emperor Wilhelm II on the occasion of the second Annual General Meeting of the KWG on October 28, 1913.[303] August von Wassermann was appointed first Director of the Institute; until then

295 Wendel (1975), 189.

296 Brocke (1990a), 153.

297 Brocke (1990a), 153, Wendel (1975), 191.

298 Brocke (1990a), 145.

299 August Oetker (1826–1918) took his doctorate in Freiburg after doing an apprenticeship as a pharmacist and studying natural sciences; he was the founder of Dr. Oetker Nährmittelfabrik and the inventor of the baking powder Backin (1891). For the history of the Oetker family and their role in the Third Reich, see Jungbluth (2004).

300 Oetker to Harnack, December 29, 1913. MPG Archives, Dept. I, Rep. 1A, 1446, sheet 2. It should be noted that a further donation of Oetker's will play an important role in the story as part of the *Goldenberg-Oetker Foundation*, cf. Chapter 5.5.

301 Ernst E. von Ihne (1848–1917), ennobled 1906, Wilhelm II's court architect. Ihne also designed the first two KWIs and the Royal Library "Unter den Linden" (today State Library of the Prussian Cultural Heritage – SBPK), the Kaiser Friedrich Museum (today Bode Museum) and the Neuer Marstall. Henning/Kazemi (2002), 21.

302 Max Guth (1859–1925), Berlin architect.

303 Henning/Kazemi (2002), 190.

he had been a Head of Department at the Institute for Infectious Diseases, directed by Robert Koch.[304] With its three-storey main building and its two stabling buildings, supplemented by a porter's lodge with apartments for the assistants,[305] the Institute was one of the most modern research establishments of its kind.[306]

The Department of Chemistry, as Head of which Wassermann proposed Carl Neuberg,[307] took up somewhat more than half of the entire floor area,[308] and was located on the second floor[309] of the main building. Wassermann described Neuberg as being "recognized as the very leading authority in the field of biological chemistry,"[310] whose appointment would be very advantageous to the KWG. In addition, he pointed out that it would not be possible to find any other scientist comparable to Neuberg for the planned salary of 6,000 marks. Neuberg's conditions for agreeing to take up the appointment were that he should be appointed a Scientific Member of the KWG and be given a ten-year contract and the position of Head of Department. In return, he waived any pension entitlement.

One of the main reasons why Neuberg was prepared to give up his job at the Royal Agricultural College and switch to the KWG was that he expected "to obtain more ample scientific success from collaboration with the undersigned [Wassermann]."[311] In addition, it has to be taken into account that professorial chairs specifically for the young subject of biochemistry were still extremely rare in Germany.[312] Furthermore, as has already been mentioned, it was uncommonly difficult for Jewish scientists to pursue academic careers at the universities of the German Empire that were commensurate with their qualifications and abilities. As Richard Willstätter, winner of the Nobel Prize in Chemistry and Head of the Department of Organic Chemistry at the KWI for Chemistry from 1912 to 1916, wrote in his autobiography:

> "It was noticeable that a disproportionately large number of Jewish or non-Aryan research workers, indeed as I see and feel it too large a proportion, were appointed to these institutes that were being created in Dahlem. They were simply available; the universities certainly did not place any obstacles in the way of their being appointed there."[313]

304 Robert Koch (1843–1910) was the first to demonstrate experimentally, by the example of anthrax, the connection between the infectious pathogen and the disease, and discovered, amongst other things, the tuberculosis and cholera germs. Nobel Prize in Medicine 1905.

305 Henning/Kazemi (2002), 190.

306 Dannenberg (1962), 93.

307 Wassermann to Harnack, April 22, 1913. MPG Archives, Dept. I, Rep. 1A, 1445, 124–125.

308 Planck (1936a), 112.

309 Wassermann to Harnack, April 22, 1913. MPG Archives, Dept. I, Rep. 1A, 1445, 124–125.

310 "[...] anerkannte allererste Kraft auf dem Gebiete der biologischen Chemie [...]": Wassermann to Harnack, April 22, 1913. MPG Archives, Dept. I, Rep. 1A, 1445, 124–125.

311 "[...] sich vom Zusammenarbeiten mit dem Unterzeichneten reicheren wissenschaftlichen Erfolg [versprach]". Wassermann to Harnack, April 22, 1913. MPG Archives, Dept. I, Rep. 1A, 1445, 124–125.

312 Krebs/Lipmann (1988), 116. See also Eulner (1970).

313 "Es war bei diesen in Dahlem entstehenden Instituten augenfällig, daß ein unverhältnismäßig großer Anteil jüdischer oder nichtarischer Forscher berufen wurde, nach meiner Ansicht und Empfindung ein zu großer. Sie waren eben zu haben, die Universitäten erschwerten ihre Berufung gewiss nicht." Willstätter (1949), 210. See also Glum (1964), 440.

But as is shown by a letter of Neuberg's written in 1955, he was himself also quite aware of this situation:

> "In October 1913, on the occasion of the inauguration of a Kaiser Wilhelm Institute in Dahlem, a conversation took place between Emperor Wilhelm II, Paul Ehrlich, Wassermann and myself. The Emperor expressed his pleasure that through the establishment of the Kaiser Wilhelm Society he was able to offer a whole number of Jewish scholars, who for whatever reason had not received the advancement they sought at the universities, worthy and lucrative positions at the new institutions."[314]

The Executive Committee of the KWG approved Neuberg's appointment,[315] and in August 1913 a contract of employment was signed between the Society and Carl Neuberg for a period of ten years, in which Neuberg, as Head of the Department of Chemistry, was granted a budget of 15,000 marks for materials and also the appointment of two assistants, a laboratory servant and one ancillary person. He was allowed free choice of his research projects and of the periodicals the results were to be published in; while he for his part undertook to pay the KWG 30 % of the proceeds of any inventions made using the Institute's resources.[316]

On October 28, 1913, the KWI for Experimental Therapy was ceremoniously inaugurated in the presence of Kaiser Wilhelm II. For conducting the Emperor round the Department of Chemistry, Neuberg was to be invested with the *Order of the Red Eagle, 4th Class* during the festivities.[317] The results of medical experiments on animal tumors, the formation of alcohol from sugar-free substances and experiments on photobiotic effects were to be demonstrated to the Emperor during his visit to the laboratories.[318] How much this occasion meant to Neuberg can be seen simply from the detail that he still had a photo of the inauguration of the Institute on display in his New York apartment in the 1950s.[319] It can also be sensed from a letter to the Archives of the KWG that Neuberg attached to a film documenting the ceremony which was in his private possession until his emigration:

> "[The film] was to be shown at my silver wedding; fate has decided otherwise.[320] If you now incorporate it into the archives of the Kaiser Wilhelm Society, I would ask to be allowed to

314 "Im Oktober 1913 fand gelegentlich der Einweihung eines Kaiser-Wilhelm-Institutes in Dahlem eine Unterredung zwischen Kaiser Wilhelm II., Paul Ehrlich, Wassermann und mir statt. Der Kaiser gab seiner Freude darüber Ausdruck, dass durch Gruendung der Kaiser-Wilhelm-Gesellschaft er einer Reihe juedischer Gelehrter, die [–] aus Gruenden welcher Art auch immer [–] an den Hochschulen nicht die gewuenschte Foerderung erhalten haben, an der neuen Stelle [...] wuerdige und auskoemmliche Position bieten koenne." Neuberg to Beitcher, March 7, 1955. CNP, Box 2: Beitcher, Kurt.

315 Harnack to Wassermann, May 7, 1913. MPG Archives, Dept. I, Rep. 1A, 1445, sheet 132.

316 Contract of employment of August 16, 1913. MPG Archives, Dept. I, Rep. 1A, 1445, sheet 160.

317 Neuberg was awarded the Order of the Red Eagle, 4th Class, by the King of Prussia (and German Emperor) on October 21, 1913. CNP, Box 10: Neuberg, Carl, Awards and Diplomas. See also the cover illustration of this volume.

318 Description of the inauguration festivities. MPG Archives, Dept. I, Rep. 1A, 1445, sheet 177–178.

319 Nachmansohn (1979), 323.

320 Neuberg's wife had died in 1929, at the age of 45. See also Chapter 5.3.

borrow it again in the improbable event of anything occurring in my life that is particularly worth celebrating."[321]

5.3. THE "GERMAN OXFORD" – NEUBERG'S LIFE IN DAHLEM

In the south-west of Berlin lies the suburb of Dahlem, which is well known today mainly for the fact that the Free University is now located there. As early as the end of the 19th century a plan was devised to relocate large parts of Berlin University to Dahlem. An area of a thousand acres forming part of the Royal Estates was to be set aside for this new utilization. The influential Minister of Education, Friedrich Althoff,[322] succeeded in getting large areas reserved for a "city of science."[323] In the first thirty years of the 20th century, the foundation of the KWG and its location in Dahlem led to the creation of a research centre that attracted the elite of both national and international science, and which the historian Fritz Stern described as a scientific location *par excellence*: "Dahlem was a kind of German Oxford, rich and often full of eccentric talents [...]."[324]

Thus still today, this place[325] is associated not only with the KWG, but also with a multiplicity of celebrated research workers, including numerous Nobel Prize winners,[326] and a whole wealth of scientific discoveries that were made there. The research centre set up in Dahlem, with its personalities and their scientific achievements, contributed very substantially to the international renown of German science. The exclusion and expulsion of countless outstanding academics during the inhuman Nazi tyranny made painful inroads not only into the KWG, but into all research establishments in Germany.[327] It must, however, be mentioned that in addition to numerous epoch-making discoveries made there even in the very first years of the KWG's existence, during the First World War poison gas weaponry was also developed in Dahlem.[328] The scientists who lived

321 "[Der Film] sollte zu meiner Silberhochzeit laufen, das Schicksal hat anders entschieden. Wenn Sie ihn nunmehr dem Archiv der Kaiser-Wilhelm-Gesellschaft einverleiben, so bitte ich um die Erlaubnis, ihn für den unwahrscheinlichen Fall noch einmal ausleihen zu dürfen, daß ein besonders feierliches Ereignis in meinem Leben eintritt." Neuberg to Cranach, January 16, 1934. MPG Archives, Dept. I, Rep. 1A, 2033, sheet 142.

322 Friedrich Theodor Althoff (1839–1908), lawyer, was called the "Bismarck of higher education." Althoff was one of the leading politicians of the Empire. Cf. also Brocke (1991).

323 Brocke (1991).

324 "Dahlem war eine Art deutsches Oxford, reich, oft voller exzentrischer Begabungen [...]." Mayer (2004).

325 See also Engel (1984). Engel describes at length the development of Dahlem from a Prussian royal estate into a suburb of superior mansions, where many industrialists, bankers, artists, architects, lawyers and academics took up residence. Cf. Engel (1990).

326 See Kazemi (2002).

327 The way the development of natural sciences in America profited from this exodus of German-Jewish scientists is described by Kennedy (2001) and Medawar/Pyke (2001).

328 On the role of the KWI for Physical Chemistry and on research into munitions, see Schmaltz (2005), cf. also Gradmann (1996).

and worked there included Albert Einstein, Fritz Haber, Otto Hahn, Lise Meitner, Otto Meyerhof, Max Planck, Otto Warburg and Richard Willstätter.[329] Carl Neuberg too, having been appointed Head of Department at the KWI for Experimental Therapy, went to live in Dahlem.[330]

As has been briefly mentioned already, Neuberg's father, Julius Neuberg, had moved from Hanover to Berlin with his wife and the two children, Carl and Anna, in 1892. On May 21, 1907,[331] Carl Neuberg married the 23-year-old Franziska Helene Lewinski. Until then he had lived with his parents, but he now moved with his young wife to the Charlottenburg district of Berlin.[332] Like his parents' home in Hanover, both the houses that Neuberg lived in before moving to Dahlem were totally destroyed in the Second World War.

Hela, as Neuberg's wife was known, was born on February 4, 1884 as the second daughter of Judicial Councillor Sigismund Lewinski and his wife Johanna in Posen, now Poznań in present-day Poland. The Lewinski family, like the Neubergs, were adherents of the Jewish faith.[333] Unfortunately, there is little information about her family. Hela's sister, Stephanie Lewinski, married the Berlin doctor Franz Nagelschmidt in 1903.[334] The rest of the family did not come to Berlin until 1904.[335] Hela's brother Karl is thought to have been killed in the First World War.[336] Her mother Johanna died in 1921[337] and her father Sigismund in 1928;[338] both were buried at the Jewish Cemetery in Weissensee, Berlin. In addition, Neuberg's granddaughter Joan Armer was able to contribute the information that Hela's family had originally, like the Neubergs, had a textile business, and must have been fairly wealthy.[339] Neuberg himself seems to have had a close relationship with his wife's parents; shortly before his death, for example, he asked a relative to send him a photo of his "revered parents-in-law."[340]

Very often, little is known of the wives of famous men. There is at least a brief description of Hela Neuberg in the memoirs of her neighbor, Charlotte Haber:[341]

329 This alphabetical list of persons is intended only as an illustration; it makes no claim to completeness and does not represent any evaluation.
330 Cf. Directors' houses: Richard Willstätter, Fritz Haber, Carl Neuberg. See Henning/Kazemi (2002), 115–121.
331 AHU, Neuberg's Personnel File, UK–N 29, Vol. 2, sheet 1. The engagement had been announced in the February. ASV, B. N. 32.
332 The address was Grolmanstrasse 12. Directories of the City of Berlin (1896–1909).
333 Civil status registers of the City of Poznań.
334 Ibid.
335 Ibid.
336 Personal communication from Joan Armer, March 11, 1998.
337 Johanna Lewinski died on February 4, 1921. Cf. Jewish Cemetery, Weissensee, Berlin: Certificate of an entry of death of February 4, 1921.
338 Sigismund Lewinski died on January 19, 1928. Cf. Jewish Cemetery, Weissensee, Berlin: Certificate of an entry of death of January 20, 1928.
339 Personal communication from Joan Armer, March 11, 1998.
340 Neuberg to Sachs, January 10, 1955. CNP, Box 14: Sachs, Helena.
341 Charlotte Haber (1889–1978), second wife of the chemist and Nobel Prize winner Fritz Haber.

"Right from the beginning of my marriage, Hela and her family formed part of the inner circle of my life. She had two extremely beautiful and gifted daughters: Irene and Marianne. Hela herself was no beauty; but her character embellished her features. Her intellectual brow, her dark but unfortunately very short-sighted eyes, her black, not very well tended hair, her warm voice, her visage that radiated goodness and intelligence enchanted her friends. The reason she neglected her external appearance can only have been that she was afraid elegant clothes would even further emphasize her disadvantageous looks."[342]

Apart from these descriptions, Charlotte Haber's recollections are remarkable above all because they show that Hela Neuberg did not see her role as being merely that of a housewife and mother, but was also apparently involved to a considerable extent in the proofreading and correction of the *Biochemische Zeitschrift,* edited by her husband:

"She became his closest collaborator, a major pillar of the 'Zeitschrift für Biochemie' that he edited. She proofread and corrected the articles that appeared in that periodical. Always ready to work for him whenever he called her."[343]

Apart from indications that Hela accompanied Carl Neuberg on numerous lecture tours abroad and went mountaineering with her husband in the Alps, there are no further mentions in his papers of the wife of whom Neuberg wrote in a letter in 1953: "There have only been two people in my life that I have really been close to: my mother […] and secondly my wife, who was carried off by leukemia at the age of 44 [sic]."[344] Hela Neuberg died on March 24, 1929, aged 45,[345] only a year after her father, Sigismund Lewinski. Neuberg initially interred his wife's ashes in the garden of their house in Dahlem.[346] But in 1936, when it became likely that Neuberg would have to sell the house, he had her urn transferred to Weissensee cemetery, where it was interred in the grave of Neuberg's parents.[347]

342 "Gleich zu Beginn meiner Ehe gehörten Hela und ihre Familie zum inneren Kreis meines Lebens. Sie hatte zwei bildschöne und begabte Töchter: Irene und Marianne. Hela selbst war zwar keine Beauté. Allein ihr Charakter verschönte ihre Züge. Ihre Denkerstirn, ihre dunklen, nur leider sehr kurzsichtigen Augen, ihr schwarzes, nicht sehr gepflegtes Haar, ihre warme Stimme, ihr durch Güte und Intelligenz beseeltes Antlitz fesselten ihre Freunde. Ihre äußere Erscheinung hat sie wohl nur darum vernachlässigt, weil sie fürchtete, daß elegante Kleidung ihr unvorteilhaftes Äußeres nur noch betonen würde." Haber (1970), 132–134.

343 "Sie wurde seine engste Mitarbeiterin, Hauptstütze der von ihm herausgegebenen 'Zeitschrift für Biochemie'. Sie redigierte und korrigierte die Aufsätze, die in der genannten Zeitschrift erschienen. Immer bereit für ihn zu arbeiten, wenn er sie rief." Haber (1970), 132–134.

344 "[...] wirklich verbunden bin ich nur zwei Menschen in meinem Leben gewesen: meiner Mutter [...] und zweitens meiner Frau, die mit 44 Jahren von einer Leukämie dahin gerafft wurde." Neuberg to Anschütz, July 12, 1953. CNP, Box 2: Anschütz, Ludwig. Armer expressed the suspicion that her leukemia had been caused by drug treatment for chronic headaches. Written communication from Joan Armer, December 14, 1997.

345 Death certificate of March 25, 1929, Jewish Cemetery, Weissensee, Berlin.

346 In a letter to Neuberg, Arnold Durig expressed his regret that after emigrating to the United States, Neuberg had no wish to return to Berlin and his house in Dahlem. Durig pointed out that the urn containing Neuberg's wife's ashes was buried in the garden of the house. Durig to Neuberg, February 2, 1948. CNP, Box 4: Durig, Arnold.

347 Interment registration of September 30, 1936. Jewish Cemetery, Weissensee, Berlin.

Carl and Hela Neuberg had two daughters: Irene Stephanie Neuberg was born in 1908 and Marianne Neuberg in 1911, both in Berlin. Neuberg took his duties as a father particularly seriously, especially after the death of his wife. Joan Armer, Irene's daughter, reports that he instructed the house servants to report to him precisely what his daughters had been doing when he was away.[348] After attending *Gymnasium*, both daughters went on to higher education. Irene, like her father, chose to study chemistry, and even took her doctorate at his institute in 1932.[349] After that, she moved to the Institut Pasteur in Paris,[350] which would later facilitate her timely emigration to the United States. The younger daughter, Marianne, started to study medicine in Heidelberg, but was not able to complete her studies due to her emigration to the United States. The early escape of both daughters was later to make things somewhat easier for Neuberg during his exile in the US, where Irene in particular played an important role in her father's life.[351]

Little is known of Neuberg's private activities; but it may be assumed that his substantial workload at the KWG, the carrying out and supervision of experiments, the publication of the results and his editorial work for the *Biochemische Zeitschrift* left him with little free time for other activities or the family. Neuberg loved classical literature[352] and historical narratives, as can be seen from his efforts from New York to regain access to his beloved library after the end of the war. He collected stamps, and was also known to be an enthusiastic mountaineer. Although he suffered from attacks of vertigo, he frequently went rambling in the mountains with his wife and friends.[353] Neuberg wrote to his friend Paul Mayer, who also began to have problems keeping his balance in old age: "The only thing that helped with this was environmental therapy, and you know that I myself made high altitude tours in the conviction that you can rest every bit as well in a glacier crevasse as in a cemetery."[354] And the investigations of the effects of high altitude climates on the human organism which Neuberg carried out on Monte

348 Written communication from Joan Armer, December 14, 1997.

349 Neuberg (1932).

350 American Men & Women of Science (1995–1996), 1385.

351 See also Chapter 7.

352 Cf. Nachmansohn (1979), 321.

353 Neuberg wrote, for example, to Helena Sachs, who had reported after the war that her hair had gone gray: "Do you remember that I have already seen [you] with white hair? It was with frost, when we were on the way from Gand through the Martelltal past the Schoentaufspitze to the Schaubachhütte. You were almost frozen, so that my wife pulled stockings over your arms." ("Entsinnen Sie sich, dass ich schon einmal mit weissem Haar gesehen habe? Es war mit Reif als wir von Gand durchs Martelltal vorbei an der Schoentaufspitze nach der Schaubachhuette gingen. Sie waren beinahe erfroren, so dass Ihnen meine Frau Struempfe ueber die Arme zog.") Neuberg to Sachs, March 6, 1946. CNP, Box 14: Sachs, Helena.

354 "Hier hat nur Umgebungstherapie geholfen und Du weisst, dass ich selbst Hochtouren gemacht habe, in der Ueberzeugung, dass man reichlich so gut in einer Gletscherspalte wie auf dem Friedhof ruht." Neuberg to Mayer, January 6, 1943. CNP, Box 9: Mayer, Paul.

Rosa on Tenerife with Nathan Zuntz, Wilhelm Caspari, Joseph Barcroft[355] and Arnold Durig in 1910 were surely not motivated only by "pure academic zeal."[356]

Neuberg's town house, which today houses the Institute of Forensic Medicine of the Free University of Berlin, is one of the more conspicuous buildings in Dahlem. The mansion, surrounded by green, park-like grounds, stands on a site bordered by Hittorfstrasse, Landoltweg and Thielallee. Still today, the exterior alone already gives an idea of how spacious the building is inside. Neuberg had purchased a plot almost 50,000 square feet in area from the *Royal Prussian Commission for the Partition of the Dahlem Estate* for the sum of 112,500 marks on March 6, 1914,[357] with the intention of having a house built on it for himself and his family. The house was designed by the architect Vincent Wisniewski, who had his office on Pariser Platz in the heart of Berlin.[358] The Neuberg family mansion was immediately adjacent to the KWI for Experimental Therapy, and later also to that for Biochemistry, and according to the architect's drawings, the lower and upper floors each consisted of twelve rooms. In addition, there were two bathrooms, a conservatory, a veranda and a big kitchen, so that the family's living area totaled around 6,500 square feet. Then there was a janitor's apartment in the basement and several attic rooms for maids.[359] How many domestic servants the Neubergs had is not known; although their granddaughter Joan Armer remembers her mother telling of at least two maids, a cook and a washerwoman who came to the house occasionally.[360]

Neuberg later stated that the cost of building the house was just as much again as what he had paid for the site.[361] The oak parquet floors, the marble stoves and the library shelves in the former study are still extant in their original condition. Although the house is presently used for other purposes, it is still possible to imagine the kind of style that the Neuberg family lived here in. Carl Neuberg had come to Dahlem to take up the position of Head of the Department of Chemistry at the KWI for Experimental Therapy, and in this capacity he had an annual salary of 6,000 marks.[362] Even with the help of his income from his previous activities, he can scarcely have been in a position to save any great sums. The fact that he

355 Joseph Barcroft (1872–1947), English physiologist, Professor at King's College, Cambridge, wrote important papers on fetal and neonatal medicine. See also Dunn (2000).

356 American Society of European Chemists and Pharmacists (1947), 6. Some of the results of this expedition are presented in Durig et al. (1910).

357 Zehlendorf District Registry, Zehlendorf Local Court [abbreviated below as ZLC]. Land registry, Berlin-Dahlem: Vol. 20, sheet 547, 9–10.

358 Application for a building permit of March 14, 1914. ZLC Zehlendorf, Zehlendorf Town Hall [abbreviated below as ZTH]. Building records, Hittorfstrasse 18, Vol I, sheet 2.

359 Construction plans of the house. ZTH. Building records, Hittorfstr. 18, Vol. I, sheets 21–23.

360 Verbal communication from Joan Armer, March 11, 1998.

361 "I got what I had paid for the land when I bought it from the Dahlem Estate. The building, which was in perfect condition, had cost just as much again." ("[...] ich erzielte den beim Erwerb von der Domäne Dahlem von mir entrichteten Bodenpreis. Das tadellose Gebäude hatte genau ebensoviel gekostet.") Carl Neuberg to Ernst Neuberg, February 6, 1948. CNP, Box 12: Neuberg, Ernst.

362 Employment contract, August 16, 1913. MPG Archives, Dept. I, Rep. 1A, No. 1445, sheets 160–161.

was, nevertheless, able to acquire this property is most easily explained by the assumption that his wife was well-to-do. Although nothing could be discovered in detail about the financial situation of the Lewinski family, both Charlotte Haber[363] and Friedrich Glum refer to the matter in their memoirs. Glum goes on to remark:

> "This case was particularly unpleasant because Neuberg as a person was still very controversial at the time; he was not very highly esteemed even by his Jewish colleagues in Dahlem. He had also committed the *faux pas* of building himself a palatial mansion on the most prominent spot in Dahlem, right alongside the institutes. He must have had a very wealthy wife. Things like that always arouse envy."[364]

From a letter of Neuberg's to the President of the KWG dating from 1929 it is possible to deduce that he may have got a loan to finance the building of the house on favorable terms through family connections with a Jewish bank.[365] How exactly this connection with Bankhaus Comes & Co. came about and from when on it existed is, however, ultimately of no consequence, since it would do no more than confirm that Neuberg had no financial problems.

The study already mentioned contained the library and an attached reading room. Neuberg was a great *amateur* and collector of literary classics.[366] In addition, his library is likely to have contained contemporary works. Although there are no documents dating from the period when Neuberg was living in Hittorfstrasse to substantiate this assumption, it may be assumed that he possessed a comprehensive collection of literature, since in his later correspondence he repeatedly mentions recent books by writers such as Thomas Mann or Stefan Zweig.[367] A large part of his library was moreover made up of his collection of international professional journals. Neuberg informed *Jewish Cultural Reconstruction*, an organization that pursued claims for compensation after the Second

363 "She [Hela Neuberg] came from a very wealthy family." ("Sie stammte aus sehr vermögender Familie.") Haber (1970), 133.

364 "Dieser Fall war besonders unangenehm, da die Persönlichkeit von Neuberg damals noch sehr umstritten war, er auch von seinen jüdischen Kollegen in Dahlem nicht sehr geschätzt wurde. Er hatte auch die Ungeschicklichkeit begangen, sich an der prominentesten Stelle in Dahlem neben den Instituten eine palastartige Villa zu bauen. Er hatte wohl eine sehr vermögende Frau. So etwas erregt immer Neid." Glum (1964), 237.

365 In this letter, Neuberg informs the President that Bankhaus Comes & Co., which a "distant relative" of his was on the Board of, would make a donation towards the construction of the Harnack House. Neuberg to Harnack, February 28, 1929. MPG Archives, Dept. I, Rep. 1A. No. 2032, sheet 52. For the construction and function of the Harnack House, see also Schroeder-Gudehus (1990).

366 Cf. Nachmansohn (1979), 321.

367 Examples of this are to be found in the correspondence with Kurt Beitcher (CNP, Box 2), Maria Kobel (CNP, Box 7) and Marianne Reinold (CNP, Box 13). He obviously had a particular regard for Thomas Mann. In addition, Neuberg's faible for short rhymes indicates that he was a great admirer of the poet Joachim Ringelnatz. One of the best examples is preserved for posterity on a postcard to Bruno Kisch, on which Neuberg reported on his vacation: "Der Schnee ist zwar ein wenig schütter / und tot ist auch mein alter Hund / man spürt das erste Lenzgewitter / man frißt und schläft sich kugelrund." ("The snow may be a bit thin, and my old dog is dead; but you can sense the first spring thunderstorm, you eat and sleep till you're as round as a ball.") Neuberg to Kisch, February 24, 1944. SBPK, Kisch papers, K3 (a–c).

World War on behalf of Jews expelled from Nazi Germany, that among the things he had lost through his emigration was a collection of periodicals consisting of over 1,000 volumes.[368] During his later years in exile in New York, Neuberg grieved particularly for this collection: "I lost not only my household belongings and my library, which I was particularly fond of, but all the property I had."[369]

There are indications that the Neuberg family accommodated foreign scholarship holders at their home from time to time. On the occasion of Neuberg's 75th birthday, for example, the Japanese scientist Masaji Tomita recalled:

> "Thirty years have already passed by since I was fortunate enough to be able to work with you in Dahlem and live with your family. I still have wonderful recollections. Once I went for a walk with Miss Irene and 'Leo' in Grunewald. Once I went to the big department store 'Weltheim' [sic, actually 'Wertheim'; Authors' note] before X-mas with two beautiful girls. Many Japanese of my acquaintance were envious that I was able to live with you so happily for such a long time."[370]

In addition, Joan Armer recalls having been told by her mother Irene that up to 27 cats lived in the house with the Neubergs at different times.[371] His affection for cats is apparent from a letter to a former assistant, Otto von Schönebeck, on the occasion of the death of the tomcat Miko, who lived with the Schönebecks:

> "On the occasion of the death of the Siamese tom, I must express my condolences not only to you but to myself as well. He was a doughty representative of a family I loved, and the last with whom I had a loose connection."[372]

368 This included the complete collections (up to 1939) of *Berichte der Deutsch-Chemischen Gesellschaft, Biochemische Zeitschrift, Chemisches Zentralblatt, Biochemical Journal, Journal of Biological Chemistry, Comptes rendus de la Societé Biologique,* Maly's *Berichte über Tierchemie, Ergebnisse der Physiologie* (Asher and Spiro), *Jahresberichte über die gesamte Physiologie* (Rona) and Hoppe-Seyler's *Zeitschrift für physiologische Chemie.* Neuberg to Jewish Cultural Reconstruction, Inc., May 3, 1951. CNP, Box 7: Jewish Cultural Reconstruction, Inc.

369 "Ich habe nicht nur meinen Hausrat, meine mir besonders am Herzen liegende Bibliothek, sondern auch jegliches Vermögen verloren." Neuberg to Butenandt, March 7, 1947. CNP, Box 3: Butenandt, Adolf. "[...] and all my possessions, including household goods and library, totally lost. I miss the latter most of all." ("[...] und allen Besitz incl. Hausrat und Bibliothek restlos verloren. Letztere fehlt mir am meisten.") Neuberg to Hofmann, February 26, 1948. CNP, Box 6: Hofmann, Eduard.

370 "30 Jahre sind schon verflossen seit ich glücklich in Berlin-Dahlem bei Ihnen arbeiten konnte und mit Ihren Familien lebte. Schöne Erinnerungen bleiben noch in meinem Gedächtnis. Einmal bin ich mit Fräulein Irene und mit 'Leo' in Grunewald spazieren gegangen. Mit zwei schönen Mädchen besuchte ich einmal vor dem X-mas das große Kaufhaus 'Weltheim'. Viele Japaner, meine Bekannten, waren beneidig, dass ich solange bei Ihnen glücklich wohnen konnte." Masaji Tomita, Professor of Biochemistry in Kobe, to Neuberg, December 22, 1953. CNP, Box 15: Tomita, Masaji. A further former pupil received into Neuberg's family circle was, as he himself reports, Bela Szemzö from Budapest. Szemzö to Neuberg, December 16, 1954. CNP, Box 15: Szemzö, Bela.

371 Written communication from Joan Armer, January 14, 1997.

372 "Zum Tode des Siamkaters spreche ich nicht nur Ihnen sondern auch mir selber Beileid aus. Er war ein wackerer und der letzte in loser Verbindung mit mir stehender Vertreter seiner von mir geliebten Familie." Neuberg to Schönebeck, April 20, 1955. CNP, Box 14: Schönebeck, Otto von.

Not much is known about any close contacts between Neuberg's family and their neighbours in Hittorfstrasse. The relationship between Hela Neuberg and Charlotte Haber, the second wife of Fritz Haber, whose house, like Richard Willstatter's, was in the immediate vicinity of Neuberg's,[373] has already been mentioned. Also very close to Neuberg's house, at Hittorfstrasse 23, was the Dahlem Protestant congregation's Church of Jesus Christ, built in 1930 and still extant, in which Martin Niemöller[374] preached at that time. Niemöller, a member of the *Pfarrernotbund* (Pastors' Emergency League)[375] and of the Confessing Church, was a symbolic figure in the resistance among Protestants to the incorporation of their churches into the Nazi system. Niemöller's sermons in Dahlem attracted great attention, and the Church of Jesus Christ was an important meeting place for large parts of the Christian community in those days.[376] Carl Neuberg himself was not untouched by these events; in a letter to his colleague Adolf Windaus he wrote:

> "I recall that shortly after the beginning of the Nazi period, a strong religious movement began, not only among confessing Christians but among Jews as well. I remember particularly vividly that some of my agricultural friends, who were not exactly atheists but had been happy to leave God to take care of himself, often attended the church in Dahlem at that time, though it was my neighbor Niemöller who was the particular attraction. I confess to you that even I, heretic though I am, was there a few times. In addition, my late wife had to do with Frau Niemöller in charity matters."[377]

Under the pressure of the general political situation and his private circumstances, Neuberg was ultimately forced to sell the house and land. Beforehand, in the summer of 1933, he had had the mansion converted into apartments.[378] His wife, as has already been mentioned, had died a few years before, and Irene too had left the family home to work at the Institut Pasteur. Neuberg himself said that he

373 Haber and Willstätter lived in Faradayweg. Cf. Henning/Kazemi (2002), 115–121.

374 Martin Niemöller (1892–1984), 1931 appointed Pastor of Dahlem parish in Berlin; as he rejected the "Aryan Law" for religious reasons, he was removed from office a short time later and prohibited from speaking, which, however, he ignored. Sent to a concentration camp from 1937 to 1945. After 1945 Niemöller was prominent in the German Protestant Church (EKD) and the World Council of Churches.

375 An association of Protestant theologians, pastors and church officials, predecessor organization to the Confessing Church. The League, set up in September 1933 in reaction to the "ecclesiastical" Aryan clauses, saw itself as being in opposition to the "German Christians," and opposed the submission of the Church to the Nazi ideology.

376 Engel (1984), 158.

377 "Ich entsinne mich, dass kurz nach dem Beginn der Nazizeit eine starke religioese Bewegung begann, nicht nur bei Bekennern christlichen Glaubens sondern auch bei Juden. Mir ist besonders lebhaft in Erinnerung, dass einige meiner agrarischen Freunde, die nie gerade Atheisten waren, aber den lieben Gott einen guten Mann sein liessen, damals vielfach in die Dahlemer Kirche kamen, wobei allerdings mein Nachbar Niemoeller einen besonderen Anziehungspunkt bildete. Ich will Ihnen gestehen, dass ich sogar auch einige Male, obgleich Heretiker, da war. Dazu kam, dass meine verstorbene Frau in caritativen Dingen mit Frau Niemoeller zu tun hatte." Neuberg to Windaus, December 23, 1947. CNP, Box 18: Windaus, Adolf.

378 A new bathroom was installed on the first floor, the salon was divided into two rooms and one of the bathrooms on the second floor converted into a kitchen. ZTH, building file Hittorfstrasse 18, Vol. 1, sheets 86, 99–100 and 104.

needed the proceeds from the sale of the house and land in order to be able to pay the "Jewish property levy" imposed by the Nazis.[379] The property was bought by the Berlin-based chemical company Aktiengesellschaft Kali-Chemie, which acquired the plot and the house for 138,000 reichsmarks (RM).[380] A further RM 8,000 was paid for the stoves, a safe and the library shelves. The handover was agreed for October 1, 1938.[381] Kali-Chemie reversed the 1933 conversion[382] and made the prestigious mansion available to its Director General Theodor Feise.[383]

In view of the fact that it was not until 1939 that Neuberg notified the authorities that he was leaving Dahlem, it must be assumed that he continued to live in the suburb after selling his house.[384] The copious correspondence that Neuberg carried on after the Second World War in an attempt to reclaim his possessions makes it clear that after moving out of Hittorfstrasse he had to put the furniture, his library and other valuable items in store in various places.[385]

After the war, Neuberg attempted to obtain compensation for the far too low price he had been paid for the house and land by Kali-Chemie. Through his cousin Ernst Neuberg, who had gone into hiding in Berlin during the war years and survived, he made contact with the company, whose headquarters were by then in Sehnde, near Hanover.[386] The company quickly indicated its readiness to acknowledge Neuberg's "claim to the difference [between] the excessively low

379 Carl Neuberg to Ernst Neuberg, February 6, 1948. CNP, Box 12: Neuberg, Ernst.

380 The "mark" (M) was the German currency from 1871 onwards. In 1923, after the great inflation, it was replaced by the "rentenmark," renamed the "reichsmark" (RM) in 1924. The "reichsmark" collapsed after the Second World War and was replaced (in West Germany) by the "Deutsche mark" (DM) in 1948. Value equivalences can only be approximate, but the following information was given by Norddeutsche Landesbank: M 100 (1914) = RM 125.60 (1938) = DM 195.70 (1950). Neuberg's assessment that the price he was paid represented only the value of the land without the house (cf. Fn. 361) is thus more or less correct; see also Fn. 391 below.

381 ZLC, land registers, Berlin-Dahlem: Vol. 20, sheets 547, 49–50.

382 ZTH, building records, Hittorfstrasse 18, Vol. 1, sheet 133.

383 Directory of the City of Berlin (1939). Theodor Heise, Honorary Doctor of Engineering, was at the same time a member of the Regional Council of Deutsche Bank for Berlin and Brandenburg. Cf. Deutsche Bank (1938), 22.

384 The form with which Neuberg gave notification that was leaving Berlin for Palestine gives his address as In der Halde 14, Dahlem. Police registration certificate of August 1, 1939. CNP, Box 10: Neuberg, Carl, Awards and Diplomas.

385 Neuberg's own account of what became of his property is as follows: furniture and large parts of the library were put in store with a haulage contractor called Fritz Stern [possibly in the Wannsee district of Berlin, Authors' note], other books in the Julius Penner foodstuffs factory in Schöneberg, where Neuberg and Theodor Sabalitschka had set up a laboratory. He attempted to send parts of his private library to the "Friends of the University of Jerusalem," but the consignment was confiscated by the Gestapo. Neuberg to Addinal, May 2, 1946. CNP, Box 1: Addinal, C. R. Neuberg left his stamp collection, china, glass, table silver and rugs in the care of his longstanding friend Felix Ehrlich in Breslau. Nothing more is known of what happened to these things, except that the stamp collection was probably auctioned by the Nazis after Ehrlich's death. Krolof-Becker to Neuberg, August 6, 1946 and October 13, 1947; Neuberg to Krolof-Becker, November 17, 1947. CNP, Box 7: Krolof-Becker, Mignon.

386 Ernst Neuberg to Carl Neuberg, September 29, 1947. CNP, Box 12: Neuberg, Ernst.

price then paid and the fair market price at the time."[387] As a result of payment difficulties, the compensation could not be paid until the company had sold the house to the Free University. It did so at a price of 200,000 Deutsche mark (DM),[388] of which a quarter was paid over to Carl Neuberg as "restitution."[389]

To conclude this matter, it must be pointed out that the information on the circumstances surrounding the sale of the house and the restitution payment after the war are all taken from the land register files and the papers left by Neuberg. According to information given on the phone, the Real Estate Department of the Hanover-based company Solvay AG, the successor to Aktiengesellschaft Kali-Chemie, also has files on the matter. However, although these documents had already been laid out in readiness, permission to examine them was refused "for reasons of data protection" when Neuberg's Jewish origin and the forced nature of the sale of the house in 1938 were mentioned.[390] Kali-Chemie doubtless benefited, when it purchased the house, from the distressed situation that Neuberg was in as a result of the political situation at the time. The purchase of the land and the building of the house had cost him around 225,000 marks in 1914; but he received a mere RM 146,000 when he sold it to the company in 1938. However, it should also be mentioned that according to the value equivalents mentioned above, the price of DM 200,000 that Kali-Chemie received when it finally resold the building in 1950 was even less than what it had paid Neuberg in 1938[391] (a reflection no doubt of the depressed economic situation at the time); that it, nevertheless, paid over DM 50,000 out of that sum to Neuberg may therefore appear fair from the company's point of view, but was still far from "restitution" of the amount that he had lost on the forced sale – although it would in any case scarcely have been possible to recompense him financially for the injustice he had suffered.

Before the Free University of Berlin had the house in Hittorfstrasse rebuilt and extended to accommodate its Institute of Forensic Medicine, it had probably also served other purposes. For example, the Director of the Institute of Forensic Medicine, Professor Volkmar Schneider, reported that the house had been used by SS units for a time.[392] Neuberg himself wrote to an American colleague in 1946:

> "My house situated in Berlin-Dahlem, Hittorfstrasse 18, in the immediate vicinity of the Kaiser-Wilhelm-Institutes, has remained undamaged and, at present, serves as officers' mess to the USA-Forces."[393]

387 Ernst Neuberg to Carl Neuberg, June 6, 1948. CNP, Box 12: Neuberg, Ernst.
388 ZLC, land registers, Berlin-Dahlem: Vol. 20, sheets 547, 81–82.
389 Höchstädt (AG Kali-Chemie) to Neuberg, November 27, 1950. CNP, Box 12: Neuberg, Ernst.
390 Telephone conversation with the Keeper of the Archives at Solvay AG, February 25, 1996.
391 According to the information given in Fn. 380, RM 146,000 (1938) would have been the equivalent of around DM 227,000 in 1950.
392 In 1996, when H. Conrads visited the Institute of Forensic Medicine, Professor Schneider reported to him that when the Free University took over the house, SS runes were found carved into the door jambs.
393 Neuberg to Adams, May 1, 1946. CNP, Box 1: Adams, Roger.

Thus the various owners and users of the house in Hittorfstrasse, taken together, provide a reflection of the highly turbulent history of Germany in the 20th century: Built in the days of the Empire as a luxurious residence for an up-and-coming academic and his family, sold under political pressure at far below value to German industry by a Jewish scientist with an international reputation forced to emigrate during the Nazi dictatorship, and for a time accommodating the SS, the mansion was used as an officers' mess by the American forces after the war. After the establishment of the Federal Republic in divided Germany, it became a university building, accessible to the public, as the Institute of Forensic Medicine of the Free University of Berlin.

5.4. THE FIRST WORLD WAR AND THE GLYCEROL SUBSTITUTES

In 1914, within a year of the opening of the KWI for Experimental Therapy, the First World War broke out, and general mobilization was ordered in Germany. As many employees and some of the directors of the institutes were called up for military service, as they had qualifications that were relevant to the war effort – although in fact most of them volunteered[394] – this had a considerable impact on the KWG and the projects being carried out at the individual institutes, some of which were financed by money from foundations or by funds contributed by industry.[395]

At the KWI for Experimental Therapy, Carl Neuberg, as Director of the Department of Chemistry, took over for a time as Acting Director of the Institute, the Director August von Wassermann having been called up.[396] Neuberg himself, as Friedrich Glum noted,[397] had been declared unfit for military service as he suffered from a pathological disorder of the sense of balance.[398] Despite the shortage of staff, Neuberg was able to continue at least a few of the projects that had already been started with the remaining personnel. At that time he was particularly concerned with investigating the biological effects of sunlight and other types of radiation on the organism, was researching into fermentation processes and the enzymes involved in them, and was also assisting with a number of projects of the Department of Bacteriology. However, these investigations were

394 Henning/Kazemi (2002), 192.

395 For the impact of the war on the KWG and its role in the First World War see Burchardt (1990).

396 Wassermann to Trendelenburg, September 25, 1914. MPG Archives, Dept. I, Rep. 1A, 1446, 34.

397 Friedrich Glum (1891–1974), studied law, from 1920 onwards Secretary General and later Director General of the Kaiser Wilhelm Society. Resigned from all political offices in 1937. See Glum (1964). Cf. also Przyrembel (2004).

398 Glum (1964), 237. In addition, a list dated 1933 of Neuberg's spells of duty at the front (confirmation by Colonel Müller-Zimmermann, Retd., of April 10, 1933) shows that he was declared unfit for military service four times. CNP, Box 10: Neuberg, Carl: Awards and Diplomas.

pushed more and more into the background, since he was increasingly asked to concern himself with research projects important to the war effort.[399]

From 1915 onwards, Neuberg's research work was directed towards developing items of equipment required by the military. He took over, for example, the direction of work on the gas protection devices for horses which the Ministry of War had commissioned from Haber's KWI for Physical Chemistry and Electrochemistry.[400] In a letter to Fritz Haber, Neuberg wrote:

> "I am making myself available to the Ministry of War for the design and testing of the equine gas masks, and will make the necessary space available in my laboratory. I shall perform this task on an honorary basis [...]."[401]

In addition, Neuberg was commissioned to write expert opinions for the military. Having been asked in 1915 to prepare "at the request of various military departments, a series of opinions on the biological effectiveness of the [weapons] used for attack and defense in the present war,"[402] he requested that the costs incurred for "necessary agents" should also be reimbursed. These "necessary agents" were probably substances for chemical weapons,[403] since a number of other scientists at the KWG were also involved in developing such munitions and investigating their effects.[404] The development and testing of means of combating infectious diseases,[405] which had now also become a matter of military importance, also fell within the ambit of the Department of Chemistry of the KWI for Experimental Therapy.

In addition, Neuberg was searching for ways of manufacturing substances that were urgently needed by the armaments industry as substitutes for various raw

399 Reports of the Department of Chemistry of the KWI for Experimental Therapy for the years 1915–1918. MPG Archives, Dept. I, Rep. 1A, 1459, sheets 46, 77, 111.

400 Meffert to Neuberg, December 28, 1916. CNP, Box 7: Kaiser-Wilhelm-Institut für Physikalische Chemie und Electrochemie.

401 "Für die Ausbildung der Pferdegasmaske und ihre Erprobung stelle ich mich dem Kriegsministerium zur Verfügung und halte den notwendigen Raum dafür in meinem Laboratorium bereit. Diese Aufgabe übernehme ich ehrenamtlich [...]." Neuberg to Haber, January 15, 1917. CNP, Box 7: Ministry of War.

402 "[...] auf Ansuchen verschiedener militärischer Ämter im Jahre 1915 eine Reihe von Gutachten über die biologische Wirksamkeit der im gegenwärtigen Kriege zum Angriff und Verteidigung." Reports of the Department of Chemistry of the KWI for Experimental Therapy for the years 1915–1918. MPG Archives, Dept. I, Rep. 1A, 1459, sheet 52.

403 Despite the 1899 Hague Convention on War on Land, under which a number of states had agreed to prohibit "the use of projectiles, the only object of which is the diffusion of asphyxiating or deleterious gases," the German army used some 150 tons of chlorine gas near Ypres in Belgium on April 22, 1915. It was Fritz Haber and his staff at the KWI for Physical Chemistry and Electrochemistry who had made it possible for this terrible weapon to be used. There followed an international arms race and a realignment in the relations between the academic and military spheres and the chemical industry. See Stiller (1995) and Schmaltz (2005).

404 Cf. Otto Hahn's powerful description of the development and testing of poison gas weapons. Hahn (1968), 111–129. See also Willstätter (1949), 237–239 and Schmaltz (2005).

405 An example of this is the development of a chemical therapy to treat typhus, reported on by Fritz Munk. Munk (1916). Cf. also Henning/Kazemi (2002), 192.

materials or for substances derived from them. Glycerol was a chemical compound that had come to be of particular importance in the First World War, being on the one hand required in the manufacture of dynamite, and on the other hand also having a wide range of applications in mechanical engineering, for example as a brake fluid. As the war had led to a shortage of fats that could be broken down to produce glycerol, it was necessary to discover new ways of manufacturing it, or substances with a similar effect. Furthermore, quite apart from the military significance, it was becoming more and more urgent to be able to resort to other source materials, since a large proportion of the fats normally used in the manufacture of glycerol came from abroad.[406] In the course of his wide-ranging research into the chemistry of fermentation, Neuberg had discovered that the natural process of the alcoholic breakdown of sugar could be "trapped," i.e. interrupted, at various points through the addition of certain chemicals.[407] These so-called trapping processes *("Abfangverfahren")* on the one hand helped to explain the sub-processes involved in the alcoholic fermentation of sugar, while on the other hand also making it possible to gain and utilize intermediate products of the process such as glycerol.[408] The process known as "glycerol fermentation," which was kept secret at the request of the German army, was subsequently implemented industrially by Wilhelm Connstein and Karl Lüdecke[409] at the *Vereinigte Chemische Werke AG* in the Charlottenburg district of Berlin.[410]

The physical properties of the substitute substances known as "Perglycerin" and "Perkaglycerin,"[411] for example their boiling points, their heat conductivities, their viscosities and the extent to which, in various degrees of dilution, they lowered the freezing point of water, were investigated by the KWI for Physical Chemistry and Electrochemistry on behalf of the Ministry of War. The results produced by the substitutes were largely the same as those produced with glycerol.[412] There followed numerous investigations of the medical and pharmaceutical applications of the substitute materials: for example, the new substances were used as an ointment base,[413] to preserve[414] tissue samples and as a lubricant[415] for catheters, cystoscopes etc. "Perkaglycerin," for example, proved itself to be "entirely equivalent to glycerol"[416] in obstinate cases of constipation in both animals[417] and humans.[418] Furthermore, most of the other investigations also

406 Connstein/Lüdecke (1919), 403.
407 Hirsch (1922).
408 Hirsch (1922) and Henning/Kazemi (2002), 192–193.
409 In 1919 Connstein and Lüdecke received the Adolf von Baeyer Medal for their research in this field.
410 Neuberg (1936), 338 and Connstein/Lüdecke (1919).
411 Neuberg to the Royal Munitions Factory in Siegburg on April 7, 1916. CNP, Box 6: Goldenberg, Geromont & Co – Chemische Fabrik.
412 CNP, Box 5: Glycerine – Misc. correspondence.
413 Joseph (1916), Nagelschmidt (1916) and Wechselmann (1916).
414 Orth (1916).
415 Posner (1916).
416 "[…] vollkommen dem Glycerin gleichwertig." Albu (1916).
417 Bickel (1916).

showed that Perglycerin or Perkaglycerin had at least equal properties to those of glycerol.

Particularly in the context of the First World War, however, the new materials were even far more significant in military technology. Perglycerin, for example, was used as a coolant for mortars and machine guns, as a brake fluid in all barrel recoil guns[419] and as a medium to keep food hot in field canteens.[420] In the view of the Ministry of War, the wide-ranging uses of the glycerol substitutes made it necessary for Neuberg, who himself, as mentioned above, was unfit for military service, to accompany the "Inspector" of the foot artillery as an expert on visits to the front:

> "As the chemical investigation of the brake fluids actually used at the front is of great importance in maintaining the artillery's fighting capability, the Ministry of War would be grateful to the Institute if it could make it possible for Professor Neuberg to join the Inspector on his visits to the front for a few weeks – starting from 9/15/1917."[421]

For the development of the substitute materials and for his services as an expert, which included instructing the troops in the field in the proper use of Perglycerin, Neuberg was awarded two Iron Crosses.[422] These military decorations meant a great deal to him.[423] Altogether, he was called upon to accompany the Inspector on at least six visits to the front.[424] On the basis of these military expeditions he

418 Boas (1916).

419 As the recoil that occurs when guns are fired can be reduced by the so-called barrel recoil, the speed of the projectile can be substantially increased.

420 Reports of the Department of Chemistry of the KWI for Experimental Therapy for the period April to September 1916. MPG Archives, Dept. I, Rep. 1A, 1459, 77–80.

421 "Da die chemische Untersuchung der an der Front tatsächlich verwendeten Bremsflüssigkeiten von grosser Bedeutung für die Erhaltung der Gefechtskraft der Artillerie ist, wäre dem Institut das Kriegsministerium dankbar, wenn die Teilnahme des Herrn Professor Dr. Neuberg an den Frontreisen des Inspizienten noch für einige Wochen vom – 15. 9. 1917 ab – ermöglicht werden könnte." Ministry of War to the KWI for Experimental Therapy, September 4, 1917. CNP, Box 7: Ministry of War.

422 These were one decoration for service at the front and one for meritorious service at home. Volz/Tönnies (1931). No evidence could be found of a third Iron Cross that Neuberg mentions in a letter after the Second World War: Neuberg to Butenandt, November 12, 1953. CNP, Box 3: Butenandt, Adolf.

423 In a letter to his later attorney, Adolf Hamburger, he asked the latter to get hold of an Iron Cross 1st Class from a junk shop in Germany, as he had lost his own. Neuberg to Hamburger, April 26, 1954. CNP, Box 6: Hamburger, Adolf.

424 The former Inspector of Foot Artillery, Colonel Müller-Zimmermann, Retd., confirmed Neuberg's service at the front on April 10, 1933. According to this document, he was active as a technical expert on various sections of the Western Front from July 1917 to October 1918 (Battle of Flanders, trench warfare on the Western Front, clashes between Arras and Albert, Battles of Soissons and Rheims, fighting before the Ypres-la-Basse front, defensive battle between Cambrai and St. Quentin, defensive battle in Flanders). CNP, Box 10: Neuberg, Carl, Awards and Diplomas.

was classified as a "front-line combatant" after Hitler's seizure of power in 1933, and so was not initially affected by the Nazis' anti-Jewish legislation.[425]

Three firms were selected to produce "Perglycerin" and "Perkaglycerin": Boehringer in Ingelheim, the Chemische Werke Dr. Heinrich Byk in Berlin and Goldenberg, Geromont & Co – Chemische Fabrik (below called Goldenberg) in Winkel. These were ordered by the Ministry of War to ensure the supply of 600 tons of glycerol substitutes.[426] All of the companies earned large profits from the widespread use of Perglycerin and Perkaglycerin.[427] Even the relatively small Goldenberg company,[428] whose General Manager, Oskar Neuberg, was a cousin of Neuberg's,[429] earned a net sum of over 750,000 marks in the 1918 business year.[430] The Goldenberg company paid these profits into a Foundation for the establishment of the KWI for Biochemistry, which was thus of great significance to Carl Neuberg's future at the KWG:

> "Since we had, with your agreement, made the entire profits from the manufacture available to the Kaiser Wilhelm Society for a new research institute, there would be no question of paying you commission on this amount."[431]

The Foundation bore the name *Goldenberg-Oetker-Stiftung*, as the Hildesheim industrialist Dr. Oetker, already mentioned above, had "displayed a lively interest in the establishment of a new Kaiser Wilhelm Institute for Biochemistry" and made a contribution to the Foundation.[432] The endowment of the Foundation amounted to a total of 15 million marks, in addition to which a further 500,000 marks was invested to earn interest that would help to cover the Institute's running costs. As Oskar Neuberg[433] explained:

425 Cf. "Act to Restore the Professional Civil Service" and "Act Amending the Act to Restore the Professional Civil Service." See also Chapter 6.1.

426 CNP, Box 7: Military chemicals.

427 In his report for the period April 1920 to March 1921 Neuberg mentions that the use of this substance had saved the government around 70 billion marks. MPG Archives, Dept. I, Rep. 1A, 1459, sheets 194–196.

428 Becke-Goering (1990), 137.

429 For the family relationships cf. Chapter 2.1.

430 Commission statement for 1918. CNP, Box 6: Goldenberg.

431 "Nachdem wir mit Ihrer Zustimmung den Gesamtgewinn an der Fabrikation der Kaiser-Wilhelm-Gesellschaft für ein neues Forschungsinstitut zur Verfügung gestellt haben, würde eine Verrechnung desselben mit Ihnen nicht in Betracht kommen." Commission statement for 1916 and 1917. Oskar Neuberg to Carl Neuberg, April 5, 1918. CNP, Box 6: Goldenberg.

432 "[…] ein lebhaftes Interesse für die Begründung eines neuen Kaiser-Wilhelm-Institutes für Biochemie an den Tag gelegt." Oskar Neuberg to the Minister of Ecclesiastical and Educational Affairs, October 16, 1917. CNP, Box 6: Goldenberg.

433 From correspondence in 1917 between Theodor Curtius (1857–1928), Professor of Chemistry in Heidelberg, and Carl Duisberg (1861–1935), chemist and industrialist, it emerges that Oskar Neuberg – whom Duisberg wrongly describes as Carl Neuberg's brother – was to be rewarded for his donation with an honorary doctorate. It was not possible to determine whether this actually occurred; the evidence only shows that Oskar Neuberg took a doctorate in Heidelberg in 1890. Cf. Neuberg (1890). In connection with his appointment as a member of the Board of Trustees he is described as Dr. phil. and Dr. med. h.c. MPG Archives, Dept.

> "We would merely associate with the establishment of this Foundation the humble request that we might be granted, for a certain number of years – 20 years would appear to us to be an appropriate length of time – influence with regard to the choice of the Director of the newly established institute. Since the funds to be allocated would for the most part be taken from our this year's earnings from the glycerol substitutes Perglycerin and Perkaglycerin, which have accrued to us as a result of the work of the Director of the Department of Chemistry of the Kaiser Wilhelm Institute for Experimental Therapy, we would be exceedingly grateful to your Excellency if we could be informed before any decision is taken of the reaction of the Ministry of Education and of the Kaiser Wilhelm Society to the plan put forward."[434]

The Senate of the KWG reacted rapidly, and resolved three days later,

> "[...] to attach such an Institute provisionally to the Kaiser Wilhelm Institute for Experimental Therapy and also to postpone the already resolved appointment of Professor Neuberg as its Director until the necessary funds for the establishment of a fully independent Institute are available."[435]

Neuberg's work during these years also attracted great attention in academic circles outside the KWG. The reactions were not always exclusively positive. Carl Duisberg, for example, remarked that he was convinced that "any able chemist who had been asked to produce substitutes for glycerol for the purposes mentioned above"[436] would have solved the problem rapidly. But for Neuberg, the general awareness of his research into glycerol fermentation represented a further academic distinction. In view of his wide ranging and diverse work in the fields of chemistry, biology and medicine, and at the suggestion of Emil Fischer, Ernst von

I, Rep. 1A, 2036, sheet 4. Later too, as a member of the Board of Trustees of the Department of Tobacco Research, he is listed as holding two doctorates. Harnack (1928), 196.

434 "Mit der Errichtung dieser Stiftung würden wir lediglich das ergebene Verlangen verknüpfen, dass uns für eine gewisse Reihe von Jahren – uns scheinen 20 Jahre eine angenehme Frist – ein Einfluß auf die Person des Direktors des neu zu gründenden Instituts vonstatten bleibt. Da die gedachten Mittel der Hauptsache nach bei uns aus dem diesjährigen Verdienst an den Glycerinersatzmitteln Per- und Perkaglycerin genommen werden müßte, der uns durch die Arbeiten des Leiters der chemischen Abteilung am Kaiser-Wilhelm-Institut für Experimentelle Therapie zugeflossen ist, so wären wir eurer Excellenz zu höchstem Danke verpflichtet, wenn vor Ablauf ein Bescheid über die Stellungnahme des Unterrichtsministeriums sowie der Kaiser-Wilhelm-Gesellschaft zu dem angeregten Plane zugehen könnte." Commission statement for 1916 and 1917. Oskar Neuberg to Carl Neuberg, April 5, 1918. CNP, Box 6: Goldenberg.

435 "[...] solches Institut vorläufig dem Kaiser-Wilhelm-Institut für Experimentelle Therapie anzugliedern und die beschlossene Ernennung des Herrn Professor Neuberg zum Direktor noch hinauszuschieben, bis die nötigen Mittel für die Errichtung des Instituts in voller Selbständigkeit vorhanden sind." Minutes of the session of the Senate, October 19, 1917 (transcript). MPG Archives, Dept. I, Rep. 1A, 1446, sheet 174.

436 "[...] jeder tüchtige Chemiker, dem die Aufgabe gestellt worden wäre, für die oben genannten Zwecke Glycerinersatzprodukte zu machen." Becke-Goering (1990), 137.

Bergmann[437] and Walther Nernst,[438] he was appointed an Associate Professor of Berlin University in February 1917.[439]

5.5. NEUBERG'S LONG STRIVING FOR AN INSTITUTE OF HIS OWN

Through his development of substitutes for glycerol, Carl Neuberg had indirectly brought himself closer to the achievement of his aim of setting up an institute that would research exclusively into biochemical issues. The fact that he was related to Oskar Neuberg and the latter's contacts to the Goldenberg company no doubt helped to expedite the realization of this objective, for the Goldenberg company had made large profits out of the industrial manufacture of glycerol substitutes. At the wish of the initiators of the Goldenberg-Oetker Foundation, Neuberg himself was earmarked to be the Director of the new KWG institute. But Germany's defeat in the First World War – the very war which had made the Foundation possible through the enormous requirement for glycerol substitutes – and the subsequent great inflation brought the plans to a standstill.[440]

At the beginning of 1919, Neuberg was offered a position at the University of Lund in Sweden.[441] The memorandum of a meeting held at the KWG to discuss this shows that Neuberg would have been very reluctant to leave Berlin. He felt himself attached to Dahlem, and he enjoyed the proximity of the institute to the house[442] he had moved into with his family only a few years before. Furthermore, he appreciated the proximity of colleagues such as Otto H. Warburg[443] and Otto Meyerhof, who were concerned with similar issues, and last but not least, his

437 Ernst von Bergmann (1836–1907) was one of the most important physicians and surgeons of his day. He succeeded Bernhard von Langenbeck as Professor of Surgery in Berlin in 1892. He was in particular involved in establishing the use of antiseptics and in the introduction of aseptic conditions and new operating procedures.

438 Walther Nernst (1864–1941) is considered to be a pioneer of physical chemistry. In 1920 he was awarded the Nobel Prize in Chemistry in recognition of his work in thermochemistry. He was involved in the establishment of the KWG and a Senator of the KWG from 1919 to 1933. Among other things, the Nernst Equation for electrochemical reaction (1888), the Nernst Distribution Law (1891) and the Nernst Heat Theorem, later known as the Third Law of Thermodynamics (1905), are named for him.

439 Request for the appointment of Carl Neuberg as an "associate professor." Faculty of Philosophy to the Prussian Minister of Ecclesiastical and Educational Affairs, December 12, 1916, and notification of the appointment by the Minister, February 13, 1917. AHU, Faculty of Philosophy, No. 1439, sheets 187–193 and 194.

440 Henning/Kazemi (2002), 194.

441 Memorandum of a meeting on February 22, 1919, relating to Professor Neuberg's appointment to the Swedish University of Lund. MPG Archives, Dept. I, Rep. 1A, 1455, sheet 6.

442 Cf. Chapter 5.3.

443 Otto Heinrich Warburg (1883–1970), cell physiologist and biochemist, 1931 Nobel Prize in Physiology and Medicine. 1914 Head of the Department of Physiology at the KWI for Biology, from 1930 Director of this KWI, from 1953 of the MPI for Cell Physiology. He had close personal contact with Neuberg which continued after the Second World War. See also Werner (1996), Höxtermann (2001). Warburg was the only Jewish director of a KWI who was able to remain in Germany throughout the "Third Reich." Macrakis (1994), 276–278.

position as editor of the *Biochemische Zeitschrift,* which was published by the Springer Verlag in Berlin,[444] presumably also played a role. In addition, Lund made it a condition that he should take Swedish citizenship, which would have been a very difficult decision for Neuberg, who was deeply attached to his native country.[445]

Although the KWG and ministry representatives who were involved in the negotiations[446] assured Neuberg that they would have greatly regretted it if he had accepted the position in Sweden, they were not able to make him any binding promises with regard to the planned institute and his future position. Instead, Neuberg was offered formal assistance with the implementation of the project if he himself was able to make all the financial arrangements:

> "The gentlemen present expressed the expectation that if Professor Neuberg were to succeed in raising such a sum for the new institute as to make it viable and allow building work to be commenced, the Senate of the Kaiser Wilhelm Society would establish the institute as a Kaiser Wilhelm Institute and propose that Professor Neuberg be appointed Director."[447]

The fact that he was able to insist on having his salary doubled,[448] which had been refused him only three months before,[449] may be viewed as a concession to a scientist who in the meantime had acquired an international reputation. His appointment as a full Professor in August 1919 may also be viewed in the same light,[450] and will have been part of the attempt to dissuade Neuberg from moving to Lund. The Minister of Science, the Arts and National Education, in a letter to the Faculty of Philosophy where Neuberg had achieved his *habilitation* in 1903, wrote:

> "As it is to be considered a matter of particular importance that Professor Neuberg should remain in his position at the Institute for Experimental Therapy, I request the Faculty to respond to this proposal at today's meeting."[451]

444 Cf. Chapter 4.

445 Memorandum of a meeting on February 22, 1919, relating to Professor Neuberg's appointment to the Swedish University of Lund. MPG Archives, Dept. I, Rep. 1A, 1455, sheet 6.

446 These were Adolf von Harnack (President of the Kaiser Wilhelm Society), Friedrich Glum (Secretary General), Emil Fischer (2nd Vice President of the Administrative Committee) and two ministry representatives.

447 "Die anwesenden Herren gaben jedoch der Erwartung Ausdruck, dass, wenn es Herrn Prof. Neuberg gelänge, für das zu gründende Institut eine solche Summe aufzubringen, dass es lebensfähig sei und mit dem Bau begonnen werden könne, der Senat der Kaiser-Wilhelm-Gesellschaft das Institut als Kaiser-Wilhelm-Institut begründen und die Ernennung von Herrn Prof. Neuberg zum Direktor desselben beantragen würde." Memorandum of February 22, 1919. MPG Archives, Dept. I, Rep. 1A, 1455, sheet 6.

448 The additional remuneration was covered by the Ministry of Education. Minutes of the meeting of the Administrative Committee of the KWG on March 21, 1919 (transcript). MPG Archives, Dept. I, Rep. 1A, 1446, sheet 240 b.

449 Harnack to Wassermann, January 22, 1919. MPG Archives, Dept. I, Rep. 1A, 1446, sheet 235.

450 Appointment as a full Professor on August 2, 1919, by the Minister of Science, the Arts and National Education. CNP, Box 10: Neuberg, Carl, Awards and Diploma.

451 "Da besonderer Wert darauf zu legen ist, dass Professor Neuberg seiner Stellung beim Institut für Experimentelle Therapie erhalten bleibt, ersuche ich die Fakultät, noch in der heutigen Sitzung zu dieser Anregung Stellung zu nehmen." Letter from the Minister of

In addition to the offer from Lund, however, Neuberg also found that he had another opportunity to make the institute that he wished to see established a reality. In June 1919 he also received an offer from the *Landwirtschaftliche Hochschule* (Agricultural College) in Poppelsdorf, Bonn.[452] Here he was offered the opportunity to establish an Institute of Biochemistry associated with the College's Department of Chemistry. It would have cost less to build an institute in Bonn than in Berlin, so that a greater proportion of the Foundation's endowment would have been available for research work proper. A letter from the Goldenberg chemical company to the President of the KWG indicates that the initiators of the Foundation were primarily concerned "with the rapid undertaking of work already to hand," and that the Foundation's interest income was in any case to go to Neuberg, irrespective of where he would be working in future.[453] However, in view of the uncertain political situation in the Rhineland after the First World War, both the members of the Foundation and Neuberg himself had second thoughts about the plan, and Neuberg asked for more time to consider.[454]

Subsequently, those responsible at the KWG discussed a variety of possible ways in which the plans for the new institute might be realized. As the documentation of these discussions is incomplete, the course they followed can only be roughly reconstructed.[455] Firstly, there was the possibility that Neuberg might move into the premises of the *Centralstelle für wissenschaftlich-technische Untersuchungen GmbH* in the Neubabelsberg district of Potsdam,[456] together with the newly established KWI for Metals Research.[457] But Neuberg was well aware of the advantages of being close to the other KWG institutes in Dahlem, and did not want to leave the "German Oxford."[458] At about the same time as the plans for Neubabelsberg, the Senate of the KWG at last formally resolved that a KWI for Biochemistry should be established with Neuberg as its Director.[459] When Hans Spemann[460] then gave up his position as Deputy Director of the KWI for Biology,

Science, the Arts and National Education to the Faculty of Philosophy, July 24, 1919: AHU, Faculty of Philosophy, No. 1468, sheet 163.

452 Minutes of the Senate, June 3, 1919 (transcript). MPG Archives, Dept. I, Rep. 1A, 1447, sheets 4–5.

453 Goldenberg to Harnack, May 17, 1919. CNP, Box 6: Goldenberg.

454 Minutes of the Senate, June 3, 1919 (transcript). MPG Archives, Dept. I, Rep. 1A, 1447, sheets 4–5. The Treaty of Versailles, which declared the Rhineland to be a demilitarized zone and placed it temporarily under allied occupation, was signed by the German delegation on June 28, 1919.

455 See also Henning/Kazemi (2002), 188–201.

456 Harnack to Neuberg, May 17, 1920. MPG Archives, Dept. I, Rep. 1A, 1447, sheet 55. The building had been used as a research institute by leading companies in the armaments industry since 1898. In 1921 it was occupied by the KWI for Metals Research. Vierhaus/Brocke (1991), 243.

457 See Henning/Kazemi (2002), 180–187.

458 Engel (1982), 12.

459 Harnack to Neuberg, May 17, 1920. MPG Archives, Dept. I, Rep. 1A, 1447, sheet 55. Report on the resolution of the Senate of the KWG of May 11, 1920.

460 Hans Spemann (1869–1941), an important "developmental physiologist" (experimental embryologist), since 1914 Head of the Department of Animal Developmental Mechanics and

which was also in Dahlem, in order to take up a position at the University of Freiburg,[461] Neuberg proposed to the Secretary General of the KWG that he himself should occupy the vacant position:

> "If you, Secretary General, would by the way also be good enough to pursue the idea that it would be very beneficial to the rapid installation of the Institute if I were, so to speak, to slip into Spemann's position, I would be sincerely grateful to you. Privy Councillor Haber, together with probably all the experts, shares my opinion that for economic reasons as well, biochemistry deserves to be paid close attention to in the near future."[462]

We can only speculate about the precise intentions behind Neuberg's proposal. Possibly he hoped that by taking over "Spemann's position" he would gradually, by roundabout ways, be able to establish an independent institute that would be concerned exclusively with research into biochemical issues. Unfortunately, we do not have any direct response from the Secretary General of the KWG to Neuberg's proposal. But Richard Goldschmidt[463] was appointed Deputy Director of the KWI for Biology, while the position of Head of the Department of Developmental Mechanics and Causal Morphology at the KWI for Biology, also previously held by Spemann, was given to Otto Mangold.[464, 465] There do not appear to have been any further efforts to establish an autonomous Institute for Biochemistry in the following two years. Despite the frustration that Neuberg must have felt at the meager outcome of his efforts, the Department of Chemistry at the KWI for Experimental Therapy continued to perform at a very high level: Between April 1920 and March 1922, for example, the Department produced 46 publications.[466]

It was only in the course of negotiations on the establishment of a Department of Tobacco Research[467] in June 1922 that renewed efforts were made to bring

Deputy Director of the KWI for Biology. 1919–1937 Professor for Zoology in Freiburg, 1935 Nobel Prize in Medicine. See also Mangold (1953), Fäßler (1997) and Kazemi (2002), 255–258.

461 Mangold (1953), 40.

462 "Wenn Sie, hochverehrter Herr Generalsekretär, nun nebenbei auch noch den Gedanken zu verfolgen die Güte hätten, dass sozusagen ein Einrücken meinerseits in die Spemann'sche Stelle die schnelle Installation des Instituts sehr fördern würde, wäre ich Ihnen aufrichtig dankbar. Herr Geheimrat Haber ist, wie wohl alle Sachverständigen, meiner Meinung, dass auch aus wirtschaftlichen Gründen für die nächste Zeit die Biochemie eine starke Berücksichtigung verdient." Neuberg to Glum, May 28, 1920 (extract). MPG Archives, Dept. I, Rep. 1A, 1447, sheet 57a.

463 Richard Goldschmidt (1878–1958) studied biology and medicine, and was an Associate Professor in Munich from 1909 onwards and Head of Department at the KWI for Biology from 1919, later also its Deputy Director. He emigrated to the USA in 1935, where he became Professor of Genetics and Cytology at the University of Berkeley. See also Satzinger (2004b).

464 Otto Mangold (1891–1962), Spemann's assistant. From 1923 Head of Department at the KWI for Biology; 1933 Professor in Erlangen and 1937 Professor of Zoology in Freiburg (as Spemann's successor), and in the years 1938–1940 Rector of the University of Freiburg.

465 Planck (1936a), 93.

466 Report on the Activities of the Department of Chemistry of the KWI for Experimental Therapy. MPG Archives, Dept. I, Rep. 1A, 1459, sheets 194–196 and sheets 252–255.

467 Cf. Chapter 5.8.

about the establishment of a KWI for Biochemistry.[468] Harnack as President warned in a report to the Minister of Education that the plans ought to be realized as soon as possible, since the already rampant inflation was likely to lead to a rapid increase in the costs of construction.[469] Although Plot 93 had been designated as the site for the biochemical institute – a plot that is today occupied by the Henry Ford Building of the Free University of Berlin – and a design for a new institute building existed,[470] the plans were not realized. The Goldenberg-Oetker Foundation's endowment of two million marks, which shortly before, on the advice of Oskar Neuberg, had been taken out of government bonds and reinvested in industrial shares,[471] was far outweighed by the 7.8 million marks[472] of the estimate for the new building – and there was no end in sight to the galloping inflation. A further alternative, that of accommodating the institute in the premises of the German Entomological Museum,[473] which was also located in Dahlem and attached to the KWG, was rejected by Neuberg in view of the far too limited amount of space that would be available there.[474]

Finally, Neuberg gave up the idea of having a physically separate institute of his own, but requested that the existence of the Institute for Biochemistry should at least be recognized by a change in the name of the KWI for Experimental Therapy.[475] Friedrich Glum, who took part in the negotiations, formulated the agreement that was reached as follows:

> "The solution I finally agreed to was to make him Director of the KW Institute for Biochemistry, as it were in exile. That is to say, he remained Head of Department in Wassermann's institute, while at the same time being appointed Director of the KW Institute for Biochemistry, which essentially only existed on paper, but for which he was able to mobilize a certain amount of funding from industrial companies he had connections to."[476]

468 In 1921, in the course of the negotiations on the establishment of the KWI for Leather Research, Neuberg was also under discussion, in response to a proposal of Haber's, as Director of this institute. Göbel (1996), 306. On the work of the KWI for Leather Research see also Schmaltz (2005), 291–325.

469 Harnack to the Minister of Science, the Arts and National Education, June 10, 1922. MPG Archives, Dept. I, Rep. 1A, 2031, sheets 3–6.

470 Preliminary draft plans for a KWI for Biochemistry by the architect Hans Jacob, August 25, 1922. MPG Archives, Dept. I, Rep. 1A, 2031, sheet 45. It is probable that Jacob had been commissioned by Neuberg himself, since the architect's address – Grolmanstrasse 14, Charlottenburg – was in the immediate vicinity of the flat Neuberg occupied before moving to Dahlem. Cf. Directories of the City of Berlin (1908–1915).

471 MPG Archives, Dept. I, Rep. 1A, 2031, sheets 30–31, 34 and 41.

472 MPG Archives, Dept. I, Rep. 1A, 2031, sheet 46.

473 The Museum was incorporated into the KWG at the beginning of October 1922. Brocke (1990b), 247–248.

474 MPG Archives, Dept. I, Rep. 1A, 2031, sheet 33.

475 Memorandum on the Negotiations concerning the Establishment of a Kaiser Wilhelm Institute for Biochemistry , September 13, 1922, in: MPG Archives, Dept. I, Rep. 1A, 2031, sheets 52–53.

476 "Der Lösung, der ich schließlich zustimmte, war, ihn zum Direktor des KW-Institutes für Biochemie gewissermaßen 'in partibus infidelium' zu machen. D. h. er blieb Abteilungsleiter im Wassermann'schen Institut und wurde nebenbei Direktor des im Wesentlichen nur auf

And indeed, the Statutes of the KWI for Experimental Therapy and Biochemistry did not provide for any changes in the hierarchical and organizational structures.[477] The Director, August von Wassermann, continued to have sole power of decision-making. Carl Neuberg, as Deputy Director under Wassermann,[478] was in future also to be involved in the budgetary planning; it may be assumed, however, that this was the way things had in fact been done in practice for years already. The fact that the "certain amount of funding" that Glum mentions as having been raised by Neuberg for his own research was in fact a substantial sum of several million marks from the funds of the Foundation seems to indicate that the Board of the KWG was not exactly being very forthcoming with its support. One real change, however, was that Oskar Neuberg, as the main sponsor of the Goldenberg-Oetker Foundation, was elected a member of the Board of Trustees of the expanded institute.[479]

As so, as Glum had correctly remarked, Neuberg's institute existed only on paper. But despite all the disappointments and reverses that Neuberg had suffered in the past years of negotiations for an institute of his own, he showed no signs of resignation. One indication of this, for example, is the way he added the words "and Biochemistry" by hand to the printed letterheads and postcards of the KWI for Experimental Therapy.[480] Neuberg was able to register a further success in his campaign in favor of the new discipline in April 1922, when one of the few Chairs of Biochemistry[481] was established at the Berlin Agricultural College, and he was appointed to it.[482] It was, however, contractually agreed with the KWG that this professorship had to be financed by the KWG itself, and that the State Government was not to incur any consequential expenses in the form of pension entitlements. It followed that the professorship was to be abolished if the KWG was unable to raise the money to fund it.[483]

The rather confused sequence of events surrounding the formal establishment of the KWI for Biochemistry may thus be summarized as follows: the institution

dem Papier vorhandenen KW-Institutes für Biochemie, für das er allerdings einige Mittel aus der ihm nahestehenden Industrie mobilisieren konnte." Glum (1964), 237–238.

477 Statutes of the KWI for Experimental Therapy and Biochemistry, December 4, 1922. MPG Archives, Dept. I, Rep. 1A, 1448, sheet 15.

478 Henning/Kazemi (2002), 194.

479 Composition of the KWI for Experimental Therapy and Biochemistry as of May 5, 1923. MPG Archives, Dept. I, Rep. 1A, 1448, sheet 1.

480 MPG Archives, Dept. I, Rep. 1A, 1448, sheet 37.

481 Deichmann (2002), 312.

482 On May 27, 1922, the Ministry of Agriculture, Estates and Forestry gave notification of the approval of the establishment of the chair with effect from April 1, 1922. AHU, Neuberg's Personnel File, UK–N 29, Vol. 3, sheet 3.

483 Agreement between the KWG and the Ministry of Agriculture, Estates and Forestry of February 21, 1922, Secs. 1 and 2. AHU, Neuberg's Personnel File, UK–N 29, Vol 3, sheet 4. A clause in the contract laid down that the appointment would not give rise to any pension entitlements either for Neuberg or for his wife. This waiver was, however, canceled again three years later. AHU, Neuberg's Personnel File, UK–N 29, Vol. 3, sheet 42. To what extent the chair was financed out of the Society's own funds and to what extent out of those of the Goldenberg-Oetker Foundation is unclear.

was set up provisionally in October 1917 and attached to the KWI for Experimental Therapy under the direction of August von Wassermann. In May 1920, it was finally resolved that it should be established as part of the KWG, but the new building planned for the institute could not be constructed for lack of money. Although Neuberg, through his wartime work on glycerol substitutes, had benefited from the major military requirements for these materials, and in roundabout ways and with the assistance of his cousin Oskar Neuberg had been able to mobilize substantial funding for the establishment of the new institute, he ironically saw these funds losing almost their entire value in the post-war years as a result of the inflation provoked by the war.[484] As Neuberg's international reputation grew, the KWG attempted to secure his loyalty through salary increases, expressions of intent and promises; but no independent Institute for Biochemistry was established. At the end of 1922, the KWI for Biochemistry, which existed only on paper, was merged with the Institute for Experimental Therapy, and Neuberg appointed Deputy Director of the KWI for Experimental Therapy and Biochemistry. De facto, however, this somewhat artificial construction did not change Neuberg's position. His aim of establishing, under the umbrella of the KWG, an internationally reputed research establishment for the discipline of biochemistry, which he himself had done so much to create, and to direct it himself, was one that Neuberg never gave up, despite all difficulties, and he firmly rejected any alternative proposals that were less than optimum. In his view, the close and intensive contact and the exchange of experience and knowledge with his colleagues in Dahlem were among the factors that were indispensable to his project, so that he was determined to have the new institute established in Dahlem and nowhere else.

When the Director of the KWI for Experimental Therapy and Biochemistry, August von Wassermann, died in Berlin at the age of 59 on March 16, 1925, Neuberg, who was in Davos doing research at the Institute of High Altitude Physiology and Tuberculosis Research,[485] broke off his stay in Switzerland and returned immediately to Dahlem. As the Deputy Director, it was of utmost importance for him to be present when the Institute's future was being discussed.[486]

One of the fundamental principles of the KWG with regard to the establishment of institutes was that an outstanding representative of an academic field inadequately represented at the state universities should be recruited and the institute built up in the way he wanted it. Thus the institutes were, in a manner of speaking, specially created for the researchers who were appointed to direct

484 Cf. an article by Haber on the scientific and technical developments fostered by the First World War, which deals among other things with the glycerol substitutes. Haber (1923).

485 Neuberg to Glum, March 10, 1925. MPG Archives, Dept. I, Rep. 1A, 1449, sheet 111.

486 Neuberg to Glum, March 21, 1925. MPG Archives, Dept. I, Rep. 1A, 1449, sheet 113. In this letter Neuberg informed Glum that he had been in Berlin again since March 18, and proposed setting up a bust of Wassermann or a memorial plaque in the Institute.

them.[487] The corollary of "Harnack's Principle,"[488] as this manner of proceeding was known, was that if an institute director died while still in office, either a new representative of the discipline who was just as outstanding as his predecessor had to be found, or the institute had to be dissolved. In the case of Wassermann's institute, however, a compromise was decided on.

Among the leadership of the KWG the view soon prevailed that now was the time to make a reality of the independence that Neuberg had always sought for his institute, which existed only on paper, and to grasp the opportunity to implement the plans for a KWI for Biochemistry which had so often fallen by the wayside.[489] The Secretary General of the KWG, Friedrich Glum, who was one of the major participants in the negotiations, stated in his memoirs that in his view the development of research into experimental therapy had already been so far advanced when the institute was set up that the discipline could have been carried on satisfactorily at the universities and the funds better invested in "up-and-coming directions, such as biology and biochemistry."[490] It was now decided that the Institutes for Experimental Therapy and for Biochemistry, which had been merged three years before, should be formally separated again.[491] Neuberg was to be Acting Director of the KWI for Experimental Therapy, which was to continue to operate on a reduced scale[492] for the time being. At that time, it was not planned to appoint anyone to Wassermann's position, although the option was to be kept open.[493] Neuberg's institute would be expanded to occupy the rooms on the second floor of the building that were thus vacated, while the Institute for Experimental Therapy would be accommodated in the basement.[494] According to Glum, Neuberg was "naturally delighted" with this arrangement.[495] The resolution on this restructuring was officially adopted in July 1925.[496]

487 Memorandum of Adolf von Harnack of November 21, 1909, on the establishment of the KWG, quoted according to Planck (1936a), 30–44.

488 Neuberg, like Max Bergmann and Otto Warburg, was one of the younger scientists who were appointed Directors after having done excellent work in KWI institutes. Laitko (1996), 600. With regard to "Harnack's Principle," see also Grau (1996) and Vierhaus (1996).

489 In a letter of April 16, 1925, Neuberg thanked the President of the Society for a letter of April 7, 1925, in which he was informed of this appointment. MPG Archives, Dept. I, Rep. 1A, 2032, sheet 2.

490 Glum (1964), 285.

491 This can be seen from the minutes of a meeting held at the Ministry of Education on May 11, 1925, to clarify Reiter's position at the Institute. MPG Archives, Dept. I, Rep. 1A, 1449, sheets 128–129.

492 MPG Archives, Dept. I, Rep. 1A, 1449, sheet 117. Of the six laboratory assistants, two secretaries and two ancillary laboratory staff, half of each category were to remain. The remaining employees (cleaner, gardeners, stable staff and stokers) were also taken over. Cf. Neuberg's personnel list of April 21, 1925. MPG Archives, Dept. I, Rep. 1A, 1449, sheet 118.

493 MPG Archives, Dept. I, Rep. 1A, 1449, sheets 128–129.

494 Addendum to the minutes of the meeting of the Senate, May 11, 1925. MPG Archives, Dept. I, Rep. 1A, 1449, sheet 130.

495 Glum (1964), 285.

496 Minutes of the 43rd meeting of the Senate of the KWG, July 7, 1925 (transcript). MPG Archives, Dept. I, Rep. 1A, 1449, sheet 188a.

The ambivalent attitude of the Kaiser Wilhelm Society in seeking on the one hand to avoid dissolving the Institute for Experimental Therapy completely while on the other hand not appointing a new director, led to some press reports that were susceptible to being misunderstood.[497] As a result, Adolf von Harnack, the President of the KWG, received letters suggesting how the funding and premises that had now become available might best be used. The Jena professor Willy Pfeiler,[498] for example, referred to his earlier discussions with August von Wassermann and recommended that the institute should be transformed into an "Institute for Animal Disease Research."[499] The cell researcher Rhoda Erdmann[500] would have liked to make use of the rooms and financial resources herself for an "Institute for Experimental Cell Research."[501] Neither suggestion was acted upon, and when only a short time later the KWG set up a Department of Cell Cultivation, but passed over Rhoda Erdmann in staffing it, she reacted with great annoyance.[502] But the negotiations relating to a possible successor to Wassermann got Neuberg himself involved in controversy.

Already in 1925, a meeting had been arranged by Friedrich Glum between Hans Reiter,[503] the Deputy Director of the Department of Bacteriology in the KWI for Experimental Therapy, and Neuberg, in order to clear up an "unfortunate matter."[504] Reiter, himself a physician, had been working at the Institute for Experimental Therapy since 1923. He had come there to deputize for the Medical Privy Councillor Martin Ficker,[505] who had been given leave of absence to

497 Newspaper cuttings on the subject (*Deutsche Allgemeine Wochenzeitung, Morgenblatt* of June 13, 1925, *Vossische Zeitung* of June 14, 1925 and *Chicago Tribune*) are to be found in the MPG Archives, Dept. I, Rep. 1A, 1449, sheets 160–163.

498 This is presumably the Willy Pfeiler (b. 1903) who headed the Institute for Animal Hygiene set up in Jena in 1919.

499 Pfeiler to Harnack, June 19, 1925. MPG Archives, Dept. I, Rep. 1A, 1449, sheet 163.

500 Rhoda Erdmann (1870–1935), the second female professor at Berlin University. From 1892 onwards she was a school teacher for ten years, then studied zoology, botany and mathematics for upper secondary teaching until 1909, having taken a doctorate under the zoologist Richard Hertwig (1850–1937) in 1908. In order to be able to do academic work she went to Yale University in New Haven, USA, not returning until 1919. In 1920 she took her *habilitation,* and in 1929 was appointed a full Professor in the Faculty of Philosophy in Berlin. Being Jewish, she was arrested on false accusations in 1933 and deprived of her positions.

501 Erdmann to Harnack, June 20, 1925. MPG Archives, Dept. I, Rep. 1A, 1449, sheet 167.

502 Erdmann to Schmidt-Ott, February 21, 1926. MPG Archives, Dept. I, Rep. 1A, 1449, sheet 219.

503 Hans Reiter (1881–1969), physician and bacteriologist, in 1916 the first to describe the oculo-urethro-synovial syndrome that is named for him. Reiter, who was a member of the Deutsche Volkspartei by 1920, joined the NSDAP (Nazi Party) in 1931 and was President of the *Reichsgesundheitsamt* (National Office of Health) from 1933–1942. He was involved in the expulsion of "Jewish" scientists, and played a part that has not yet been investigated in the human experiments in the Nazi concentration camps, which he was demonstrably aware of. Keitel (2003), (2004a) and (2004b). See also Stürzbecher (1983) and Maitra (2001).

504 Neuberg to Glum, April 1, 1925, Reiter to Glum, April 3, 1925, in: MPG Archives, Dept. I, Rep. 1A, 1449, sheets 114–115.

505 (Philipp) Martin Ficker (1868–1950), hygienist and bacteriologist, was involved in building up the Department of Microbiology at the *Instituto Oswaldo Cruz* in Rio de Janeiro.

conduct microbiological research in São Paulo.[506] Reiter was concerned to prevent the threatened reduction in staff, his own position being among those under threat. He told the KWG management that Wassermann had promised him he would be able to stay at the Institute, even in the event of Ficker, for whom he was deputizing, returning from Brazil.[507] In addition, the mere fact that he had deputized for the sick director seems to have led him to consider himself a suitable successor.[508] However, both the KWG management and the representatives of the Ministry of Education made it unmistakably clear to Reiter that the Society had not entered into any commitments towards him with regard to employment in the longer term, and that if Ficker were to return from Brazil, he would have to expect to be released.[509] In the subsequent period, Reiter conducted himself most peculiarly, much to the displeasure of the Society. The Director General of the KWG, for example, complained to Neuberg that certain staff of the Institute were not working long enough hours;[510] all of whom, according to the information provided by Neuberg, were working under Reiter's direction.[511] When asked to comment, the latter maintained with great vehemence that investigations had proved that any activity lasting more that seven hours was to be considered injurious to health, and furthermore detracted from the enjoyment of the job.[512] For the management of the Society, it was totally unacceptable for staff to be managed in such a way, and all the more so in times when the budgetary situation was so precarious. It insisted on the legally prescribed working hours being adhered to, and refused to countenance any exceptions.[513] Moreover, Reiter did not appear to be prepared to accept the reductions in the personnel and materials budgets of the Institute for Experimental Therapy after its reduction in size. He attempted, for example, to push ahead with research and commercial development projects with the assistance of private donations,[514] this apparently being the main reason why the KWG decided to part with Reiter.

> "We are thus of the opinion, which is shared by Mr. Krüß, Head of Department in the Ministry, that continued cooperation with you would not be fruitful. [...] The Kaiser Wilhelm Society does not wish to see the Institute supported by private funds, since this could easily lead to its becoming dependent upon donors, and to the true character of the Institute as a

506 Henning/Kazemi (1990), 921–922.

507 This is indicated by the minutes of a meeting held at the Ministry of Education on May 11, 1925, to clarify Reiter's position at the Institute. MPG Archives, Dept. I, Rep. 1A, 1449, sheets 128–129.

508 Glum (1964), 285–286.

509 MPG Archives, Dept. I, Rep. 1A, 1449, sheets 128–129.

510 Glum to Neuberg, June 20, 1925. MPG Archives, Dept. I, Rep. 1A, 1449, sheet 166.

511 Neuberg to Glum, June 21, 1925. MPG Archives, Dept. I, Rep. 1A, 1449, sheet 170.

512 Reiter to Neuberg, June 22, 1925. MPG Archives, Dept. I, Rep. 1A, 1449, sheets 175–176.

513 Glum to Neuberg, June 29, 1925. MPG Archives, Dept. I, Rep. 1A, 1449, sheet 177.

514 For example, in a letter to Neuberg dated June 30, 1925, Reiter requested with regard to a laboratory assistant whose work was financed by a pharmaceutical company and who was due to be dismissed on the following day that her employment should be prolonged so that certain experiments could be completed. MPG Archives, Dept. I, Rep. 1A, 1449, sheet 182.

research establishment being obscured. For the same reason, the Society does not wish to see ancillary staff employed who are remunerated out of private funds."[515]

Reiter's peculiar behavior apparently resulted from his hopeless expectation of being appointed to a leading position in the KWG. The disputes about people not working their full number of hours appear either to have been motivated by injured pride, or else were intended to show the Acting Director, Neuberg, in a bad light. Reiter regarded Neuberg as a competitor, and held him partly to blame for the fact that things had not turned out favorably for him. With regard to the donations from private business, it may be assumed that he was desperately attempting to enhance his own status and to make himself indispensable by developing promising drugs. But his plans went awry, among other things because the management of the KWG had already decided to use the funding formerly allocated to the KWI for Experimental Therapy for other purposes. Hans Reiter was finally dismissed with effect from April 1, 1926;[516] but the frustration of his plans was to have an ominous sequel for Neuberg after the Nazis had come to power.[517]

In the middle of the 1920s, the KWI for Experimental Therapy was finally broken up into several parts, and continued to exist on paper for legal reasons only, as the KWG still wished to keep open the option of appointing a new director. The Department of Chemistry in Wassermann's institute had given birth to the KWI for Biochemistry, now completely independent with Neuberg as its Director. Furthermore, Wassermann's last pupil, Felix Klopstock,[518] had remained as a "guest scientist," working in two rooms[519] on serological issues, particularly with regard to tuberculosis and syphilis. In 1934 Klopstock's post was attached to the Department of Immunochemistry in the KWI for Biochemistry,[520] but this was officially closed down again the following year when Neuberg was forced to resign.[521] Martin Ficker, who was now working in Brazil, had been employed as a

515 "Wir sehen daher im Einvernehmen mit Herrn Ministerialdirektor Krüß ein längeres Zusammenarbeiten mit Ihnen für nicht ersprießlich an. [...] Der Kaiser-Wilhelm-Gesellschaft ist eine Unterstützung des Institutes von privater Seite infolge der damit leicht verbundenen Abhängigkeit von Geldgebern und einer Verwischung des eigentlichen Forschungscharakters des Instituts nicht erwünscht. Aus dem gleichen Grunde wünscht die Gesellschaft auch nicht die Beschäftigung von Hilfskräften, die von privater Seite bezahlt werden." The President of the KWG to Reiter, July 3, 1925. MPG Archives, Dept. I, Rep. 1A, 1449, sheet 184.

516 This is shown amongst other things by a letter from Glum to the Chairman of the Board of Trustees of the KWI dated July 6, 1925, in which further decisions on the future of the Institute were also explained. MPG Archives, Dept. I, Rep. 1A, 1449, sheet 188.

517 See Chapter 6.1.

518 Klopstock, about whose life and work hardly anything is known, was for a time listed in the Institute's activity reports as a scholarship holder of the *Notgemeinschaft der deutschen Wissenschaft*. Cf. Harnack (1928), 198.

519 Planck (1936a), 112.

520 The KWI for Experimental Therapy and the KWI for Biochemistry now had a joint budget. Henning/Kazemi (2002), 195.

521 Accounts of the KWI for Biochemistry for 1937. MPG Archives, Dept. I, Rep. 1A, 1463a, unnumbered sheet. The Department was officially closed at the end of October 1936; however, Klopstock already had to vacate the premises on August 1, 1936. Glum to Klopstock, June 11, 1936. MPG Archives, Dept. I, Rep. 1A, 2033, sheet 209.

scientific member of the KWI for Experimental Therapy since 1923, mainly in the field of diagnostics and therapy for infectious tropical diseases, in particular leprosy. This "microbiological research unit," which in terms of manpower consisted at any given time of Ficker and one or two assistants, was attached administratively and for budgetary purposes to the KWI for Experimental Therapy from 1926 to 1934; but despite this, the department continued to exist until Ficker's death in 1950.[522] Furthermore, the KWG was attempting to expand its international academic relations by planning, as its Secretary General wrote,

> "[...] initially to give a series of young researchers the opportunity to work in the Institute for a few years, and to apply the existing funds budgeted for the KWI for Experimental Therapy to this purpose."[523]

A first step in this direction was the appointment of the Danish cell researcher Albert Fischer.[524] Friedrich Glum had looked him up in Copenhagen on the recommendation of Otto Warburg, and had invited him to come to the KWG in Dahlem as a guest scientist.[525] Neuberg too found Fischer's academic qualifications convincing,[526] and so the Senate of the KWG decided that Fischer should be employed for a period of initially three years starting in April 1926.[527] It was expressly and officially pointed out that this appointment, which Rhoda Erdmann had hoped she would be given, was also to be regarded as a favor in return "for the hospitality so frequently shown to German scholars at foreign research institutes."[528] Organizationally, the "Dr. Albert Fischer Guest Department for Cell Culture" was attached to the KWI for Biology, and within it to the Department of General Physiology under Otto Warburg.[529]

Fischer and his staff were concerned with physiological and pathological processes in the cultivation of tissue, above all in relation to the investigation of malignant tumors and of therapy for them.[530] His work was funded out of the residual budget of the KWI for Experimental Therapy,[531] some of whose rooms

522 Henning/Kazemi (1990), 921–922.

523 "[...] zunächst einer Reihe von jüngeren Forschern Gelegenheit zu geben, für einige Jahre im Institut arbeiten zu können und wir die für das KWI für Experimentelle Therapie vorhandenen Etatsmittel dafür verwenden." Glum to Weinberg, July 6, 1925. MPG Archives, Dept. I, Rep. 1A, 1449, sheet 188.

524 Albert Fischer (1891–1956) conducted research in the Guest Department for Cell Culture in the so-called "Gatehouse" from 1925–1932.

525 Glum (1964), 338–339. Otto Warburg had apparently developed a particular interest in tissue cultures in the course of his cancer research work. Cf. Florkin et al. (1975), 267.

526 Neuberg to Glum, June 29, 1925. MPG Archives, Dept. I, Rep. 1A, 1449, sheet 178.

527 Minutes of the meeting of the Senate of December 7, 1925 (transcript). MPG Archives, Dept. I, Rep. 1A, 1449, sheet 214.

528 "[...] für die vielfach erfolgte Inanspruchnahme von Gastfreundschaft durch deutsche Gelehrte in ausländischen Forschungsstätten." Harnack to Gustav Stresemann, January 18, 1926. Brocke (1990b), 319, Fn. 214.

529 Planck (1936a), 92.

530 See Fischer's report in: Planck (1936b), 274–275.

531 This is indicated by Neuberg's budgets for the 1928/29 fiscal year, in which he remarks that Fischer's department consumed so much material that it had to be paid for partly out of the Institute for Biochemistry's budget. MPG Archives, Dept. I, Rep. 1A, 2038, sheets 20–21.

Fischer's department also occupied.[532] As from the beginning of April 1930, the work was financed by funds from the Carlsberg Foundation[533] in Copenhagen.[534] The Guest Department existed until the end of March 1932,[535] when Albert Fischer was given an institute of his own in Copenhagen, and left the KWG.

5.6. THE KAISER WILHELM INSTITUTE FOR BIOCHEMISTRY

The foundation and final actual establishment of the KWI for Biochemistry was a long drawn out process lasting almost eight years. This period was marked by Neuberg's high degree of involvement in both his academic and his private affairs, by reverses caused by political and economic factors, and also by personal feuds. For Neuberg, it was a matter of particular pride that the KWI for Biochemistry was the first independent research establishment of its kind. Through his unceasing efforts on behalf of his still young subject of biochemistry, which was concerned with issues in the border area between chemistry, biology and medicine, Neuberg contributed in large measure to its establishment as an independent discipline.[536] The following years up to Neuberg's forced resignation as Director of the Institute in 1936 were characterized by wide-ranging research activities and the great diligence of the Institute's staff in getting their work published.

Now that the Institute had achieved its independent existence, it had to be provided, like every other KW Institute, with a Board of Trustees. This had the task of taking personnel and budgetary decisions.[537] Its members were proposed by Carl Neuberg[538] and appointed by the Senate of the KWG for a period of five years. Some of the Trustees, for example Oskar Neuberg, who has already been mentioned several times, had previously been Trustees of the earlier KWI for Experimental Therapy and Biochemistry. In addition, one representative each from the Reich (i.e. national) Ministry of the Interior and the Prussian (i.e. state) Ministry of Science were appointed members of the Board of Trustees, which consisted initially of eight,[539] and from 1931 onwards of seven[540] persons. The

532 This is indicated by a remark in the auditors' report on the accounts of the Kaiser Wilhelm Institute for Biochemistry, dated March 31, 1934: "whereas the Gatehouse, formerly occupied by Fischer's Guest Department, is standing empty." ("[...] während das früher von der Gastabteilung Fischer bewohnte Torhaus leer steht.") MPG Archives, Dept. I, Rep. 1A, 2040, sheet 1.

533 This Foundation was set up by the Danish brewery Carlsberg in 1876, in order to promote not only culture (the *Ny Carlsberg Glyptothek* Museum in Copenhagen) but also the sciences in general and the natural sciences in particular.

534 Administrative Headquarters to Neuberg, April 7, 1930. MPG Archives, Dept. I, Rep. 1A, 2039, sheet 44.

535 Henning/Kazemi (1990), 915.

536 Harnack (1928), 107.

537 With regard to the structure and organization of the KWG cf. Brocke (1999a) and (1999b).

538 Neuberg's proposal for appointments to the Board of Trustees. Memorandum of February 24, 1926. MPG Archives, Dept. I, Rep. 1A, 2036, 4.

539 Harnack (1928), 196–197.

Institute's Board of Trustees was composed as follows: The KWG was represented by its President, Adolf von Harnack, who was at the same time Chairman of the Board of Trustees, and by the Secretary General Friedrich Glum. The Ministries delegated Minister of State Carl Friedrich Schmidt-Ott, Max Donnevert, *Ministerialrat* (Counselor) in the *Reich* Interior Ministry, Werner Richter, *Ministerialdirektor* (Head of Department) in the Prussian Ministry of Science, the Arts and National Education, and the chemist and Privy Councillor Nikodem Caro. Industry was represented by Carl Duisberg (chemist and chairman of the National Federation of German Industry, from Leverkusen) and Oskar Neuberg (factory director and Carl Neuberg's second cousin, from Wiesbaden).

The activities of Neuberg's institute over the following ten years can be traced in the annual reports of the KWI for Biochemistry, which appeared in *Naturwissenschaften*.[541] Between October 1924 and September 1935 more than 80 researchers are known to have worked at the Institute and to have published,[542] mainly in the *Biochemische Zeitschrift*.[543] However, these annual reports also demonstrate that the number of scientists involved in the research was even greater.

Year of report	No. of personnel	People publishing	Difference
1925	11	No details available	–
1926	16	22	6
1927	20	28	8
1928	20	24	4
1929	20	27	7
1930	15	20	5
1931	16	18	2
1932	15	16	1
1933	15	17	2
1934	15	15	0
1935	11	12	1

Although Neuberg, "out of regard for the Institute's character as a research establishment and in accordance with its rules, never took more than a small number of graduate students,"[544] he did supervise the writing of a number of doctoral theses at his various research establishments.[545] His daughter Irene, for example, was one of those who wrote their theses at the KWI for Biochemistry.[546] Altogether, the Kaiser Wilhelm Institute for Biochemistry produced more than 400 publications

540 Oskar Neuberg was then no longer a Trustee. Adolf von Harnack was succeeded on his death by the retired Minister of State Friedrich Saemisch. Minutes of the meeting of the Senate of May 12, 1931 (transcript). MPG Archives, Dept. I, Rep. 1A, 2037.

541 From 1924 onwards, the periodical *Naturwissenschaften* was the KWG's vehicle for public announcements and official publications.

542 See Chapter 11.3.

543 Cf. Chapter 4.

544 "[...] in Rücksicht auf den Charakter der Anstalt als einer Forschungsstätte bestimmungsgemäß stets nur eine kleine Zahl von Doktoranden annahm [...]" Planck (1936a), 114.

545 See Chapter 11.2.

546 Neuberg (1932).

during the period 1924–1935, Neuberg himself being involved in over half of them.

Year of report	Publications	With Neuberg	Neuberg alone
1925	31	23	3
1926	34	23	2
1927	57	39	2
1928	53	32	2
1929	37	9	0
1930	34	22	4
1931	36	12	1
1932	37	18	2
1933	33	13	0
1934	37	13	1
1935	20	12	1
Total	409	216	18

The Institute's ten-year research report, published in the KWG's yearbook for 1936, even speaks of 670 different publications. Neuberg, who is likely to have written the article himself, probably included publications from the years 1922 to 1925, when the KWI for Biochemistry was still part of the KWI for Experimental Therapy and Biochemistry.

Neuberg identified powerfully and quite emotionally with *his* institute. This may have been because of the long and not always easy path he had had to travel before he was able to realize his dream of a separate institute for biochemistry. His feeling of responsibility towards the Institute went so far that he had a separate telephone line installed to connect the Institute with his house, which was only a few streets away. Neuberg wrote in a letter on this subject:

> "In order to allow me to be contacted should any danger arise, or indeed at any time, there is a telephone line linking the Kaiser Wilhelm Institute for Biochemistry, Thielallee 69, and my private residence, Hittorfstrasse 18. The telephone is beside my bed."[547]

Neuberg's intense personal commitment to the Institute is demonstrated, for example, by his written exchanges with the architect who had been commissioned to fit out the laboratories:

> "In view of the small amount of experience you have in building laboratories, I have willingly sacrificed my time over a period of weeks to supervise the work of installing the plumbing, and contrary to the usual custom, have had to concern myself with every gas tap, water tap, socket and pipe, indeed with even the smallest pieces of joinery work. You owe it to this

547 "Um bei Gefahr und überhaupt jeder Zeit erreichbar zu sein, besteht eine telefonische Verbindung zwischen dem Kaiser-Wilhelm-Institut für Biochemie, Thielallee 69 und meiner Privatwohnung, Hittorfstr. 18. Der Apparat steht an meinem Bett." Letter from Neuberg to Steglitz-Zehlendorf Police Station, December 31, 1933. MPG Archives, Dept. I, Rep. 1A, 2033, sheet 131.

reckless expenditure of my time that I was able to eradicate most of the errors you made. In the case of the basins, however, you faced me with a *fait accompli*."[548]

The personnel establishment varied from year to year, the costs being distributed, according to requirements, between the budgets of the KWI for Biochemistry, the KWI for Experimental Therapy and the Department of Tobacco Research. In 1932, for example, the KWI for Biochemistry employed seven scientific assistants, two technical assistants, three laboratory assistants, two secretaries, a mechanic and an animal keeper.[549]

There are number of factors demonstrating that there must have been an excellent spirit of collegiality at the Institute. In the first place, the numerous publications indicate an intensive and constructive working atmosphere. Further evidence may be seen in the efforts of the staff on Neuberg's behalf from 1933 onwards[550] and the correspondence between Neuberg and his former staff after his emigration.[551] The following poem, "The Incorrigibles of the Kaiser Wilhelm Institute for Biochemistry," which Neuberg composed on the occasion of his 50th birthday, illustrates the informal, relaxed manner that was customary at the Institute:

"Wenn ein Mensch mit 50 Jahren	"When a man, at the age of 50
Noch auf dieser Erde wohnt	is still dwelling upon this earth
Und vergaß dorthin zu fahren,	and has forgotten to wend his way
Wo der Fürst der Hölle thront,	to the realm of the Prince of Darkness,
Glauben die an Jahren jünger,	those who are younger in years
Gut sei die Gelegenheit,	think it is a good opportunity
Vor dem grossen Allbezwinger	to come before the all-conquering one
Zu Beginn der Greisenzeit	at the beginning of the dotage
Des Objekts zu demonstriern	of the said object, and demonstrate;
Gut gemeint ist euer Plan,	your plan is well-intentioned
Und ich will bei euch probiern	and I am glad to be with you as I try
Abwehr gegen Altersnah'n."[552]	to fend off approaching old age."

Furthermore, Neuberg as Director of the Institute undertook steps to promote the interests of his staff in many ways. For example, he was vehement in his efforts to persuade the President of the KWG that Felix Klopstock, who had already been a

548 "Wegen Ihrer geringen Erfahrungen auf dem Gebiete des Laboratoriumsbaus habe ich willig wochenlang meine Zeit auf die Beaufsichtigung der Installationsarbeiten verwendet, und mich entgegen der sonstigen Gepflogenheiten um jeden Gashahn, Wasserhahn, um jede Steckdose und jedes Rohr, ja um die kleinste Tischlerarbeit kümmern müssen. Diesem Raubbau an meiner Zeit haben Sie es zu verdanken, daß ich die meisten Fehler, die Sie begangen haben, noch ausmerzen konnte. Im Fall der Becken haben Sie mich jedoch vor ein Fait accompli gestellt." Neuberg to the builder Laternser, June 30, 1926. MPG Archives, Dept. I, Rep. 1A, 2046/1.

549 List relating to the personnel budget of March 11, 1932. MPG Archives, Dept. I, Rep. 1A, 2039, 83.

550 See Chapter 6.1.

551 Despite intensive investigations, it was unfortunately not possible to find any member of Neuberg's staff to interview personally.

552 Printed poem of Carl Neuberg's, "Die Unverbesserlichen des Kaiser-Wilhelm-Instituts für Biochemie," July 26, 1927. CNP, Box 7: KWI for Biochemistry.

research worker in the Department of Chemistry at the KWI for Experimental Therapy, should be given the opportunity to take his *habilitation*.[553] When members of staff were ill and had money problems as a result, Neuberg would apply to the management of the KWG to approve extraordinary expenditure,[554] and he also sought to obtain salary increases for staff if he thought they were appropriate – staff whose efforts he was always appreciative of.[555]

Thanks to Neuberg's international reputation and his work as the editor of the *Biochemische Zeitschrift*, the KWI for Biochemistry was an attractive place for foreign biochemists to do research at. In addition to those scientists whose research was published while they were in Berlin, the Institute's records contain repeated references to foreign scholarship holders. At times, the demand was so great that "as a result of the throng of foreigners […] even German candidates have had to take a back seat."[556] Altogether, scholarship holders from 18 different countries worked at the KWI for Biochemistry.[557] The information in the files does not allow a detailed overview of Neuberg's foreign pupils and scientific assistants to be compiled, but as an example, there follows a list[558] of Japanese scientists who spent some time working at the Dahlem Institute:

553 Neuberg to Harnack, February 28, 1929. MPG Archives, Dept. I, Rep. 1A, 2032, 52.

554 Neuberg requested, for example, the approval of an emergency grant out of the Institute's funds to assist the sick assistant Joseph Burkard. Neuberg to Administrative Headquarters, June 1, 1933. MPG Archives, Dept. I, Rep. 1A, 2039/3. Neuberg also sought an additional grant to Maria Kobel, when she was ill after inhaling bromine vapors. Neuberg to Administrative Headquarters, November 24, 1935. MPG Archives, Dept. I, Rep. 1A, 2033, 190.

555 Neuberg's request on behalf of the laboratory servant Klatt may be quoted as an example: "[…] that Klatt, who up to now has had to support his otherwise completely destitute mother, and now wishes to marry and naturally has not been able to accumulate anything in the way of reserves, should be granted the same basic salary as Noske [animal keeper at the Institute, Authors' note], since Klatt has not only worked here for longer but has also performed much more arduous duties than Noske, is untiringly hard-working and often busy at the Institute well on into the night." ("[...] Klatt, der bisher seine völlig mittellose Mutter hat unterhalten müssen, jetzt heiraten will und natürlich nicht die geringste Rücklage hat machen können, das Grundgehalt von Noske zu gewähren, denn Klatt hat nicht nur eine längere sondern auch viel schwerere Dienstzeit als Noske, ist unermüdlich fleißig und oft des nachts im Institut tätig.") Neuberg to Administrative Headquarters, March 22, 1927. MPG Archives, Dept. I, Rep. 1A, 2038/4.

556 "[…] infolge des großen Andrangs von Ausländern, [...] selbst einheimische Interessenten haben zurückstehen müssen [...]." Neuberg to Glum, August 23, 1929, in response to an inquiry from the Uruguayan Embassy to the KWG with regard to the acceptance of Dr. Miguel Rubino. MPG Archives, Dept. I, Rep. 1A, 2032, 56 and 57.

557 Scientists from the following countries were at the KWI for Biochemistry at different times: Austria, Czechoslovakia, Denmark, France, Great Britain, Greece, Hungary, India, Italy, Japan, Latvia, Poland, Romania, Spain, Sweden, Turkey, Uruguay and USA. Planck (1936a), 114.

558 Sent to Neuberg on the occasion of his 75th birthday, when Honorary Memberships of the Japanese Biochemical Society and the Imperial Society of Agriculture of Japan, honors specially created for him, were conferred upon him. CNP, Box 11: Neuberg, Carl, Birthday Congratulations. The list does not, however, appear to be complete, as other names are to be found elsewhere. Cf. Neuberg to Kimura, March 10, 1954. CNP, Box 7: Kimura, T.

Akamtsa, Dr. Shigeru	Professor of Biochemistry, Chiba
Fujise, Dr. Shin-Ichiro	Professor of Biochemistry, Tohoku, Sendai
Imanishi, Dr. Yoshio	Director of Internal Medicine, Kakogawa, Hyogo
Iwatsuru, Dr. Ryuzo	Professor of Internal Medicine, Wakayama
Kageura, Dr. Naom	Professor of Internal Medicine, Nagasaki
Kumagawa, Dr. Hachiro	Director of a newspaper, Kawasaki
Kurono, Dr. Kanrok	Industrial Director, Kita-Ku, Tokyo
Kuroya, Dr. Mashahiko	Professor of Bacteriology, Tohoku, Sendai
Nagai, Dr. Kazuw	Professor, University of Hokkaido, Minaminijo, Sapporo
Nagayama, Dr. Takeyosh	Professor Emeritus, Minato-Ku, Tokyo
Ochiai, Dr. Eij	Professor of Pharmaceutical Chemistry, Tokyo
Ohta, Dr. Kosh	Pediatrician, Atami, Kanagawa
Soda, Dr. Tokur	Professor of Biochemistry, Tokyo
Tomita, Dr. Masaji	Professor of Biochemistry, Kobe
Yamakawa, Dr. Ichir	Court Physician, Atami, Kanagawa
Yamasaki, Dr. Izue	Professor of Agricultural Chemistry, Kiushu, Fukuoka

The fact that the working climate in the Dahlem Institute was so extraordinarily good, from both a human and an academic point of view, was no doubt one reason why so many foreign colleagues and former guest scientists of Neuberg's institute assisted him during the time of his political persecution and emigration, and why his own staff proved to be so loyal.[559] Neuberg himself, in addition to his research work and his editing of the *Biochemische Zeitschrift*, now also had to devote himself increasingly to the administrative tasks incumbent upon the Director of the KWI for Biochemistry and Acting Director of the KWI for Experimental Therapy. His international reputation led to his being invited to attend numerous congresses and undertake many lecture tours both in Germany and abroad.[560] It was also during this period that Neuberg received most of the numerous honors and decorations that were conferred upon him.[561]

But in addition to his duties at the KWG, Neuberg was also involved in the preparations for the establishment of *Notgemeinschaft der Deutschen Wissenschaft*, the "Emergency Association of German Science," which had been set up in 1920. This organization, established jointly by all the academies and colleges of science in Germany together with the KWG and other organizations, had made it its task to acquire funding for science in Germany and to distribute it to the participating institutions. In order to make a proper selection of research projects that were worthy of support, specialist committees were formed for each individual branch of science, in which between two and ten academics, acting on

559 See Chapter 6.5.

560 In the files of the Max Planck Society relating to the KWI for Biochemistry there are numerous mentions of Neuberg traveling to lecture or to attend congresses in Germany and abroad.

561 See Chapter 8.

an honorary basis, decided how the funds should be awarded.[562] Since 1922, Neuberg had been a member of the Chemistry Committee and Commissioner for the fields of biochemistry and colloid chemistry.[563] It remains an open question whether he used his membership of the committee to promote work in any particular fields. In a letter to the Secretary General of the KWG, in which Neuberg requests an extraordinary grant for additional materials, he makes clear that he has not used his position to obtain funding preferentially for his institute.

> "I would allow myself to add that I do not want to approach the *Notgemeinschaft der Deutschen Wissenschaft* in this matter, because I would not consider it right, as a member of the Chemistry Committee, to make any claims on that body in my own interests."[564]

5.7. MARIA KOBEL AND PAUL MAYER – COLLEAGUES AND FRIENDS OF NEUBERG'S

In the comprehensive list of Carl Neuberg's publications, it is noticeable that there were always a considerable number of women involved in joint publications originating from the KWI for Biochemistry. Furthermore, it can be seen that compared with the other institutes of the KWG, a relatively large number of female academics were employed at Neuberg's institute. Vogt[565] also points out that the number of female students taking doctorates at the KWI for Biochemistry was unusually large for those days; among them was Marthe Vogt,[566] who was later to become a famous scientist. The question as to what prompted Neuberg to give so much assistance to women and promote their academic careers – something that was also not customary at the time – remains open, however; even Ines Mandl, who worked with Neuberg in New York in later years, was unable to give any explanation. In her view, Carl Neuberg made no distinction as to whether an applicant was a man or a woman. In deciding who to employ, he only took qualifications and achievements into account.[567]

562 Brocke (1990b), 228–233, and the publications in the series *Beiträge zur Geschichte der Deutschen Forschungsgemeinschaft* ("Contributions on the History of the German Research Foundation"), edited by Rüdiger vom Bruch and Ulrich Herbert, which have been appearing since 2006.

563 Cf. Notgemeinschaft der deutschen Wissenschaft (1928), 192.

564 "Hinzuzufügen erlaube ich mir noch, daß ich die Notgemeinschaft der Deutschen Wissenschaft deshalb nicht in Anspruch nehmen möchte, weil ich es für richtig halte, als Mitglied des Fachausschusses für Chemie keinerlei Anforderungen in eigenem Interesse geltend zu machen." Neuberg to Glum, March 20, 1926. MPG Archives, Dept. I, Rep. 1A, 2032, 24.

565 Vogt (1996).

566 Marthe Vogt (1903–2003), daughter of the notable research scientists Oskar and Cecilie Vogt and sister of Marguerite Vogt (b. 1912), molecular biologist at the Salk Institute in San Diego. Marthe Vogt left Germany with her sister for political reasons in 1935. Through her work on neurotransmitters she became one of Britain's most important pharmacologists and was the second woman to be elected a Fellow of the Royal Society (in 1952). See also Vogt (1999), Muscholl (2003), Cuthbert (2005).

567 Ines Mandl, New York, personal communication on October 25, 1996.

It was no doubt also Neuberg's lack of prejudice with regard to women scientists that made it possible for the Department of Tobacco Research at the KWI for Biochemistry to be headed by Maria Kobel, one of the few women to be appointed a Head of Department by the KWG. It is appropriate to take her as a particularly outstanding example of all the excellent researchers of both sexes who were associated with Neuberg, since she was a companion on his way through all the stages of his life, including flight and emigration, from the beginning of their collaboration in 1924 until his death in 1956.

Maria Kobel, born on August 5, 1897 in Liegnitz, Silesia, as the eldest of ten children, studied chemistry in Breslau (now Wrocław in Poland), and was appointed to the KWI for Biochemistry by Neuberg in 1924.[568] She was chosen over the heads of all the male applicants, presumably on the basis of her doctoral thesis, which was considered one of the best of her year.[569] Her mentor Neuberg furthered the career of the "most excellent Maria Kobel"[570] to the best of his ability. After only three years, he appointed her Head of the new Department of Tobacco Research at the KWI for Biochemistry.[571] In order to prevent Maria Kobel from moving to a job in the chemical industry, he also campaigned for a life insurance to be taken out for her, financed from institute funds.[572] Together with her he published a series of papers, including several concerned with the biochemistry of tobacco. Although many of his former staff were dismissed after the Nazis had come to power and Butenandt had been appointed to the Chair, he succeeded, even after his forced resignation in 1936, in finding her a job at the German Chemistry Society's Beilstein publishing house.[573] After the Second World War the Beilstein office moved from Berlin to Frankfurt, where Maria Kobel worked until she retired in 1975. But although, like her former boss Neuberg, she never again managed to occupy a position corresponding to her academic abilities after the war, it was only with great reluctance that she retired at the age of 78.[574]

She continued to visit Neuberg regularly in his laboratory until his emigration,[575] and remained in contact with him even after his escape. After a reunion in 1951, Neuberg described the close friendship that had developed out of their working together as follows:

568 Vogt (1997), 43.
569 Ibid.
570 "[…] der sehr ausgezeichneten Maria Kobel." Neuberg to Glum, December 4, 1947. CNP, Box 5: Glum, Friedrich.
571 See Chapter 5.8.
572 Minutes of the meeting of the Board of Trustees of the Department of Tobacco Research, June 26, 1930. MPG Archives, Dept. I, Rep. 1A, 2054, 7.
573 Neuberg to Glum, December 4, 1947. CNP, Box 5: Glum, Friedrich.
574 Personal communication from Bernd Wollny, Beilstein-Informationssysteme, January 14, 1997.
575 Vogt (1997), 43.

> "The friendship with you is one of the few things that means something in my life. I am even more grateful to you for your affection than for the attentiveness that you have displayed towards me and the time that you have sacrificed for me."[576]

One of the reasons for this particular closeness was no doubt Maria Kobel's personal efforts on behalf of his wife Hela when she was seriously ill; Maria Kobel had devised preparations for the treatment of her liver disease.[577] The regular correspondence between Neuberg and Maria Kobel, which is infused with mutual respect and friendship, naturally deals not only with private matters but also with scientific topics, for example the correct nomenclature of chemicals. Shortly before her death in 1986, Maria Kobel described the years in which she had worked with Neuberg as the happiest time of her life.[578] And much the same was true for Carl Neuberg:

> "Dear Miss Kobel, it may be that I remember the time in Dahlem with an even much greater and more joyful sense of satisfaction than you do, and firmly anchored in this recollection is the feeling of gratitude I owe to you as the one among my staff with whom an ideal relationship of confidence and comprehension existed in every respect."[579]

Among the male colleagues with whom Neuberg also cultivated relationships of friendship were Paul Mayer and Hans von Euler-Chelpin. Shortly after the opening of the KWI for Biochemistry in May 1925, Neuberg had proposed these two scientists to the President of the KWG for appointment as external members.[580] Neuberg's proposal was accepted. Both were appointed External Scientific Members of the Kaiser Wilhelm Institute for Biochemistry at the meeting of the Senate of the KWG on August 7, 1925.[581] That Euler-Chelpin,[582] to whom Neuberg had been bound by close personal and academic ties since the beginning of the 20th century, should have been appointed such a member may not appear at all surprising, in view of the fact that he was awarded the Nobel Prize in Chemistry, jointly with Arthur Harden,[583] in 1929. Paul Mayer,[584] on the other hand, who was one of Neuberg's few faithful and lifelong male friends, has been almost completely forgotten, so that it is appropriate to say more about him at this point.

576 "Die Freundschaft mit Ihnen gehört zu den wenigen Dingen, die in meinem Leben etwas bedeuten. Dankbar bin ich Ihnen mehr noch für Ihre Zuneigung, als für die Aufmerksamkeiten, die Sie mir erwiesen und die Opfer an Zeit, die Sie mir gebracht haben." Neuberg to Kobel, November 20, 1951. CNP, Box 7: Kobel, Maria.

577 Neuberg to Kobel, August 17, 1948. CNP, Box 7: Kobel, Maria.

578 Vogt (1997), 46.

579 "Liebes Fräulein Kobel, vielleicht denke ich mit noch viel groesserer und freudigerer Genugtuung als Sie an die Dahlemer Zeit, und in dieser Erinnerung ist verankert das Gefühl der Dankbarkeit, die ich Ihnen schulde, als demjenigen Mitarbeiter, mit dem ein ideales Verhältnis des Vertrauens und Verstehens in jeglicher Hinsicht obwaltete." Neuberg to Kobel, July 28, 1954. CNP, Box 7: Kobel, Maria.

580 Neuberg to Harnack, May 22, 1925. MPG Archives, Dept. I, Rep. 1A, 2032.

581 Transcript of the minutes of the meeting of the Senate of August 7, 1925. MPG Archives, Dept. I, Rep. 1A, 2032.

582 Cf. Schulte (2001b).

583 See also Chapter 8.2.

584 An initial biographic appreciation of Mayer is to be found in Engel/Conrads (1998), from which – unless otherwise stated – the following information is also taken.

Paul Mayer was born in Berlin on October 29, 1872, as the son of Jacques Mayer, a doctor in Karlsbad. Paul Mayer himself stated that he was an adherent of the Jewish faith. He took his school leaving examination at the French Lycée in Berlin, then studied medicine in Vienna and took his doctorate there in 1897. For the next four years he worked as a medical assistant at the Institute of Pathology at the Charité Hospital in Berlin, while at the same time carrying out physiological and chemical experiments in Ernst Salkowski's laboratory. It was there that he got to know the young chemist Carl Neuberg in 1898. When Mayer took over his father's practice in the famous spa Karlsbad, his friend Carl Neuberg made it possible for him to carry on doing experimental work in his institute. Initially, Mayer was able to continue his experiments at the Institute of Chemistry of the Charité. From 1909 onwards he worked at the Institute of Animal Physiology of the Royal Agricultural College in Berlin, where Neuberg had in the meantime become Head of Department. But even when Neuberg moved to the KWI for Experimental Therapy in 1913, a place in the laboratory there was made available to Mayer, as was also the case later at the KWI for Biochemistry. During the winter months, Mayer regularly took advantage of this opportunity to do experimental work. He published articles in the fields of balneology and metabolic physiology, presumably in connection with his work in Karlsbad. In the field of biochemistry, he was particularly interested in glucuronic acids, which Neuberg also investigated, as well as in the optical qualities of various organic substances.

But his friendship with Neuberg went far beyond merely a relationship between colleagues. Ines Mandl remembered that Neuberg occasionally mentioned expeditions to the mountains which he had undertaken together with Mayer.[585] In addition to his honorary membership of Neuberg's institute, Mayer was also awarded an honorary doctorate in agriculture by the Berlin Agricultural College[586] in 1929, and three years later, at Neuberg's suggestion, was honored by being elected a Member of the Imperial Academy of Sciences Leopoldina. In the other direction, it was on the initiative of his friend Paul Mayer, through a former of pupil of Neuberg's, Kurt P. Jacobsohn, that Neuberg was appointed an honorary member of the Portuguese Academy of Sciences in 1934.[587]

Like Neuberg, Paul Mayer too was forced to flee Nazi Germany. He emigrated to England in 1938, but remained in correspondence with his friend Neuberg even after the war. Mayer's congratulations to Neuberg on the occasion of his 70th birthday, which the latter celebrated in exile in America, convey an impression of their profound friendship: "[...] honored by your fellow scientists and loved by your friends, of whom I am proud to be the oldest."[588] Paul Mayer died in London on November 6, 1949, at the age of 75.

585 Personal communication from Ines Mandl, October 25, 1996.

586 Neuberg held a Chair of Biochemistry there from 1922 onwards.

587 Extensive correspondence on this matter is to be found in CNP, Box 7: Jacobsohn, Kurt.

588 "[...] geehrt von Deinen Fachgenossen und geliebt von Deinen Freunden, von welchen letzteren ich stolz bin, Dein ältester Freund zu sein." Engel/Conrads (1998), 85.

5.8. THE DEPARTMENT OF TOBACCO RESEARCH AT THE KAISER WILHELM INSTITUTE FOR BIOCHEMISTRY

The first conference to explore what interest there was in the establishment of a Department of Tobacco Research took place on June 1, 1922.[589] This meeting, chaired by the Nobel Prize winner Fritz Haber, was attended by Carl Neuberg and by representatives of the German tobacco farmers, the tobacco processing industry and the *Reich* Ministry of Economic Affairs, who unanimously expressed considerable interest in such an establishment. The cigarette manufacturers and also in particular the cigar producers hoped that the department's work would lead to improvements in the cultivation of tobacco plants on German soil, in order to make the country independent of expensive imports – a matter of great economic significance to post-war Germany with its lack of foreign exchange. The fact that the biochemist Neuberg was involved in the planning right from the beginning can be explained by the fact that his core academic concern, namely research into fermentation processes, also played a major role in the processing of tobacco. Neuberg summarized the motives for setting up the department as follows:

> "As the quality of the tobaccos grown in Germany has, in the judgment of connoisseurs, deteriorated very substantially, it is an attractive proposition both from a scientific and from an economic point of view, once suitable varieties have been bred, to place the processing of tobacco, which is often empirically rough and ready, on a scientific basis. The government is interested in this for tax reasons, and industry from a commercial point of view."[590]

The President of the KWG having indicated his fundamental agreement to such a department being attached to the KWG,[591] Neuberg wrote a memorandum on the establishment of a research facility of this kind.[592] The aim of the collaboration was to be the achievement of an improvement in the quantity and quality of the tobacco produced in Germany, with the aid of a systematic and close exchange between academic research and the experience of tobacco cultivators and processors. The processes of the growth, ripening and fermentative transformation of the harvested material were to be analyzed in detail, possibilities identified for improving the quality during growth, and experiments carried out on the cultivation of tobacco varieties not previously grown in Germany. It was planned to divide the tasks of the research establishment into three organizational units: Fundamental scientific research was to take place in a department of the KWI for Biochemistry under Neuberg's direction, the agricultural experiments were to be carried

589 Report by Haber on the meeting of June 2, 1922. MPG Archives, Dept. I, Rep. 1A, 2049, sheets 1–6.

590 "Da nach dem Urteil der Sachkenner die Qualität der in Deutschland gewachsenen Tabake sich ganz erheblich verschlechtert hat, ist es vom wissenschaftlichen wie auch vom volkswirtschaftlichen Standpunkt aus eine reizvolle Aufgabe, nach der Kultivierung geeigneter Sorten die vielfach empirisch-rohe Zubereitung des Tabaks auf eine wissenschaftliche Grundlage zu heben. Hieran ist aus steuerlichen Gründen das Reich, von ökonomischen Gesichtspunkten aus die Industrie interessiert." Neuberg (1928), 110.

591 Harnack to Neuberg, June 10, 1922. MPG Archives, Dept. I, Rep. 1A, 2049, sheet 7.

592 Memo of Neuberg, June 13, 1922. MPG Archives, Dept. I, Rep. 1A, 2049, sheets 13–15.

out by the Chamber of Agriculture of the State of Baden in Forchheim, near Karlsruhe,[593] and the breeding experiments on commercial agricultural land in Bavaria and present-day Baden-Württemberg. As far as the funding of the project was concerned, various representatives at the conference indicated that they were prepared to provide annual funds totaling some 7 million marks.[594] After this initial meeting, Neuberg showed great commitment and skill in his efforts to arrange a further meeting of the main institutions with an interest in the project.[595] This meeting, attended by individual members of the Reichstag and representatives of various ministries and associations, finally took place in December 1922.[596] Apart from a draft bill for the "Promotion of German Tobacco Research,"[597] this meeting brought no essential progress. Nevertheless, all sides indicated that they were still interested in the creation of such a research establishment.

It was not until two years later that there were further signals from the government side and from industry indicating that the existing plans might be realized.[598] The Senate of the KWG announced that it agreed to the establishment of the department and the necessary extension of Neuberg's institute.[599] Work was started, but had to be interrupted again, since the level of the ongoing government grant had not been determined,[600] which meant that the permanent existence of the Department of Tobacco Research did not appear to be adequately safeguarded.[601] After further tedious negotiations, agreement was finally reached at a meeting at the National Ministry of Food in April 1926, which placed the research projects on a solid financial basis. The Ministry contributed a sum of 40,000 marks, to be

593 After the Second World War, the Federal Institute for Tobacco Research was established in Forchheim; its results were published in the "Veröffentlichungen der Bundesanstalt für Tabakforschung, Forchheim bei Karlsruhe."

594 This was made up of an annual payment of 4.5 million marks out of tobacco tax revenues and a contribution of 5 marks per zentner [50 kilos – approx. a hundredweight; Translator's note] of German tobacco from the tobacco planters and processors. In addition, the representatives of the cigarette industry announced that they were prepared to make a one-off donation of 500,000 marks. Report by Haber on the meeting of June 2, 1922. MPG Archives, Dept. I, Rep. 1A, 2049, sheets 1–6.

595 Neuberg to Glum, November 28, 1922. MPG Archives, Dept. I, Rep. 1A, 2049, sheet 23.

596 Notes on a meeting on December 8, 1922. MPG Archives, Dept. I, Rep. 1A, 2049, sheets 49–53.

597 MPG Archives, Dept. I, Rep. 1A, 2049, sheets 67–72. It was not possible to determine whether this bill was ever brought before parliament; it is, however, certain that no such law was ever passed.

598 Glum informed Neuberg on August 22, 1924 that the sum of 20,000 marks was now available to set up the research department. These funds came from the Ministry of Agriculture and from the industry. MPG Archives, Dept. I, Rep. 1A, 2049, sheet 84.

599 Minutes of the meeting of the Senate on December 16, 1924 (transcript). MPG Archives, Dept. I, Rep. 1A, 2049, sheet 95a Top 15.

600 Neuberg to the *Reich* Minister of Food, January 30, 1925. MPG Archives, Dept. I, Rep. 1A, 2049, sheet 107.

601 Glum to the Tobacco Farmers' Association, March 6, 1925. MPG Archives, Dept. I, Rep. 1A, 2049, sheet 116.

used principally by Neuberg for research purposes.[602] The KWG provided a garden area, laboratory and secretarial rooms, equipment and ancillary staff to a total value of 20,000 marks.[603] In the investigations that now got under way, the research issues at the forefront of Neuberg's concern[604] were the analysis of the fermentation conditions of the tobacco leaf, the problem of the so-called quick fermentation and the examination of the possibilities of transferring procedures used in other countries to production in Germany.[605] Furthermore, what happens when the tobacco leaf is processed was to be investigated in more detail:

> "There are two views which are fundamentally opposed to each other. According to the one, it is predominantly purely chemical changes that make the fresh tobacco leaf fit for consumption; while according to the other, it is predominantly the influence of microorganisms. There is no agreement as to what is changed during the transformation, of whatever kind, of the fresh tobacco leaf by technical measures, and what factors primarily influence the conversion of the leaf, which would otherwise decay, into the final product ready for use."[606]

A separate Board of Trustees was appointed for the Department of Tobacco Research; considering the small budget, it was a relatively large body. In addition to representatives of various ministries both of the *Reich* government and of the States of Prussia, Baden and Bavaria, it also included delegates from the tobacco growers' associations and tobacco product manufacturers.[607] Maria Kobel, who had been on Neuberg's staff for many years, was put in charge of the Department, initially as his deputy,[608] but later with complete responsibility in her own right.[609]

Despite the continued uncertainty with regard to the funding of the research, which compelled the Department to make do with less and less money,[610] a remarkable number of publications appeared within five years:

602 Notes on a meeting at the *Reich* Ministry of Food on April 28, 1926. MPG Archives, Dept. I, Rep. 1A, 2049, 167.

603 MPG Archives, Dept. I, Rep. 1A, 2055, sheet 6.

604 Neuberg's overview of the work to be done, May 12, 1926. MPG Archives, Dept. I, Rep. 1A, 2049, sheets 172–174.

605 An indication of a fact-finding journey is to be found in a note from Neuberg to Harnack on October 20, 1927. MPG Archives, Dept. I, Rep. 1A, 2050, sheet 219. According to this, Neuberg planned to visit the Italian tobacco research station Scafati, where there were plans for Maria Kobel to work for a time.

606 "Zwei grundsätzlich verschiedene Anschauungen stehen einander gegenüber. Nach der einen sind es vornehmlich rein chemische Veränderungen, die das frische Tabakblatt genußfähig machen, nach der anderen handelt es sich vornehmlich um die Wirkung von Mikroorganismen. Es ist keine Einigkeit darüber vorhanden, was bei den wie auch immer gearteten Umformungen des frischen Tabakblattes durch die technischen Maßnahmen geändert wird, und welche Faktoren in erster Linie die Überführung des zersetzlichen Blattes in das verwendungsbereite Enderzeugnis bedingen." Neuberg/Kobel (1926).

607 The Board of Trustees consisted of a total of 18 members; that of the KWI for Biochemistry, by contrast, had only 8 members. Harnack (1928), 196–197.

608 Harnack (1928), 197.

609 KWG (1929).

610 In the 1929/30 fiscal year the Department's budget amounted to a mere M 26,500, drawn almost entirely from the national budget. MPG Archives, Dept. I, Rep. 1A, 2055, sheet 36.

Neuberg, Carl/Kobel, Maria (1926): Zur Biochemie des Tabaks. Die Naturwissenschaften 14 (1926), 1182–1184.
Neuberg, Carl/Kobel, Maria (1926): Über die Vorgänge im frischen und getrockneten Tabakblatt vor und während der Fermentation. Biochemische Zeitschrift 179 (1926), 459–490.
Neuberg, Carl/Ottenstein, Berta (1927): Übertritt von Methylalkohol in den Tabakrauch. Biochemische Zeitschrift 188 (1927), 217–226.
Neuberg, Carl/Kobel, Maria (1927): Über die enzymatische Abspaltung von Methylalkohol aus Pektin durch ein Ferment des Tabaks. Biochemische Zeitschrift 190 (1927), 232–240.
Neuberg, Carl/Ottenstein, Berta (1928): Bildung von Methylalkohol bei Autolyse frischer Tabakblätter. Biochemische Zeitschrift 197 (1928), 491–501.
Andreadis, Thales (1929): Über den Nitratgehalt des Tabaks und über eine Fehlerquelle der bisherigen Bestimmungsmethode. Biochemische Zeitschrift 204 (1929), 484–492.
Neuberg, Carl/Kobel, Maria (1929): Isolierung von Methylalkohol aus Tabakrauch. Biochemische Zeitschrift 206 (1929), 240–244.
Andreadis, Thales (1929): Untersuchungen über die Vorgänge bei der Tabakfermentation. Biochemische Zeitschrift 211 (1929), 378–394.
Kobel, Maria/Scheuer, Max (1929): Über den Kohlenhydratumsatz im Tabakblatt. Nachweis von Methylglyoxal als Zwischenprodukt im Stoffwechsel grüner Blätter. Biochemische Zeitschrift 216 (1929), 216–223.
Schwaibold, Julius (1929): Über den Jodgehalt frischer, dachreifer und fermentierter Tabakblätter. Biochemische Zeitschrift 218 (1929), 318–320.
Neuberg, Carl/Kobel, Maria (1930): Studien an Tabakenzymen. Biochemische Zeitschrift 229 (1930), 455–463.
Neuberg, Carl/Ostendorf, Clara (1930): Über ein Modell für die Pectase. Biochemische Zeitschrift 229 (1930), 464–466.
Neuberg, Carl/Burkard, Joseph (1931): Über neue stickstofffreie Bestandteile des Tabakrauches. Biochemische Zeitschrift 243 (1931), 472–484.
Neuberg, Carl/Scheuer, Max (1931): Pektin des Tabaks. Biochemische Zeitschrift 243 (1931), 461–471.

According to Neuberg's reports for the years 1930–1932, the Department of Tobacco Research worked in close collaboration with the research unit of the Chamber of Agriculture of the State of Baden at Forchheim, near Karlsruhe.[611] To judge by the reports and publications listed above, its work was largely concerned with the constituent substances of the tobacco leaf and tobacco smoke, the conditions of fermentation, and experiments on the assimilation of plants and the rapid drying of tobacco.

Although even in 1934 there were proud reports of the sensational cultivation of nicotine-free tobacco at the National Institute for Tobacco Research,[612] in 1935 – probably in connection with the National Socialists' ambivalent anti-tobacco campaign,[613] but also no doubt as a move in the direction of forcing Neuberg's retirement – the Ministry of Food withdrew the last funds available to support the Department's work.[614] Neuberg was instructed to continue working as long as the

611 Neuberg's activity reports, 1930–32. MPG Archives, Dept. I, Rep. 1A, 2054, sheets 77–81.
612 Zähringer (2004).
613 Zähringer (2004). See also Proctor (1999), especially Chapter 6, 173–247.
614 Planck to the Chairman of the Board of Trustees, Saemisch, July 9, 1935. MPG Archives, Dept. I, Rep. 1A, 2057, 28.

money lasted, and then to close the department and dismiss the staff.[615] Like the Department of Immunochemistry already mentioned above, the Department of Tobacco Research was dissolved as of October 31, 1936.[616]

The circumstances leading to the formation of the Department of Tobacco Research provide a good example of Carl Neuberg's attempts to further reinforce his position within the KWG. In his efforts to set up the new department, Neuberg was apparently pursuing the objective, despite the generally poor financial situation, of expanding his institute's area of work, but of doing so by mobilizing the necessary funds outside the KWG through collaboration with government and industry. With skill and persistence he succeeded, as he had already done during the First World War, in putting the results of his research to practical use, while utilizing the funds he was able to raise as a result to pursue his own scientific interests. Even in his later years, Neuberg, himself a heavy smoker,[617] seems still to have retained an interest in the topic of research into tobacco plants[618] – which he had spent eight years working on, alongside his regular work as Director of the KWI for Biochemistry, together with his close colleague Maria Kobel.

615 Planck to Neuberg, July 9, 1935. MPG Archives, Dept. I, Rep. 1A, 2057, 29.

616 Accounts for the 1937 fiscal year. MPG Archives, Dept. I, Rep. 1A, 1463a.

617 Carl Neuberg to Ernst Neuberg, November 20, 1947. CNP, Box 12: Neuberg, Ernst.

618 During his time in the United States, Neuberg collected numerous newspaper cuttings reporting on processes serving to reduce the nicotine content of tobacco leaves by fermentation methods. CNP, Box 15: Tobacco-Misc.

6. FROM BERLIN TO NEW YORK – NEUBERG'S ESCAPE FROM HITLER'S GERMANY

6.1. NEUBERG'S DISMISSAL

The history of the KWG during the National Socialist reign of terror has been and still is the subject of numerous investigations.[619] All that will be done here, in order to provide a better understanding of the processes that affected Carl Neuberg, is to give a brief overview of the dramatic changes associated with the coming to power of Hitler and the *National Socialist German Workers' Party*.[620]

From a historical perspective, the time of the Weimar Republic (1918–1933) may be regarded as a distinctly science-friendly period, even if, during its later years, economic problems led to repeated cuts in the state's share of the funding provided for science. In principle, political circles were well aware of the economic importance of basic research. It was considered part of Germany's cultural mission to promote it, no doubt with the aim, among others, of strengthening Germany's reputation as a nation of science.

There was a clear caesura in this respect when Hitler came to power. The new Chancellor had a completely antiquated understanding of technology, and thanks to his fundamental ignorance, nurtured a deep mistrust of science in general. Achievements in academic fields, especially if they were the work of *Jewish* members of the community, were regarded by Hitler simply as a tissue of lies, produced with the sole objective of weakening the German nation. The dictator described cultivated people, scientists and artists as "smart-aleck weaklings."[621] Until the time later on when the Nazi leadership came to recognize the importance of basic research in the preparations for war, the predominant aim was to introduce the *"führerprinzip,"* the principle of unquestioning obedience to ideological authority, and to force Jews and politically undesirable persons out of their positions in order to replace them with faithful party members, irrespective of how qualified they were.[622] Apart from the countless unknown persons who were persecuted for religious, racial or political reasons, the Nazi tyranny led to an

619 More detailed descriptions are to be found in Albrecht/Hermann (1990), Macrakis (1994), Schüring (2003) and Beyler (2004). See also the series on the history of the KWG during the National Socialist period [www.mpiwg-berlin.mpg.de/KWG/index.htm, 03/02/2006].

620 The following information is based, where not otherwise stated, on the account by Albrecht/Hermann (1990).

621 "geistreiche Schwächlinge".

622 Cf. the critical analyses regarding the German universities during the Nazi period that have appeared over recent years, e.g. van den Bussche (1989), Jakobi et al. (1989), Szabó (2000), Remy (2002), Hoßfeld et al. (2003), Bruch/Schaarschmidt (2005) and Lohff (2005).

exodus of leading personalities from the fields of science and culture,[623] the consequences of which continue to be felt right down to today.[624]

The KWG was not immune to these processes. The new regime had no coherent policy on the future of the KWG, since this research society, unlike the state institutions such as the universities, occupied a special position as a "private association." Despite increasing state participation in the individual KW Institutes, the KWG continued to draw funding from the private sector, too. As only a semi-public-sector organization, therefore, it was able to avoid official interventions, at least to a certain extent.[625] The new people in power and their adherents within the KWG were united only in the view that in the interests of the "national revolution" the "Jewish influence" had to be reduced. During the first phase of this so-called revolution, it had been attempted, with the assistance of Nazi cells within companies and organizations and a network of informers always ready to intrigue and plot, to eliminate communist and social democratic influences.[626] In the second phase, from April 1933 onwards, the attack on "Jewish" scientists began, whose "influence" these party members viewed with hostility as being detrimental to the "Aryan nation." The Nazi regime and its supporters sought to eliminate this influence by bringing about the dismissal and expulsion of Jewish scholars in favor of academics who were faithful to the party line.[627] In considering how this affected individual scientists working at the KWG, account must be taken of the extent to which the particular institute at which a Jewish academic was employed was financed by public funds. A distinction also has to be made between the university positions of the scientists concerned and their employment at the KWG.

On April 7, 1933, as one of the first pieces of legislation adopted by the Hitler administration, the *Gesetz zur Wiederherstellung des Berufsbeamtentums* – "Act for the Restoration of the Professional Civil Service" – was promulgated.[628] Appearing at first sight to be a measure to simplify public administration, it revealed its political perfidy in its Sections 3 and 4, which made it possible for "non-Aryan" or politically unreliable civil servants to be dismissed.[629] Of the total of 1,052 KWG staff on full contracts of employment, 52 fell under the provisions of the "Aryan Law," as it came to be known. Of these, 8 were the directors of

623 Cf. Macrakis (1994), Schüring (2003) and Beyler (2004).
624 Cf.. Medawar/Pyke (2001).
625 The public sector contribution (*Reich* and Prussia) was some 46% in 1932; or according to other figures, even over 50%. In addition, it must be taken into account that nine out of the total of 29 directors' positions of KWIs in Germany were financed by the State of Prussia, and that the individual KWIs were supported by state and private funding in varying proportions, i.e. were subject to state influence to differing extents. See Albrecht/Hermann (1990), 359.
626 Cf. Schüring (2003), 157–160.
627 Albrecht/Hermann (1990), 360.
628 With regard to the exodus of scientists from the KWG see Macrakis (1994), Schüring (2003) and Beyler (2004).
629 *Reichsgesetzblatt* [Legislative Gazette] 34 (1933).

institutes, 9 were heads of department and 21 were scientific assistants.[630] This does not, however, take account of those members of staff who also lost their jobs as a result of the way the vacant directorships were filled.

Although Neuberg counted as a "Jew" within the meaning of Section 3, the law in its original form could not be applied to him, since he was able to claim exemption in two respects under the provisions of Section 3 (2).[631] But the subsequent *Act Amending the Act for the Restoration of the Professional Civil Service*[632] did apply to him, and it was under its provisions that he was deprived of his university positions at the beginning of April 1934 and forced into retirement. The Prussian (and soon also national) Minister of Education, Bernhard Rust,[633] who like Neuberg came from Hanover,[634] informed him on March 28, 1934:

> "Pursuant to Section 6 of the Act for the Restoration of the Professional Civil Service of 4.7.1933, you are hereby retired from your position as Professor at the Agricultural College. [...] At the same time, I hereby withdraw your authorization to teach as an Honorary Professor of the University of Berlin."[635]

The Section 6 of the Act mentioned here provided, in its revised version of June 1933, that even honorary civil servants could be retired if this was considered necessary "for the simplification of administration or in the interests of the service."[636] This section of the Act thus gave very broad leeway for interpretation, and allowed government officials to get rid of many long-serving civil servants who were undesirable for political and/or racial reasons.

Subsequent to his compulsory retirement from his university chair, for which he had not even drawn a salary since his appointment to the KWG, Neuberg attempted to get the measure amended to the extent that he would retain the title of Professor Emeritus; but his application was rejected. A letter addressed to the

630 Cf. Table 1 in Albrecht/Hermann (1990), 361.

631 According to Sec. 3 (2) of the Act, civil servants who had entered the service before August 1, 1914, or who had been front-line combatants in the First World War - both of which applied to Neuberg – were exempted from its provisions.

632 *Reichsgesetzblatt* [Legislative Gazette] 68 (1933).

633 Bernhard Rust (1883–1945) studied German and philosophy for the teaching profession, joined the Nazi party in 1924 and became *gauleiter* of Hanover North as early as 1928. After the Nazis had come to power, he was appointed Minister of Science, the Arts and National Education of the State of Prussia in April 1933, and of the *Reich* in April 1934. He saw to it that all areas of education, science and culture were permeated by the Nazi ideology, and that thousands of Jewish intellectuals were immediately dismissed. In 1945 Rust evaded all responsibility by committing suicide. Weiß (1998), 392. Rust's cousin, Agnes Linke, was Neuberg's secretary at the KWI for Biochemistry.

634 Höffkes (1986), 277.

635 "Aufgrund § 6 des Gesetzes zur Wiederherstellung des Berufsbeamtentums vom 7.4.1933 werden Sie hiermit als ordentlicher Professor der Landwirtschaftlichen Hochschule in den Ruhestand versetzt. [...] Gleichzeitig entziehe ich Ihnen die Lehrbefugnis als Honorar-Professor an der Universität Berlin." Prussian Minister of Science, the Arts and National Education to Neuberg, March 28, 1934. AHU, Neuberg's Personnel File, UK–N 29, Vol. 1, sheet 8.

636 *Reichsgesetzblatt* [Legislative Gazette] 68 (1933).

management of the Agricultural College[637] shows that in requesting such an amendment, Neuberg was concerned not with material but rather with intangible values. He was seeking to avoid any appearance of having been dishonorably discharged from office. This letter impressively demonstrates how his sense of duty and his loyalty to the state still prevailed, even though the new rulers were threatening to destroy everything he had labored at and built up over the preceding decades.[638] But it also serves to demonstrate how bitter this time must have been for Neuberg, as he saw his academic career prematurely and forcibly ended:

> "What I would like to achieve is no more than simply to remain a Professor Emeritus, while waiving any pecuniary advantage or the fulfillment of any academic tasks, instead of being pensioned off. The reason is that while I am quite prepared to disappear from the stage, I would not want to become involved in any unfortunate discussions in Germany or abroad; for if I remain officially listed among the Professors Emeriti, nobody will know whether I vacated my position voluntary, or for reasons of health, or because I had reached the age limit after having been in public service for 36 years, since 1898. The fact that I would also prefer such a solution for reasons of personal honor is one I do not need to explain. Far be it from me to comment in any way upon any measures that may be necessary for reasons of state; but everybody knows that I have always attempted to do my duty both in peacetime and in wartime, and have not been guilty of any kind of offense."[639]

But Neuberg's efforts failed.[640] His length of service was determined at around 37 years, and he was awarded an annual pension of some RM 10,800, to be paid by the KWG.[641]

637 After receiving the Ministry's rejection of his request, Neuberg wrote the following letter in the hope that the Rector of the Agricultural College might still be able to achieve something for him by interceding with the Ministry.

638 See also Chapter 6.4.

639 "Was ich erreichen möchte ist nichts anderes, als daß ich unter Verzicht auf jede pecuniäre Besserstellung oder Ausübung akademischer Tätigkeit emeritiert statt pensioniert werde. Der Grund dafür ist, daß ich sang- und klanglos abtreten, aber nicht in irgendwelche unerwünschten Diskussionen des In- und Auslandes gezogen werden möchte; denn wenn in den offiziellen Verzeichnissen mein Name unter den Emeritierten geführt wird, weiß niemand, ob ich freiwillig oder aus Gesundheitsgründen oder wegen Erreichen der Altersgrenze ausgeschieden bin, nachdem ich 36 Jahre, seit 1898, im Staatsdienst tätig war. Daß ich auch vom Ehrenstandpunkt eine solche Regelung vorziehen würde, bedarf keiner Begründung. Mir liegt es fern, über irgendwelche Maßnahmen der Staatsnotwendigkeit überhaupt zu sprechen, aber jeder weiß, daß ich im Frieden wie im Kriege meine Pflicht zu tun mich bemüht und keinerlei Schuld auf mich geladen habe." Neuberg to the College Administration, July 7, 1934. AHU, Neuberg's Personnel File, UK–N 29, Vol. 3, sheet 128.

640 "Neuberg's wish, presented verbally, for his removal from office to be converted into emeritus status, is not granted." ("Dem mündlich vorgetragenen Wunsch Neubergs, die Zurücksetzung in eine Emeritierung umzuwandeln, wird nicht entsprochen.") Minister of Science, the Arts and National Education to Neuberg, July 6, 1934. AHU, Neuberg's Personnel File, UK–N 29, Vol. 3, sheet 129. See also Neuberg to the Rector of the Agricultural College, July 17, 1934. AHU, Neuberg's Personnel File, UK–N 29, Vol. 3, sheet 129.

641 Ministry of Science, the Arts and National Education to the Rector of the Agricultural College, August 23, 1934. AHU, Neuberg's Personnel File, UK–N 29, Vol. 3, sheet 149. According to this, Neuberg's pension was set at RM 10,810 p.a. and his period of service determined at 37 years and 138 days.

Neuberg's position as the director of an institute of the semi-public KWG must, however, be considered separately from his university activities. For those KWG employees who were subject to the provisions of the *Act for the Restoration of the Professional Civil Service*, the factors of decisive importance to their future were the extent to which the KWI employing them was financed by public funds, and the position they held in it. As early as June 1933 Max Planck[642] had presented a report to the Ministry on how the new Act was to be implemented at institutes that received more than 50% of their financing from public funds: Nineteen dismissals were recommended, and three doubtful cases and five cases of hardship listed. By the middle of July, all the "non-Aryan" employees mentioned in the report had been dismissed with the exception of two cases of hardship that were still under examination. The equivalent report in respect of the institutes that were predominantly privately financed, on the other hand, was not presented until September 1933. Furthermore, the KWG's Administrative Headquarters only provided figures in this report, and did not recommend any dismissals.[643] The Administrative Headquarters' personnel policy, later described by Friedrich Glum as *"Selbstgleichschaltung"* – "voluntary conformity" and submission to the dictates of Nazi ideology[644] – was characterized by the fact that the KWG attempted wherever possible to keep its elite researchers, while applying the new provisions directly and, with few exceptions, consistently in the lower and middle tiers.[645] Some historians describe Max Planck's actions in implementing the political exclusion of Jewish colleagues as being as vacillating "as trees in the wind."[646] This is not the place to investigate whether Planck's method of proceeding could be considered to be "resistance" to the political directives. From the present day point of view, it is not always possible to determine what his intentions were when he decided, as President of the KWG, when and how he would stand up for a particular scientist: Was he concerned with German science in general, were his actions directed entirely towards safeguarding the interests of the KWG, or was he, in some individual cases, also driven by personal motives?[647] A distinct measure of reserve was perceptible among those scientists at the KWG not affected by the discriminatory laws towards those of their colleagues who were threatened with dismissal, even though they had worked closely together for years. As Albert Einstein remarked to the physicist Max Born:

642 Max Planck (1867–1947), born in Kiel, from 1892 Professor of Theoretical Physics in Berlin. Formulated the quantum theory in 1900 (Nobel Prize 1918). During the politically critical period he was President of the KWG from 1930–1937 and in 1945; in 1949 it was reinstated under the name Max Planck Society *(Max-Planck-Gesellschaft)*. Cf. also Pufendorf (2006).

643 Albrecht/Hermann (1990), 366–367.

644 Cf. Glum (1964), 443 and Albrecht/Hermann (1990), 369–371.

645 Albrecht/Hermann (1990), 366.

646 Planck to Ernst Telschow (1899–1988), Secretary General of the KWG from 1937. Cf. Albrecht/Hermann (1990), 372. See also Przyrembel (2004) and Pufendorf (2006).

647 Cf. the account in Albrecht/Hermann (1990), 372–375. See also Beyler (2004) and Przyrembel (2004).

> "As you know, I have never had a particularly favorable view of the Germans (in a moral or political respect) [...]. But I must admit that I have nevertheless been to some extent surprised by the degree of their brutality and cowardice."[648]

In addition to pursuing all the legal avenues open to him in order to keep Jewish directors of institutes in their positions, Planck also attempted to persuade scientists who had been dismissed and were considering fleeing Germany to remain in the country, in order to mitigate the "damage" their emigration would cause to German science in general and the KWG in particular. Carl Neuberg was among those that Planck attempted to keep in Germany; he advised him to stay, and he did. It was a decision Neuberg would greatly regret a few years later.[649]

The events that led to Neuberg's dismissal from his position as Director of the Institute have already been outlined by Arnold Nordwig.[650] But the evaluation of documents in the archives of the MPI and the APS[651] has brought aspects to light that had previously been only inadequately considered, or even escaped notice altogether. These are dealt with in the following account: Carl Neuberg's dismissal as Director of the KWI for Biochemistry was largely a result of the machinations of a laboratory servant and mechanic at the institute, Kurt Delatrée-Wegner.[652] It was his calumnies that triggered a tug-of-war lasting over a year between the KWG, Nazi party organizations and three ministries, at the end of which Neuberg's position was so weakened that he was forced into retirement at the end of September 1934.

Delatrée-Wegner, who a short time before had tended to be more susceptible to communist than to Nazi ideas, had joined the Nazi party a few months after Hitler's seizure of power, and was appointed *"Obmann"* – foreman or shop steward – at the Kaiser Wilhelm Institute for Biochemistry by the party's "organization of industrial cells" (*"Nationalsozialistische Betriebszellenorganisation"* – NSBO).[653] Thanks to his new position as a representative of the ruling party, he felt himself authorized to cause unrest at his place of work,[654] and according to

648 "Du weißt, daß ich nie besonders günstig über die Deutschen dachte (in moralischer und politischer Beziehung) [...]. Ich muß aber gestehen, daß sie mich doch einigermaßen überrascht haben durch den Grad ihrer Brutalität und Feigheit." Einstein to Max Born (1872–1970), the Danish theoretical physicist, May 30, 1933. Quoted according to Albrecht/Hermann (1990), 368–369, note 27.

649 Neuberg to Butenandt, March 7, 1947, MPG Archives, Dept. III, Rep. 84/1, sheet 608. Neuberg did not take Walther Nernst's advice to leave the country.

650 Nordwig (1983).

651 Conrads (2002), Schüring (2004).

652 In some documents Delatrée-Wegner is called Wegner, in others Wegener.

653 A statement of Neuberg's, dated March 6, 1934, gives information on Delatrée-Wegner's party membership: "Wegner told me on May 2, 1933, that he had become a National Socialist on the previous day [...]." MPG Archives, Dept. I, Rep. 1A, 2035, sheets 94–95. For Wegner's appointment as an "Obmann" in the NSBO, see MPG Archives, Dept. I, Rep. 1A, 2035, sheet 51.

654 Cf. statements by institute workers Otto von Schönebeck and Eduard Hofmann regarding Delatrée-Wegner's political attitude. MPG Archives, Dept. I, Rep. 1A, 2035, sheet 106.

Neuberg, saw it as his task to "cleanse the institute of Jews."[655] The first written complaint about Delatrée-Wegner was submitted in May; Neuberg forwarded the letter to the Administration, pointing out in his covering letter that all the signatories were "Aryans."[656] The complaints became more concrete a short time later, and the staff asked for the mechanic to be dismissed; they accused him of lying, calumny, theft and arrogance, in particular towards foreign guests,[657] as a result of which Neuberg suspended Delatrée-Wegner the next day.[658] In the days that followed, however, Neuberg was able to persuade the staff to withdraw their request for Delatrée-Wegner's dismissal, so that his suspension was revoked and he was able to return to work.[659]

The matter escalated, however, when Delatrée-Wegner physically attacked the scientific assistant Werner Schuchardt, who was one of those who had signed the letter complaining about him; whereupon Neuberg dismissed him without notice.[660] However appropriate and understandable Neuberg's reaction may have been, his decision was, nevertheless, naturally a hazardous one at that time and one that was easily twisted to represent an affront to the Nazi leadership. Just when Jews were losing their jobs all over the country under the provisions of the "Aryan Law," here was an institute director of Jewish origin dismissing a card-carrying member of the NSDAP.

It was of course impossible for this action, which was doubtless perceived as a provocation and an open declaration of war, to remain without consequences. On July 20, 1933, Neuberg was summoned to appear before the so-called "Brandenburg District Office for Industrial Peace," a meeting which the Office Manager of the KWG's Administrative Headquarters, Fritz Arndt,[661] also attended at Neuberg's request. In the course of this meeting, which Delatrée-Wegner and the SA *Sturmbannführer* of Dahlem were also present at, the party officials who were there demanded the reinstatement of the mechanic in the most unpleasant terms, even threatening to send Neuberg to a concentration camp.[662] But despite these massive threats, Neuberg refused to sign the record and thereby revoke the dismissal. It later turned out that this "commission," headed by a certain Neufind,

655 Statement by Neuberg. MPG Archives, Dept. I, Rep. 1A, 2035, sheets 94–95.

656 Letter of complaint of May 10, 1933, and Neuberg's covering letter of the following day. The letter was signed by the assistants Auhagen, Collatz, Kobel, von Schönebeck and Schuchardt and the technical assistants Kühl and Reymann. MPG Archives, Dept. I, Rep. 1A, 2035, sheets 1–2.

657 MPG Archives, Dept. I, Rep. 1A, 2035, sheets 7–10.

658 The documents are not completely unambiguous, in that some describe the complaint as having led to the "suspension," others to the "dismissal" of Delatrée-Wegner. However, whether it was suspension or dismissal proved to be irrelevant in the ensuing events. MPG Archives, Dept. I, Rep. 1A, 2035, sheets 35 and 100.

659 Neuberg's instruction to Delatrée-Wegner to return to work, May 23, 1933. MPG Archives, Dept. I, Rep. 1A, 2035, sheet 15.

660 Record of the occurrence of June 3, 1933, and Neuberg's letter of dismissal to Delatrée-Wegner of June 11, 1933. MPG Archives, Dept. I, Rep. 1A, 2035, sheets 19–21.

661 A member of the administrative staff (1884–1968). See also Brocke (1990b), 267.

662 MPG Archives, Dept. I, Rep. 1A, 2035, 32 f., sheets 35–36.

was probably self-appointed, and that the meeting had taken place without the knowledge or authorization of the responsible ministry or of any higher ranking party officials.[663] On the same evening, Neuberg received threatening telephone calls at home:

> "At 11 p.m. on July 20, Dr. Collatz received a call in Professor Neuberg's private apartment from somebody whom from the voice he took to be Mr. Wegener. Professor Neuberg himself was rung up at 2.15 in the same night by a caller who said: 'We just wanted to make sure you haven't cleared off yet, sweetie.'"[664]

But neither Neuberg himself nor the KWG's Administrative Headquarters allowed themselves to be intimidated by these threats, so that Delatrée-Wegner continued to be refused reinstatement.[665] The party organization now started to look for other ways to drive the Jewish institute director out of his position. It was claimed that Delatrée-Wegner had reported Neuberg to the chairman of the KWI Staff Council two days after his first suspension in May 1933, accusing him of insulting the *führer*.[666] The alleged insult was said to consist of Neuberg's having called Hitler a "bull in a china shop." However, there were no witnesses,[667] and in view of the short time between Delatrée-Wegner's suspension and this denuncia-

663 This can be seen from a letter from the Secretary General of the KWG to the *Reich* Minister of the Interior, September 16, 1933. MPG Archives, Dept. I, Rep. 1A, 2035, sheets 51–54.

664 "Am 20. Juli um 11 Uhr abends wurde in der Privatwohnung Prof. Neubergs Dr. Collatz angerufen von jemand, den er nach der Stimme für Herrn Wegener hielt. Prof. Neuberg selbst wurde um $^1/_4$ 3 Uhr in der selben Nacht angerufen mit den Worten: 'Wir wollten uns nur überzeugen, Süsser, ob Du noch nicht ausgerückt bist'." MPG Archives, Dept. I, Rep. 1A, 2035, sheets 35–36.

665 Memo from Telschow to Glum, July 27, 1933. MPG Archives, Dept. I, Rep. 1A, 2035, sheet 39. After Wegner's dismissal, Neuberg wrote him the customary certificate of employment, which, however, contained the remark "Dismissed on account of contentions." The District Head of the NSBO, Hecker, demanded that Neuberg remove this remark, which he finally did. MPG Archives, Dept. I, Rep. 1A, 2035, sheets 77–80.

666 With regard to this incident and the following confrontation, see MPG Archives, Dept. I, Rep. 1A, 2035, sheet 45. At this point, doubt may be cast upon whether Delatrée-Wegner had really reported Neuberg as early as May. The accusations were taken up by Wegner's fellow party members, and they were not put forward until the attempts to get Delatrée-Wegner reinstated at the Institute after his dismissal had been going on for months. Neuberg commented on this as follows: "W.'s whole lack of credibility is shown by the circumstance that neither he nor his henchmen, who had practically begged for his reinstatement, had ever raised the accusation against me that he then invoked 10 months later. Surely it would have been irreconcilable with his own dignity to have wanted to work under a man accused of such a thing." ("Die ganze Unglaubwürdigkeit des W. geht doch auch aus dem Umstand hervor, daß weder er noch seine Mittelsmänner, die um seine Wiedereinstellung geradezu bettelten, niemals die nach 10 Monaten von ihm beschworene Beschuldigung gegen mich erhoben haben. Es muß doch als unvereinbar mit der Würde erscheinen, unter Leitung eines derartig beschuldigten Mannes arbeiten zu wollen.") Neuberg to Glum, March 6, 1934. MPG Archives, Dept. I, Rep. 1A, 2035, sheet 100.

667 Delatrée-Wegner received support from two party members working at the KWG. Georg Graue, scientific assistant at the KWI for Chemistry, and the manager of the Harnack House, Giersch, both declared that Wegner had told them of Neuberg's alleged remark on the day it occurred. MPG Archives, Dept. I, Rep. 1A, 2035, sheet 60.

tion it is more than probable that it was viewed as an act of revenge on the part of the laboratory servant. Despite this, there was an official confrontation between Neuberg and his accuser. This meeting was initiated by the Society's Secretary General, Friedrich Glum, who thereby caused Neuberg to harbor a long-standing grudge against him. After the war Neuberg wrote to Glum:

> "I too am reluctant to recall that embarrassing episode in the past. I know only too well that as things were at the time it would have been difficult for you to handle the Wegner case in any other way. But what I found unpleasant was that contrary to what I had agreed with you, I was confronted with him in your office. This allowed him to claim, as he did at every opportunity, that the Kaiser Wilhelm Society supported him, even though at the time even Frick[668] decided in my favor, as my police files show."[669]

Neuberg, probably rightly, emphatically denied ever having used the alleged expression or anything like it.[670] The way things were, it was certainly highly improbable that Neuberg, as a "non-Aryan," should have expressed himself in such a disrespectful manner about Hitler after the Nazis had come to power and above all to a member of the Party; and this was how the KWG Administration saw the matter, too.[671] Despite this, in February 1934 members of the Institute were questioned about Neuberg's political attitudes, causing the latter, also in February, to write to the President of the KWG, Max Planck, and ask for his support.[672]

Despite all this, at the beginning of March 1934 the Neuberg case was handed over to the courts, and investigations into a charge of insulting the Chancellor were initiated.[673] Although the KWG's Administrative Headquarters forwarded a letter written by Neuberg in his defense to the Ministry a few days later, the case appeared to have already been decided at this level, since Ministerial Counselor Max Donnevert informed the Deputy General Manager of the KWG, Ernst Telschow, in April that it would be even easier at this point in time to accept a request from Neuberg to be given emeritus status and thus to "vacate his position with honor."[674]

During this period when he was under threat, Neuberg did not feel that he was being given adequate support by the KWG's Administrative Headquarters. Accor-

668 Wilhelm Frick (1877–1946), *Reich* Minister of the Interior 1933–1943.

669 "An die peinliche Episode der Vergangenheit denke auch ich ungern. Genau weiss ich, dass Sie unter den damaligen Verhältnissen, den Fall Wegner nur schwerlich anders behandeln konnten. Unangenehm habe ich nur empfunden, daß ich gegen Abrede mit Ihnen, in Ihrem Dienstzimmer mit ihm confrontiert wurde. Dies gab ihm zu der überall ausgestreuten Behauptung Anlass, dass die Kaiser-Wilhelm-Gesellschaft ihn stütze, obgleich damals sogar Frick zu meinen Gunsten entschied, wie meine Polizeiacten ausweisen." Neuberg to Glum, March 2, 1948. CNP, Box 5: Glum, Friedrich.

670 MPG Archives, Dept. I, Rep. 1A, 2035, sheet 45.

671 Glum to Buttmann, a senior civil servant in the Ministry of the Interior, September 16, 1933. MPG Archives, Dept. I, Rep. 1A, 2035, sheet 54.

672 Cf. Schüring (2004), 350.

673 Memo of Telschow, March 3, 1934. MPG Archives, Dept. I, Rep. 1A, 2035, sheets 91–92.

674 Memo of Telschow, April 20, 1934. MPG Archives, Dept. I, Rep. 1A, 2035, sheet 118. This shows that Neuberg had already submitted such a request to the President of the Society at some earlier time, which no reference can be found to in the documents now available.

ding to his attorney, the Ministry of Education had not been sent any detailed documents by the Society in Neuberg's defense. His letter to the KWG's Administrative Headquarters on the matter has overtones of despair, but also of the will to resist:

> "For me, this is a struggle for my honor, which in this case is also the honor of the Kaiser Wilhelm Society; at the same time, I am fighting for the honor of all those who work in the institute I head. [...] It seems to be a very strange situation when the Administrative Headquarters requests me to keep them informed of how things are proceeding, instead of the other way round. If they have withdrawn their confidence from this Institute and from myself, one of their oldest officers, then I do not think that I am being immodest, I who have otherwise never bothered the Administrative Headquarters, if I ask them to make a simple statement to that effect. If this fear I have expressed is unfounded, then I believe all the more that I and my staff have a claim to be given powerful support."[675]

Neuberg did not know that the Administrative Headquarters and the President of the KWG, Max Planck, had indeed spoken up in his favor;[676] but either their efforts were too late, or else they were in any case, in view of the political situation, doomed to failure. Neuberg's fate at the KWG was sealed on the political level, and all that appears to have remained was to find a suitable way of removing him from his directorship at the KWG.

Both Neuberg and the Administrative Headquarters having recognized the signs of the times, it was already being considered how matters at the KWI should proceed after Neuberg's resignation as Director. In these negotiations, Neuberg demanded among other things that all his assistants should keep their jobs after his departure, and that he himself should be given a modest opportunity to continue with his work in the little Gatehouse on the Institute site where the Danish cell biologist Albert Fischer had worked.[677] In neither case was he able to get his way.

The charge of political unreliability[678] against Neuberg had to be dropped,[679] however, since even at that time, the patent lack of credibility of the mechanic Delatrée-Wegner did not allow any other decision. As Neuberg was classified as a First World War front-line combatant, he could not at first be removed from his

675 "Es handelt sich für mich um einen Kampf um meine Ehre, die im vorliegenden Fall auch die Ehre der Kaiser-Wilhelm-Gesellschaft ist; ich kämpfe zugleich um die Ehre aller Angestellten des von mir geleiteten Instituts. [...] Es erscheint als eine seltsame Situation, daß die Generalverwaltung mich ersucht, sie über den Fortgang der Ereignisse auf dem Laufenden zu halten, statt umgekehrt zu verfahren. Falls sie dem hiesigen Institut und mir, einem ihrer ältesten Beamten, das Vertrauen entzogen hat, so glaube ich, der ich niemals die Generalverwaltung behelligt habe, nicht unbescheiden zu sein, wenn ich eine einfache Erklärung darüber erbitte. Sollte diese von mir geäußerte Befürchtung grundlos sein, so denke ich mit meinen Mitarbeitern wohl erst recht Anspruch auf eine tatkräftige Unterstützung zu haben." Neuberg to the KWG's Administrative Headquarters, March 21, 1934. MPG Archives, Dept. I, Rep. 1A, 2035, sheet 104.

676 MPG Archives, Dept. I, Rep. 1A, 2035, sheets 102 and 112.

677 Memo of Telschow, April 20, 1934. MPG Archives, Dept. I, Rep. 1A, 2035, sheet 118.

678 Act for the Restoration of the Professional Civil Service, Dept. 4. *Reichsgesetzblatt* (Legislative Gazette) 34 (1933).

679 Pelckmann to Neuberg, April 20, 1934. MPG Archives, Dept. I, Rep. 1A, 2035, sheet 122.

positions either on racial or on political[680] grounds. However, the amendment of the "Aryan Law" soon made it possible for him to be straightforwardly dismissed from his university positions in Berlin. Although the calumnies of the mechanic Delatrée-Wegner did not officially lead to his dismissal from the KWG, it can only be assumed that the political sensitivity of the whole affair contributed to Neuberg's finally being deprived of his position as Director. His forced retirement from the KWG was in all probability implemented with effect from September 30, 1934, having been carried through, like his compulsory retirement from the University in March 1934, on the basis of Section 6 of the amended Act for the Restoration of the Professional Civil Service.[681]

Max Planck did attempt to intervene with the Ministry to prevent this dishonorable dismissal,[682] but his primary concern appears to have been to avoid any damage being inflicted on the Society. In the first place he was afraid that there would be general unrest amongst the KWG scientists; secondly, he feared that the reputation of German science would suffer damage abroad; and last but not least, he wanted to prevent the Rockefeller Foundation,[683] which had already agreed to give the Society generous financial assistance, from withdrawing from this commitment on account of Neuberg's dismissal. Furthermore, he was concerned that Neuberg's dishonorable forced retirement would jeopardize the appointment of Hans von Euler-Chelpin as his successor:[684]

> "At present I have to consider it as being certain that Professor von Euler, in view of the extraordinarily sensitive mood in Sweden, would refuse the Directorship of the Kaiser Wilhelm Institute for Biochemistry, being concerned that he might be reproached with disloyalty to Professor Neuberg, with whom he has been in contact for many years and of whose Institute he is an external scientific member."[685]

680 Cf. Act for the Restoration of the Professional Civil Service, Secs. 3 and 4. *Reichsgesetzblatt* [Legislative Gazette] 34 (1933).

681 Cf. Planck (1936a), 114. Karlson (1990), 99, states that Neuberg submitted his resignation himself; Nordwig (1983) says that he was probably compulsorily retired pursuant to Sec. 6 of the Act; Schüring (2004) writes that Neuberg officially lost his position on October 1. Macrakis (1994), 279, indicates that he was forced to resign; in Table 3 in Albrecht/Hermann (1990), 364, his resignation is even stated to have been "voluntary" and to have occurred in 1936.

682 Max Planck to the *Reich* Minister for Science, the Arts and Education, Bernhard Rust, July 24, 1934. Schottlaender (1988), 92.

683 On the role of the Rockefeller Foundation and the funding it provided to the KWG, cf. Kay (1993).

684 See also Macrakis (1994), 279–280.

685 "Zur Zeit muß ich als sicher annehmen, daß Herr Prof. v. Euler angesichts der außerordentlich empfindlichen Stimmung in Schweden die Annahme der Direktorstelle am Kaiser-Wilhelm-Institut für Biochemie in der Sorge, daß ihm unkollegiales Verhalten gegen Herrn Prof. Neuberg, mit dem er seit langen Jahren in Verbindung steht und dessen Institut er als auswärtiges wissenschaftliches Mitglied angehört, vorgeworfen werden könnte, ablehnen würde." Schottlaender (1988), 92.

Planck, in his letter to the Minister of Science, the Arts and National Education, Bernhard Rust, pursued a strategy in the interests of science and of the KWG[686] whereby Neuberg would initially remain in office until such time as the Rockefeller Foundation had finally approved its financial support,[687] after which Neuberg would submit a request to be given emeritus status, so that negotiations could then be taken up again with Professor von Euler-Chelpin as his successor.

In later negotiations concerning his pension, Neuberg stated that he had submitted his request for emeritus status to Planck on August 11, 1934, but that Planck had apparently "not forwarded this letter, or done so only after a delay."[688] The Ministry was not prepared to convert the forced retirement into emeritus status; but Neuberg was allowed to continue to direct the Institute on an acting basis for the time being. His work for the KWG finally ended on September 30, 1936, when Adolf Butenandt was appointed the Institute's new Director.[689]

Neuberg's assistants, and indeed the other staff of the KWI for Biochemistry too, displayed extreme loyalty to him despite the uncertainty of their own positions. Almost all the employees demonstrated this through their willingness to testify on behalf of their Director. In the correspondence concerning the Delatrée-Wegner case that is preserved in the archives, it is pointed out again and again that all these witnesses were of "Aryan origin," were sympathetic towards the "national revolution,"[690] and in some cases were even members of Party organizations.[691] Even the new employees appointed in 1933 to replace other staff dismissed on grounds of racial ideology attempted to save Neuberg from his threatened suspension as Director of the Institute.

After the war, Neuberg wrote about the attitude of his staff, including his secretary Agnes Linke:[692]

> "Those who were dismissed in May 1933 were attached to me, as were all my staff, even in the time of conflict, as you will remember. They stayed with me right up until my dismissal, however much I advised them not to. Almost all of them have written to me now, including Dr. Hofmann.[693] My last secretary was a cousin of Rust, the Minister of Education. She put

686 For the description of Planck's policy as being like "trees in the wind" ("Bäume im Wind"), cf. Albrecht/Hermann (1990), 369–375 and Macrakis (1994).

687 It was a matter of RM 1.5 million in convertible currency.

688 Neuberg to Otto Hahn, October 1, 1947. CNP, Box 6: Hahn, Otto. This point was important to Neuberg since it represented a diminution of 25% in his later remuneration.

689 Glum to Neuberg, August 20, 1936. CNP, Box 7: Kaiser Wilhelm Gesellschaft.

690 MPG Archives, Dept. I, Rep. 1A, 2035, sheets 1 and 3.

691 The assistants Eduard Hofmann and Ernst Auhagen were members of the SA, Otto von Schönebeck was listed as a member of the SS association. Maria Kobel and Werner Schuchardt described themselves as non-political, and Herbert Collatz was a member of the DNVP (German National People's Party) until its dissolution. Cf. evidence given by Attorney Sack. MPG Archives, Dept. I, Rep. 1A, 2035, sheet 105.

692 The secretary Agnes Linke succeeded Margot Selberg, who was dismissed for racial reasons. Bernhard Rust headed the Ministry which finally forced Neuberg into retirement. MPG Archives, Dept. I, Rep. 1A, 2033, sheets 121 and 123; 2040, sheet 125. See also Neuberg to Linke, January 26, 1948. CNP, Box 8: Linke, Agnes.

693 Neuberg's papers do indeed contain quite a substantial amount of correspondence with many of his former assistants, secretaries or technical assistants.

the portrait of Hitler and 'Mein Kampf' in a room I never went into, she searched the files to find documents I had long forgotten which showed that I had managed, through a Jewish doctor, to get [Delatrée-]Wegner accommodated in a sanatorium free of charge, after he had thrown a burning kerosene lamp at his mother, and the poor woman had come to me in tears, asking for my help."[694]

As such upright conduct was unfortunately not to be taken for granted at that time, it is appropriate to list those members of staff who are mentioned in the files as having made statements in Neuberg's favor, or as having officially taken his part in some other way:

Dr. Ernst Auhagen, Assistant
Millie Berlak, Secretary
Dr. Herbert Collatz, Assistant
Stefan Grzycki, Scholarship holder from Poland
Dr. Eduard Hofmann, Assistant
Charlotte Kühl, Technical Assistant
Agnes Linke, Secretary
Neumann, Mechanic
M. Reymann, Technical Assistant
Otto von Schönebeck, Assistant
Werner Schuchardt, Assistant
Margot Selberg, Secretary

The affair relating to the laboratory assistant Delatrée-Wegner, who was allegedly mentally ill,[695] was, however, not the only threat to Neuberg's person and position. Neuberg himself and his institute were also exposed to the enhanced political influence under the Nazi regime of Hans Reiter, with whom he had

694 "Die im Mai 1933 Entlassenen hingen an mir, wie es alle meine Mitarbeiter – dessen werden Sie sich erinnern – selbst in der Conflictzeit taten. Sie blieben bis zu meiner Entlassung bei mir, so sehr ich ihnen auch abriet. Fast alle haben mir jetzt geschrieben, auch Dr. Hofmann. Die letzte Sekretärin war die Cousine des Kultusministers Rust. Sie brachte Hitlerbild und 'Mein Kampf' in einem Raum unter, den ich nie betrat, sie suchte aus den Acten die von mir längst vergessenen Belege heraus, dass ich Wegner durch einen jüdischen Arzt kostenlosen Aufenthalt in einem Sanatorium verschafft hatte, als er seiner Mutter eine brennende Petroleumlampe an den Kopf geworfen und die arme Frau weinend um meine Hilfe gebeten hatte." Neuberg to Glum, March 2, 1948. CNP, Box 5: Glum, Friedrich.

695 Medical certificate stating that Wegner was suffering from severe mental illness. MPG Archives, Dept. I, Rep. 1A, 2035, sheet 63. As to the subsequent fate of the mechanic Delatrée-Wegner, Otto von Schönebeck informed his former chief Neuberg: "[...] here I have to report to you that I came across my old adversary, Wegner, who told me (would you believe it?) that he was 'a victim of political persecution.' Such creatures are called chameleons. He had gone to Knoll in Ludwigshafen and I had two clashes with him in public in 1936. My wife was afraid he would report me to the police, but in the end he was too much of a coward." ("[...] hier muß ich Ihnen doch berichten, daß mir mein alter Widersacher, Wegner, über den Weg gelaufen ist und mir mitteilte, man staune, 'politisch Verfolgter' sei. Chamäleon nennt man solche Tiere. Er war nach Ludwigshafen zu Knoll gekommen u. ich hatte 1936 zweimal Zusammenstöße in der Öffentlichkeit mit ihm. Meine Frau fürchtete er würde mich anzeigen, aber dazu war er dann doch zu feige.") Schönebeck to Neuberg, September 20, 1948. CNP, Box 14: Schönebeck, Otto von.

already crossed swords in the days of the KWI for Experimental Therapy.[696] As has already been mentioned elsewhere, Reiter had hoped to succeed Wassermann as Director of the Institute after the latter's death. But instead of being appointed Director, Reiter was dismissed from the KWG. In the meantime, he had risen to high office in the state: As an expert in racial hygiene[697] he had been appointed President of the *Reichsgesundheitsamt*, the National Office of Health, in 1933.[698] It is impossible to avoid the impression that he wished to take revenge on Carl Neuberg, whom he held to be among those to blame for his earlier humiliation.[699] Friedrich Glum remembers the following incident:

> "One day a former secretary[700] of Professor Neuberg's came to me and showed me a letter from the President of the *Reichsgesundheitsamt*, Professor Reiter, in which she was called upon to hold herself in readiness to make a statement that Professor Neuberg had attempted acts of indecency with his secretaries."[701]

Reiter's attacks were, however, not confined to the Director of the Institute, but appeared to encompass the other members of staff as well. In a letter written after the war, for example, Neuberg states:

> "He [Reiter, Authors' note] sought to exploit the situation to such an extent that he even placed hindrances in the paths of all my Aryan assistants, in particular of the most excellent Maria Kobel, who finally, thanks to my old connections, just managed with great difficulty to find a job at the Beilstein publishing house."[702]

Furthermore, the *Reichsgesundheitsamt* had plans to take the KWI for Biochemistry under its own wing and thus place it under the direction of its President, Hans Reiter.[703] The reason given for this plan was the alleged importance of the research activities being carried out at the *Reichsgesundheitsamt*, and the lack of space there. This proposal, which was not directed simply against Neuberg, whose dismissal was imminent in any case, but also targeted the independence of the KWG, was, however, not pursued by the responsible ministries.

To conclude: The KWI for Biochemistry serves as a clear example of how in that "time of conflict" it was not only Jewish scientists and employees who were victims of Nazi policy. In addition to the personnel already mentioned who lost

696 Cf. Chapter 5.5.

697 Since 1933, Reiter had also been the University Officer for Rostock of the National Socialist Teachers' Association. The subjects he lectured on included racial hygiene, and he introduced so-called "social hygiene study tours." See Heiber (1991) and Keitel (2004a).

698 Keitel (2004a), 245–246.

699 Cf. Glum (1964), 286.

700 This refers to Agnes Linke, the cousin of Education Minister Rust.

701 "Dann kam eines Tages eine frühere Sekretärin von Professor Neuberg zu mir, die mir einen Brief des Präsidenten des Reichsgesundheitsamtes, Professor Reiter, zeigte, in dem sie aufgefordert wurde, sich für eine Aussage bereit zu halten, daß Professor Neuberg Versuche gemacht hätte, sich an seinen Sekretärinnen zu vergehen." Glum (1964), 446.

702 "Seine Konjuncturpolitik ging so weit, daß er meinen sämtlich arischen Assistenten das Fortkommen erschwerte, namentlich der sehr ausgezeichneten Maria Kobel, die schließlich noch mit knapper Mühe und Not durch meine alten Beziehungen in der Beilstein Redaktion unterkam." Neuberg to Glum, December 4, 1947. CNP, Box 5: Glum, Friedrich.

703 Nordwig (1983), 52–53.

their jobs on the basis of the so-called "Aryan Law," the politically motivated forced retirement of the Director also led to the dismissal of "Aryan" employees. Disregarding those members of staff who had already sought and found new positions before Neuberg's final suspension,[704] five assistants and three laboratory workers at the KWI for Biochemistry were affected in this way.[705] Thus at Neuberg's institute alone, at least 17 people lost their jobs through no fault of their own, due to the discriminatory legislation of the National Socialists.

6.2. ADOLF BUTENANDT: NEUBERG'S SUCCESSOR

Shortly after Neuberg had been "retired," and this had become known at the KWG, a number of scientists applied for his position as Director of the KWI for Biochemistry. In a letter to the Nazi Minister of Education, Bernhard Rust, Max Planck refers to a chemist named Lüdicke,[706] who was said to have applied for the position of Director of the Institute with a recommendation from the "Faculty Association of the Kaiser Wilhelm Society." [707] Planck expressed his suspicion that this application, received by the *Reich* Ministry of Science, could be connected with the calumnies raised against Neuberg, and pointed out that there was no such thing as the "Faculty Association of the Kaiser Wilhelm Society."[708] There was merely an academic council, but this body had not yet discussed Neuberg's replacement. In November 1935 Professor Houben[709] expressed his interest in the Institute's premises. He suggested to Planck that he might take over the direction of the Institute and work there with his assistants, but would vacate the premises again in good time if a new appointment was made.[710] Planck refused this request, pointing out that the procedure of making a new appointment was pending.[711]

704 Otto von Schönebeck moved to the company IG Farben in July 1935. MPG Archives, Dept. I, Rep. 1A, 2040, sheet 181. Ernst Auhagen had already done the same in December 1933. Bayer-AG (1967). When two technical assistants (v. Vitzum and Reymann) left the Institute, and what became of them, is unknown.

705 Only the secretaries employed to replace Jewish predecessors in 1933, the two cleaners and three further lower level employees were taken over by the new Director, Adolf Butenandt. MPG Archives, Dept. I, Rep. 1A, 2041, sheet 12.

706 It must be left open whether this might have been Karl Lüdecke, who was involved in the industrial manufacture of the glycerol substitutes. According to information given personally by Dr. Kazemi of the MPG Archives on May 5, 2006, no reference to a "Lüdicke" could be found in the archived documents.

707 "Dozentenschaft der Kaiser-Wilhelm-Gesellschaft." Max Planck to the Minister of Science, the Arts and National Education, Bernhard Rust, July 24, 1934. Schottlaender (1988), 94.

708 With regard to the role played by the *Nationalsozialistische Deutsche Dozentenschaft* (National Socialist German University Faculty Association – NSD) cf. Lohff (2005), 221–223.

709 Josef Houben (1875–1940), chemist, from 1921 Head of the *Chemische Reichsanstalt* (National Institute for Chemistry) in Dahlem, Berlin, and a professor at the University. He laid down both offices in 1933. Houben worked on the chemistry of terpenes and camphors.

710 Houben to Planck, November 4, 1935. MPG Archives, Dept. I, Rep. 1A, 2033, sheet 199.

711 Planck to Houben, November 6, 1935. MPG Archives, Dept. I, Rep. 1A, 2033, sheet 200.

Despite the pain of Neuberg's forced retirement, scientific work continued at the institute under his acting directorship, as is demonstrated by the publications dating from this period.[712] In the meantime, the KWG's efforts to persuade Hans von Euler-Chelpin or Adolf Windaus[713] to take over Neuberg's position at the KWI for Biochemistry had failed. It was not until May 1936 that a successor for the position of Director was found: the chemist Adolf Butenandt,[714] who was appointed on May 1, 1936. Butenandt had taken over the Chair of Organic Chemistry at the *Technische Hochschule* (College of Advanced Technology) in Danzig (now Gdańsk in Poland) from Alfred Wohl, a Jewish professor dismissed in 1933.[715] Although Butenandt's biographer and former subordinate Peter Karlson mentions "political reservations" against Butenandt's appointment in the Ministry of Science,[716] these appear to have been quickly dispelled.[717]

During his three years in Danzig, Butenandt had been engaged in successful research into human sexual hormones.[718] After a visit to America, Butenandt was offered an appointment at Harvard University in Boston[719] in the fall of 1935,

712 Cf. the first volume of the periodical *Enzymologia Acta Biocatalytica* edited by Carl Oppenheimer in 1936/37. Oppenheimer (1874–1941) taught biochemistry and economic chemistry at Berlin Agricultural College from 1908 to 1936. In 1932 he was appointed a member of the Leopoldina, was forcibly retired (with emeritus status) in 1936 and emigrated to The Hague. He published many biochemical works, was the founder of the *Biochemisches Zentralblatt*, and influenced the rapidly developing field of enzymology for several decades. His periodical *Enzymologica* afforded many emigrant research workers the opportunity to publish their work. It was renamed *Molecular and Cellular Biochemistry* in 1972.

713 Adolf Otto Reinhold Windaus (1876–1976), Professor of Chemistry in Freiburg, Innsbruck and Göttingen, is considered to be the founder of modern vitamin research. He discovered vitamin D, and described the role of cholesterol in the human organism, for which he was awarded the Nobel Prize in Chemistry in 1928.

714 Karlson (1990), 99–103. In addition to Butenandt, Professor Fritz Kögl (1897–1959) from Utrecht was also under discussion as a possible successor; he had worked as an assistant to Butenandt in Göttingen before being appointed to a Chair in Utrecht in 1932. The theory of cancer that he developed was later found to have been based on falsified results. See Karlson (1990), 299 and cf. also Trunk (2004), 258–259.

715 Alfred Wohl (1863–1939) was appointed a full Professor at the College of Advanced Technology in Danzig and Head of the Organic Chemistry Laboratory in 1904. He was accepted as a full member of the German Academy of Sciences Leopoldina in 1932. In the following year, he was dismissed as being politically unreliable. One year before his death, Wohl emigrated to Stockholm. Oddly, Schieder/Trunk (2004) make no reference to Butenandt's predecessor in Danzig. (After the First World War Danzig had been detached from Germany and made a Free City under League of Nations protection; the Nazis gained power there in 1933.)

716 Minutes of a meeting of the Executive Committee of the KWG, April 9, 1935. Quoted according to Karlson (1990), 100.

717 In view of Butenandt's fundamentally anti-Semitic and nationalistic attitude, nothing much stood in the way of his acceptance of the Nazi political philosophy. See also Schieder (2004).

718 Butenandt (1981), 2, see also Satzinger (2004a).

719 Schieder (2004), 35–37.

which forced the KWG to pursue its attempts to recruit him with greater urgency and probably also helped to dissipate any doubts about his political reliability.[720]

The Ministry's resistance to the appointment of somebody from outside the *Reich* was dispelled not least by Butenandt's application to join the Nazi party[721] in the spring of 1936.[722] Without directly mentioning his own application to join the party, Butenandt, in the transfiguring light of old age, gave the following account of his reservations with regard to the Nazi party and his "assessment" of the political situation in the "German Reich" at that time:

> "There were a number of people in Danzig who joined the NSDAP only because they saw this as being an expression of their commitment to Germany. It was only after I had moved to Berlin that I was able to make a proper appraisal of how things had developed in the Reich in the meantime."[723]

In mid-May 1936, about a month after Butenandt had joined the Party on April 7, the KWG entered into negotiations with him about his appointment. In the course of the discussions[724] it was agreed, among other things, that the Institute buildings should be converted in accordance with Butenandt's wishes, and that a house should be built for him and his family on the Institute site.[725] His predecessor was also mentioned in these discussions:

> "Professor Butenandt requested that Professor Neuberg should have no further connection with the Institute in future. This is also the wish of the Ministry. Consequently, Professor Neuberg will have to be informed that it is not possible for him to be allowed any special facilities to work in the Institute or in any of the Institute's buildings."[726]

This makes it more than clear that Butenandt wanted to make a complete break, to know that all contacts with his predecessor would be terminated immediately. The mention of "facilities to work" refers to the laboratory rooms in the Gatehouse on

720 Karlson (1990), 100. The Ministry representative, Rudolf Mentzel, had previously headed the Nazi party organization in Göttingen, and was now Head of Research at the Ministry.

721 Deichmann (1993).

722 A document drawn up by the *Reich* Ministry of Propaganda in 1942 is interesting in this respect, in which among other things academics are listed who appear suitable, from a political and ideological point of view, to be called upon to give lectures. The list is divided into three headings, with Adolf Butenandt appearing in the first category as "unrestrictedly positive." Heiber emphasizes, however, that even this first category "does not include any very wild people." Heiber (1991), 376.

723 "Mancher in Danzig bekannte sich zur NSDAP, nur weil er meinte, es sei ein Bekenntnis zu Deutschland. Erst nach meiner Übersiedelung nach Berlin konnte ich die Entwicklung, die sich inzwischen im Reich vollzogen hatte, richtig beurteilen." Butenandt (1981), 3.

724 Notes on a meeting between Professor Butenandt, Professor Mentzel, Dr. Telschow and Friedrich Glum, May 12, 1936. MPG Archives, Dept. I, Rep. 1A, 2041, sheet 4. According to this, Butenandt's house was to cost around RM 70,000.

725 Karlson (1990), 100–103.

726 "Prof. Butenandt hat darum gebeten, daß Prof. Neuberg in Zukunft in keiner Beziehung zum Institut mehr stehen soll. Dies entspricht auch den Wünschen des Ministeriums. Herrn Prof. Neuberg wird infolgedessen zu eröffnen sein, daß besondere Arbeitsmöglichkeiten im Institut bezw. im Gebäude des Instituts ihm nicht eingeräumt werden können." MPG Archives, Dept. I, Rep. 1A, 2041, sheet 4a.

the Institute site, which Neuberg would have liked to set himself up in.[727] This was the building that the Guest Department for Cell Culture had been accommodated in until its dissolution, the department which had mainly housed the Danish biologist Albert Fischer from 1925 to 1932.[728] Contrary to what he said on May 12, 1936, as quoted above, Butenandt in his autobiography presents the matter in the light that he, "as a young scholar newly arrived, did not want to have to burden [himself] with the decision" as to whether Neuberg should be allowed to work in the Gatehouse or not.[729] The fact that Butenandt had his way and that Neuberg was denied the possibility of working in the Gatehouse is documented by a letter of Eduard Hofmann's:

> "You know that he [Neuberg, Authors' note] would have been happy if he had merely been able to go on working in the porter's lodge of our institute in Dahlem, and that he was not even allowed this possibility."[730]

It is worth noting what Neuberg himself felt about the rejection of his request. He recounted his successor's behavior to Heinrich Otto Wieland,[731] in his own characteristically pointed but all the more impressive manner, as follows:

> "According to the KWG, Butenandt said at the time that he would find it 'intolerable' for me to remain in two little rooms in the Gatehouse of my former Institute, where Planck wanted to accommodate me, and made his acceptance of the position dependant upon my being removed without trace. Well, that would have been in accordance with the spirit of the time, and I think of it without bitterness, whenever I am reminded of it in any way."[732]

When Neuberg set up a laboratory in the Schöneberg district of Berlin at his own expense,[733] he did receive a limited amount of support from his successor.[734] Butenandt emphasized in his obituary for Carl Neuberg that "every imaginable kind

727 Memo of Telschow, April 20, 1934. MPG Archives, Dept. I, Rep. 1A, 2035, sheet 118.

728 This information is to be found in an audit report on the accounts as of March 31, 1934. MPG Archives, Dept. I, Rep. 1A, 2040.

729 MPG Archives, Dept. I, Rep. 1A, 2033, sheets 218–223.

730 "Sie wissen ja, dass er glücklich gewesen wäre, wenn er nur in einem Pförtnerhäuschen unseres Dahlemer Institutes hätte arbeiten können und dass man ihm nicht einmal diese Möglichkeit gegönnt hat." Hofmann to Hahn, February 11, 1949. MPG Archives, Dept. II, Rep. 1A, Neuberg's Personnel File.

731 Heinrich Otto Wieland (1872–1957) studied chemistry under Adolph Baeyer, worked in military research at the KWI for Chemistry for a short time during the First World War, and received the Nobel Prize in Chemistry in 1927 for his work on the structures of triterpenes. Martin (1988), 94 and Schulte (2001a).

732 "Nach Aussage der K.W.G. hat Butenandt s. Z. mein Verbleiben in 2 kleinen Raeumen des Torhaeuschens meines ehemaligen Instituts, wo mich Planck unterbringen wollte, fuer ‚untragbar' gehalten und die Annahme seiner Berufung von meiner restlosen Beseitigung abhaengig gemacht. Nun, das wuerde dem Zeitgeist entsprochen haben und ich denke ohne Bitterkeit daran, wenn ich irgendwie daran erinnert werde." Neuberg to Wieland, December 16, 1948. CNP, Box 18: Wieland, Heinrich O.

733 See also Chapter 6.3.

734 For example, books to the value of RM 5,000 owned by Neuberg privately were purchased off him by the Institute. MPG Archives, Dept. I, Rep. 1A, 1463.

of apparatus" had been "secretly moved to a bakery in Steglitz."[735] But this needs to be taken with quite a large pinch of salt, since the operation can hardly have been as "secret" as Butenandt presents it. The files of the KWI for Biochemistry contain numerous indications that Butenandt asked KWG Headquarters if he was authorized to lend out to his predecessor apparatus that he did not require himself.[736] Schüring suspects that not only Butenandt but Max Planck and Ernst Telschow too were involved in Neuberg's final expulsion from the KWG environment; for attempts were made, by delaying tactics, by not answering letters and by giving contradictory instructions, to get rid of Neuberg as quickly as possible in order to protect the KWG against some supposed harm.[737] After the Second World War, Neuberg himself expressed the view that it had not been Butenandt alone that had been responsible for the events, which he summarizes in a manner that expresses his disappointment in his own very characteristic way:

> "I laughed at the euphemistic use of the expression 'retirement.' I experienced it as a metathesis, a kick in the behind; and it was certainly not Butenandt that administered it."[738]

After the war, Neuberg and Butenandt conducted regular correspondence, which has already been the subject of an evaluation by Ute Deichmann.[739] The two researchers attempted, to their mutual benefit, to return to apparently normal relations as between academic colleagues. Butenandt subsequently supported Neuberg in his efforts to obtain an adequate pension, while Neuberg in return granted Butenandt partial absolution, as it were, by conducting this correspondence.

During the 1950s, moreover, Butenandt maintained that he had offered Neuberg a position at his old institute immediately after the war:

> "We were pleased that Carl Neuberg, although he had been forced to quit his position in 1935 under the pressure of the legislation of the time, maintained his friendship towards and interest in our Institute; and we accepted sympathetically the fact that he was unwilling to accept my offer to return to us in 1945."[740]

735 Butenandt (1956), 226.

736 Otherwise, Butenandt would have put this equipment in store in the Institute's loft. MPG Archives, Dept. I, Rep. 1A, 2033, sheets 218–223. Request for permission to hand over the equipment. MPG Archives, Dept. I, Rep. 1A, 2033, sheet 223. In the Institute's accounts for 1938, there is a remark that there is no lending slip for a piece of equipment being used by Neuberg. The apparatus was either to be brought back to the Institute or sold to Neuberg. MPG Archives, Dept. I, Rep. 1A, 1463.

737 Schüring (2004), 352–354.

738 "Gelacht habe ich über den euphemistischen Ausdruck meines ‚Ruecktritts'. Ich habe ihn als Metathesis, als Tritt in den Ruecken empfunden; diesen hat Butenandt sicher nicht ausgeteilt." Neuberg to Wieland, February 2, 1949. CNP, Box 18: Wieland Heinrich O.

739 Deichmann (1993).

740 "Daß Carl Neuberg, obwohl er 1935 unter dem Druck der damaligen Gesetzgebung seine Arbeitstätte verlassen mußte, unserem Institut seine Freundschaft und sein Interesse bewahrte, hat uns beglückt; daß er 1945 mein Angebot, zu uns zurückzukehren, nicht mehr annahm, haben wir verstanden." From a speech given at the inauguration of the new University Institute of Physiology and Chemistry and the new MPI for Biochemistry building in Munich, 1957. Butenandt (1981), 458.

In his obituary for Neuberg, Butenandt even remarked that Neuberg could have had "a position in accordance with his own wishes."[741] But these are astonishing statements, since he did not renew contact with Neuberg until 1947, doing so in a letter in which no such offer is even hinted at.[742] Two more years later, Neuberg is still writing that he cannot say anything about a possible return to Germany, since he has had no concrete offers: "Where am I supposed to return to? Where would I be able to work? Who has made me an offer?"[743]

Butenandt himself reveals the nature of the "position in accordance with his own wishes" offered to Neuberg in a letter to the then President of the Max Planck Society, Otto Hahn, in which he suggested that his predecessor should be proposed for the Cross of Merit of the Federal Republic of Germany.[744] This letter indicates that his invitations to Neuberg during his later visits to Germany were merely to spend some time at his old institute as a "guest."

To summarize, it may be stated that Butenandt's behavior towards his predecessor, the founder of the KWI for Biochemistry, was not as noble as he himself later presented it, or as others have done elsewhere. Butenandt, undisputedly one of the outstanding researchers and leading figures in the academic politics of his day, profited substantially from the dismissal of the Jewish scientist Neuberg, just as he had previously inherited Wohl's chair. During the negotiations on his appointment to the KWG, he took no identifiable steps to support his predecessor after he had been "retired," but on the contrary, requested that Neuberg should have no further connection with the institute he had founded.

Once Butenandt had secured his professional and political position, he did make efforts to assist Neuberg in building up tolerable conditions in which to work, and after the war he was successful in his efforts to present his "assistance" as a courageous and disinterested act. In view of the documentary sources, however, this representation appears untenable.

The post-war correspondence between the two scientists and the "deal" it represented were of benefit to both sides, as Deichmann also has shown. Neuberg obtained his pension.[745] Through the contact with his predecessor, living in exile in America, Butenandt was in a way partially exculpated for occupying the position Neuberg had been forced to vacate. In the matter of Carl Neuberg's return to Germany, the highly respected Butenandt appears, to formulate it benevolently, to have undergone a revision of his recollections after Neuberg's death, so that

741 Butenandt (1956), 227.

742 Butenandt to Neuberg, January 29, 1947. CNP, Box 3: Butenandt, Adolf.

743 Neuberg to Anschütz, February 1, 1949. CNP, Box 2: Anschütz, Ludwig. He also wrote in a similar vein to Eduard Hofmann: "I cannot answer your inquiry as to whether I would return to Germany with a joyful 'Yes' until someone tells me to what position, under what conditions and to where I should return." ("Ihre Anfrage, ob ich nach Deutschland zurückkehren würde, kann ich mit einem freudigen Ja erst beantworten, wenn man mir sagt, in welche Stellung, unter welchen Bedingungen u. wohin ich zurückkehren sollte.") Neuberg to Hofmann, February 20, 1949. CNP, Box 6: Hofmann, Eduard.

744 Butenandt to Hahn, November 26, 1953. MPG Archives, Dept. II, Rep. 1A, Neuberg's Personnel File.

745 Cf. also Chapter 7.3.

having been appointed President of the MPG in 1960, four years after Neuberg's death, he was able to appear as a committed proponent of "restitution," and in no way responsible for the injustice that had been done.[746]

6.3. NEUBERG'S LAST YEARS BEFORE HIS EMIGRATION (1936–1939)

Little is known about the laboratory, mentioned above, in which Neuberg went on working after his forced dismissal and expulsion from the KWI for Biochemistry. This *Biologisch-Chemische Forschungsanstalt* was established by Carl Neuberg and the Berlin pharmacologist Theodor Sabalitschka.[747, 748] In all probability, the laboratory was located in the Schöneberg district of Berlin, and not, as Butenandt believed, in Steglitz; there is one letter among Neuberg's papers that appears to indicate this.[749] Other post-war letters lead to the conclusion that the laboratory was accommodated in part of the foodstuffs factory *Nahrungsmittelfabrik Penner AG*. This assumption is supported by the fact that the Penner company had been selling the preservative p-oxybenzoic acid ester (*Nipaester*), which Sabalitschka had discovered, since 1925.[750] Only from Butenandt's recollections is it possible to draw conclusions as to how the rooms were equipped. In his obituary for Neuberg, Butenandt describes the working conditions as follows:

> "In 1936, we secretly moved every imaginable kind of apparatus from the Kaiser Wilhelm Institute for Biochemistry to a bakery in Steglitz: even an entire laboratory table and a fume cupboard with all the necessary apparatus, together with a centrifuge, Warburg's apparatus, distillation and evaporation equipment, went over there – and it gradually developed into a small but well-equipped laboratory where Neuberg could work and where he felt happy, despite all the injury that had been done to him."[751]

It may well be assumed that in the three years between Neuberg's final expulsion from the KWG in 1936 and his emigration in 1939, the atmosphere in this laboratory was not quite as romantic as it is here made to appear.

746 Butenandt (1981), 4–9.

747 Theodor Sabalitschka (1881–1971), a Catholic Christian, studied pharmacy and took his *habilitation* at Berlin University in 1922. In 1936 his license to teach was withdrawn for political reasons and he was dismissed in 1940. In 1945 he took over the directorship of the Berlin Pharmaceutical Institute, but got involved in political controversies and left the university service in 1949. Cf. Engel (2005).

748 Ammon/Dirscherl (1956).

749 The printed letterhead of a letter to Kurt Jacobsohn in 1937 gives the address Belziger Strasse 71 in Schöneberg. Neuberg to Jacobsohn, March 12, 1937. CNP, Box 7: Jacobsohn, Kurt.

750 Engel (2005), 320.

751 "Heimlich haben wir 1936 aus dem 'Kaiser-Wilhelm-Institut für Biochemie' alles Mögliche an apparativen Hilfsmitteln in eine Steglitzer Bäckerei geschafft; selbst einen ganzen Laboratoriumstisch und ein Digestorium mit allen Installationseinrichtungen neben Zentrifuge, Warburg-Apparatur, Destillations- und Verdampfungsgeräten, wanderten hinüber – und es entstand allmählich ein kleines gut eingerichtetes Laboratorium, in dem Neuberg arbeiten konnte und wo er sich – trotz allem was man ihm angetan – glücklich fühlte." Butenandt (1956), 226.

Neuberg's publications during this period are principally concerned with topics he had already published on before his suspension from the KWG. It should, however, be noted that the names of the co-authors include some people that Neuberg had not worked with before. It remains open whether all the publications on which his internationally celebrated name appears really were produced in collaboration with him, or at the very least authorized by him.

Neuberg's correspondence shows that he had been able to deposit a large part of his private scientific library[752] in rooms at the Penner factory. When he fled to the Netherlands in 1939, he attempted to take it with him, but this was apparently prevented by an employee of the company.[753] Neuberg, who in addition to his books had also deposited some of his furniture and his collection of specimens in the Penner factory, attempted after the Second World War to get back at least part of the library, which was very dear to his heart. But he soon had to acknowledge that this was a hopeless undertaking. His household goods were apparently destroyed during the war in an American air raid.[754] It was not possible to determine whether Neuberg's valuable books and his collection of specimens really fell victims to the bombing or whether they ended up in other hands.[755] All that is certain is that Neuberg himself never got them back.

More than six years had passed since the Nazis had come to power before Neuberg finally decided to escape from Germany. For the remaining members of the Jewish community within the *Reich,* conditions had progressively deteriorated, and external circumstances were becoming more and more threatening. During this period there had been calls to boycott Jewish businesses, professors, teachers, lawyers and doctors. A large number of discriminatory laws and ordinances which had no other purpose than to press ahead with the exclusion and denigration of the Jewish population passed through the submissive Reichstag unhindered.[756] In November 1938, the harassment of Jews by the German Nazis reached its first shocking climax in the pogroms of the so-called *"Reichskristall-*

752 Cf. Chapter 5.3.

753 Böhm to Neuberg, May 24, 1951. CNP, Box 3: Böhm, Ernest.

754 Neuberg to Sabalitschka, October 8, 1947. CNP, Box 14: Sabalitschka, Theodor.

755 There is an indication that Neuberg suspected Sabalitschka, his colleague at the time, of possibly not being totally innocent with regard to the disappearance of the books. He wrote to Rudolf Ziffer, with whom he had worked in Schöneberg and who was living in Montevideo, Uruguay, after the war: "From Warburg, who is here, I hear that Sabalitschka is doing well. He naturally refuses to give anything back, although his house and belongings are undamaged and he is now a big earner as Director of the Pharmaceutical Institute." ("Von Warburg, der hier ist, hoerte ich, dass es Sabalitschka gut geht. Er weigert sich natuerlich, irgendetwas zurueckzugeben, obgleich sein Haus und Hausrat unbeschaedigt und er jetzt Direktor des pharmazeutischen Instituts und Grossverdiener ist.") Neuberg to Ziffer, January 14, 1949. CNP, Box 18: Ziffer, Rudolf.

756 In particular, the so-called "Nuremberg Laws" *("Nürnberger Gesetze")* of September 15, 1935, "legitimized" the most brutal interference with the lives of Jewish citizens. The *Reichsbürgergesetz* (*Reich* Citizenship Act) of 1935 laid down that only "nationals of German or related blood" could be citizens of the *Reich.* In consequence, no Jew was permitted to occupy an official position any more, and Jewish civil servants had to leave the service by December 31, 1935.

nacht." The subsequent "atonement payments," by which members of the Jewish community were compelled to pay themselves for the damage caused by German mobs, robbed many of their livelihoods.

These circumstances have to be taken into account in attempting to imagine the conditions that Neuberg lived under in the last years before his flight. The processes that had led to Neuberg's dismissal as Director of the Institute, and the growing influence that State and Party were gaining over both the universities and the KWG, also make it clear that academia in Germany was in no way an ivory tower. It is astonishing in these circumstances that Neuberg had not turned his back on Germany long before. There follows a consideration of the reasons that caused him to stay.

Neuberg was equally proud of his Jewish origin and of his belonging to the German nation; he had never denied his Jewish descent. Nachmansohn wrote: "Neuberg's deep devotion and genuine love of Germany did not present a conflict for him in stressing his Jewish identity."[757] This inner conviction and his political attitude lead to the conclusion that Neuberg, as an admirer of German culture, would have felt an early flight from the Nazis to represent surrender to a gang of criminals. But like so many others, he underestimated the dangers that were to emanate from the regime. Nachmansohn describes Neuberg's attitude shortly after Hitler's appointment as Chancellor in 1933 as follows:

> "Jews should not be intimidated; they should fight. The regime could not last long. He predicted that the violence and anti-Semitic propaganda would die down within a couple of months. He had lived through several such dangerous-looking periods."[758]

In addition, Neuberg probably felt himself secure against attack, both as a combatant who had done his country service in the First World War, and in his capacity as a long-serving civil servant. Furthermore, in the confrontation with the mechanic Delatrée-Wegner described above, his innocence and his irreproachable conduct towards the state had even been officially confirmed by a court. Neuberg was certainly not the only one to make such a false assessment. The following remarks may give an impression of his attitude, and provide an example of what many others among those affected thought:

> "Why should I emigrate? Nothing is eaten as hot as it is cooked. After all, we live under the rule of law. What can happen to me? I was a soldier, I fought on the western front for four years, was an NCO and hold the Iron Cross First Class."[759]

757 Nachmansohn (1979), 325.

758 Ibid., 326–327.

759 According to the works of reference, however, Neuberg's two decorations were Class II Iron Crosses, and only certificates for these medals are to be found among his papers. Cf. Fn. 766. "Weshalb soll ich auswandern, es wird nichts so heiß gegessen wie es gekocht wird. Wir leben schließlich in einem Rechtsstaat. Was kann mir schon passieren, ich bin Frontkämpfer, habe vier Jahre lang an der Westfront gekämpft, war Unteroffizier und habe das Eiserne Kreuz erster Klasse." Richarz (1982), 339. "Nothing is eaten as hot as it is cooked" is a German proverb, meaning that apparently drastic measures are always less extreme in their practical application than what was originally proposed.

Neuberg's loyalty to Germany and his efforts to secure himself a legal right to stay, or at least to ensure that his remaining would be tolerated, by demonstrating his membership of the German nation went so far that he even offered his services to the Nazi government. In a strictly confidential letter to the Director General of the KWG, Friedrich Glum, Neuberg, who had already been forcibly retired from his university posts but was still at the time Director of the KWI, drew attention to the fact that through his longstanding academic contacts it would be possible for him to visit a factory in Strasbourg in order to "[...] gain an insight into a technical process that is of great importance, both economically and militarily. It relates to the production of butyl alcohol and acetone – the latter being a raw material for artificial rubber and gas munitions."[760]

Glum forwarded this letter to the Ministry of Defense.[761] At the same time, he pointed out both Neuberg's non-Aryan descent and the difficulties that Neuberg had had inflicted on him by government offices and party organizations.[762] No evidence can be found as to whether such a visit to the factory concerned ever took place. Neuberg's offer shows that he was keen to avoid offering the new regime any pretext for denouncing him as an "enemy of the people," but it also reveals that he was even prepared to exert himself on behalf of this new Germany, despite all the shameful treatment he had already been subjected to.

Planck, who, as has already been mentioned, had spoken up for Neuberg in the interests of science,[763] appears to have advised him to stay in Germany. This is made clear by a letter[764] that Neuberg wrote to Karl Thomas[765] after the war, in which he stated that owing to his activities as a technical expert during the First World War, he was for a time forbidden to leave Germany. However, no official documents have been found confirming that Neuberg really was classified as a bearer of military secrets, and therefore prohibited from leaving the country.[766]

It may also have been the case that Neuberg's personal and family situation initially kept him from embracing the uncertain existence of an emigrant and asylum-seeker. At the time of his final suspension as Director of the Institute, he was almost 60 years old, an age at which it would have appeared difficult, indeed almost hopeless, to attempt to attain to anything like a suitable position on which to build up a new existence in another country. Neuberg's beloved wife Hela had died in 1929, and both his daughters were married. Irene was living in Paris, later

760 Neuberg to Glum, July 4, 1934. MPG Archives, Dept. I, Rep. 1A, 2033, sheet 168.
761 Glum to Major-General Liese, July 7, 1934. MPG Archives, Dept. I, Rep. 1A, 2033, sheet 170.
762 Cf. Chapter 6.1.
763 This support declined markedly after the appointment of Butenandt in May 1936. See above and cf. Schüring (2004), 352–353.
764 Neuberg to Thomas, December 21, 1949. CNP, Box 15: Thomas, Karl.
765 Karl Thomas (1883–1969) was Professor of Physiological Chemistry at the University of Leipzig from 1921, Head of Department at the KWI for Physiology of Effort from 1945, and from 1947 onwards Director of the Medical Research Institute of the MPG in Göttingen.
766 It is thus certainly possible that Neuberg may have at least not wholly unconsciously erred in the question of a prohibition on his leaving the country, in order to enhance the significance of his services during the First World War; cf. the question of his Iron Crosses, Fn. 759.

in Istanbul, and Marianne had moved with her husband to Vienna.[767] Only his sister Anna was still living in Germany. As both of his daughters were already safely abroad, there did not appear to be any sufficient reason for him to emigrate. Even before his forced retirement from the KWG, Neuberg had had the opportunity to be appointed Director of a research institute that the *Hoffmann-LaRoche* company was planning to set up at Grenzach[-Wyhlen], Baden, in 1936.[768] It was not possible to find out in any detail why this plan, which was also supported by the management of the KWG, did not come to fruition. Neuberg's disinclination to work for industrial companies[769] may have played a decisive part in his rejection of the offer, or else his hope that there would continue to be room for him in the semi-public-sector KWG and the general hope that the Nazi nightmare would soon come to an end.[770] Three years later, Neuberg seems to have become aware of the seriousness of the situation. One passage in a letter to his former pupil Kurt Jacobsohn in Lisbon is almost a cry for help:

> "I grieve to note that you have still not found your old teacher a position as a laboratory servant. His desire for such a position is a thoroughly serious one, as things are not going all that well for him here."[771]

6.4. NEUBERG'S ATTITUDE TO JEWISH IDENTITY AND HIS DECISION TO EMIGRATE

Neuberg's loyalty and feeling of belonging to Germany have been described in the previous chapter. From his exile in America, Neuberg, whose political attitude was described by Kisch as that of a "German nationalist,"[772] summarized his relationship to Germany as follows in a letter to the Nobel Prize winner Heinrich O. Wieland, who was teaching in Göttingen:

767 Verbal communication from Joan Armer, June 14, 1998.

768 Glum to Neuberg, June 17, 1936. CNP, Box 7: Kaiser-Wilhelm-Gesellschaft.

769 Cf. Chapter 7.2.

770 Even after his compulsory retirement as Director, Neuberg, who was still Acting Director of the KWI for Biochemistry, attempted to obtain funds from the Rockefeller Foundation, which he hoped would allow him to continue to work at the Institute free of state interference. Schüring (2004), 351, and Macrakis (1994).

771 "Trauernd wird bemerkt, daß Sie Ihrem alten Lehrer immer noch nicht eine Laboratoriumsdienerstelle verschafft haben. Der Wunsch nach einer ist durchaus ernst, da es ihm hier nicht zum besten geht." Neuberg to Jacobsohn, April 19, 1939. CNP, Box 7: Jacobsohn, Kurt. At this point it may be remarked that Jacobsohn, who had been at Neuberg's Institute from 1926 to 1929 and had subsequently worked, *inter alia*, at the University in Lisbon, always did his best on his teacher's behalf. From his correspondence with Neuberg, which he added to Neuberg's papers after the latter's death and which represent one of the few original sources of remarks of Neuberg's in the late 1930s, it can be seen that Jacobsohn repeatedly attempted to obtain recognition for Neuberg abroad. Among other things, he was behind Neuberg's appointment as an Honorary Member of the Portuguese Academy of Sciences in 1934, which can only have strengthened Neuberg's position in that time of conflict.

772 Kisch (1966), 329.

> "I have no legitimization to come bearing Christmas gifts, since I still hold to the formulation originating from the political economist Franz Oppenheimer, the brother of Carl Oppenheimer of the enzymes: 'German national feeling, Jewish awareness of descent and pride in his home country west of the Elbe.' (My ancestors, on both my father's and my mother's side, can be shown to have lived in the State of Hanover for three hundred years)."[773]

According to Ines Mandel, Neuberg was no orthodox representative of his faith; he had a detailed knowledge of Jewish traditions and the scriptures, but in religious questions his attitudes were liberal and he did not follow Talmudic precepts in his daily routines or in his way of life.[774] His interest in history in general and his efforts to uphold a symbiosis or co-existence of Jewishness and German nationality may perhaps be illustrated by the following example, in which he tells a former member of his staff about an interesting article he has read in a magazine:

> "It was concerned with the fact that the Hilde sagas of the *Gudrunlied* appeared in old Yiddish literature 130 years before the Ambras manuscript. This discovery shows that cultural relations have existed between the Jewish community and its Teutonic environs since almost the very earliest times. Even so, the fact that the Jews incorporated the most important Germanic heroic epic into their treasury of song earlier than their neighbors did is remarkable."[775]

On the basis of his conviction that peaceful co-existence between the two faiths was desirable, Neuberg was committed to promoting an integration of the Jewish population in Germany that would, nevertheless, allow them to maintain their own cultural and religious identity. This can be deduced amongst other things from his membership of the *Centralverein deutscher Staatsbürger jüdischen Glaubens* ("Central Association of German Citizens of the Jewish Faith").[776] This association was devoted in particular to combating the anti-Semitism that was rampant and growing in Germany, and to promoting the recognition of Jews as equal German citizens.[777] There are indications that for a short time Neuberg was a

773 "Als Christkindlbringer habe ich nicht die richtige Legitimation, denn ich halte es immer nach Art der vom Nationalökonomen Franz Oppenheimer, dem Bruder des enzymatischen Carl Oppenheimer, herrührenden Formulierung: 'Deutsches Nationalgefühl, jüdisches Stammesbewußtsein und westelbischer Heimatstolz.' (Im Lande Hannover saßen nachweislich meine Ahnen väterlicher- und mütterlicherseits 300 Jahre)." Neuberg to Wieland, February 3, 1948. CNP, Box 18: Wieland, Heinrich O.

774 Kisch (1966), 329, and personal communication from Ines Mandl, October 25, 1996.

775 "Es handelt sich darum, dass die Hilde-Sagen des Gudrunliedes 130 Jahre vor der Ambraser Niederschrift in der altjuedischen Literatur niedergeschrieben sind. Aus dieser Entdeckung folgt, dass kulturelle Beziehungen zwischen germanischer Umwelt und Judentum seit fast Urzeiten bestanden haben. Dass aber die Juden den bedeutendsten germanischen Heldengesang frueher als die Umwelt ihrem Liederschatz einverleibt haben, ist bemerkenswert." Neuberg to Hofmann, March 10, 1954. CNP, Box 6: Hofmann, Eduard.

776 A variety of sources witness to this fact. Among other things, Neuberg's name is to be found in the list of members of the Greater Berlin State Association of the *Centralverein* as of October 24, 1931. Leo Baeck Institute Library, membership list of the Greater Berlin Association of the Centralverein, MfW 12–III.

777 Cf. also Barkai (2002).

member of the association's Central Committee.[778] Although no unambiguous proof of this could be found, it is certain that Neuberg wrote at least one article for the Central Association's magazine dealing with the role and importance of Jewish scientists in biochemical research.[779] In this article, Neuberg refuted the frequently raised anti-Semitic insinuation that the Jews are predominantly involved in purely commercial and mercantile activities. He pointed out the extraordinary large number of Jewish scientists who had acquired international recognition in the field of biochemistry, which Neuberg himself represented. The reasons why so many Jewish scientists were attempting to make their way in precisely this still very young discipline were described by Neuberg as follows:

> "Under the prevailing conditions, in the majority of cases there remained only one possibility for Jewish researchers in the natural sciences to work in accordance with their abilities and pursue, often under great external hindrances and serious deprivations, what they perceived to be the right way. This possibility was to seek new areas of research outside the officially recognized ones that were jealously guarded by existing institutions, and so to establish an uncertain and anxious existence in scientific fringe areas, just as in everyday life they were relegated to medieval exclusion on the peripheries of the cities."[780]

Even in his later years, Neuberg remained faithful to his feeling of having an obligation, in an unorthodox and undogmatic manner, towards the Jewish faith and its traditions. Even after his expulsion, however, he continued to feel himself to be a German, socialized in the days of the Empire, who was proud of the achievements and virtues of his people, which he valued highly. Despite his feeling of belonging to Jewish culture and his expulsion from Germany precisely because of that background, he expressed himself in an astonishing manner with regard to the prejudices against people of Jewish origin:

> "The Jewish organizations are indeed characterized by those inferiorities that weigh upon our race. If the collectors – long live the spirit of America! – allow 50-55% of the money collected for Israel to disappear into their own pockets and are light-years away from the idea that one might do anything without pecuniary reward, this simply demonstrates the depths of the Jewish mentality. I'm too old to get worked up about it, but still, the fact that my Jewish staff members, wherever in the world they may be living, only write when they want something, whereas the Christians write without ever asking anything of me, gives me food for thought. I don't believe I treated my Jewish staff any worse than the non-Jewish ones. In

778 Leo Baeck Institute Library, structure of the State Association of the Centralverein, November 14, 1947, MfW 12–III.

779 Neuberg (1927).

780 "Unter den obwaltenden Umständen, ist den jüdischen Naturforschern in der Mehrzahl der Fälle nur eine einzige Möglichkeit offengeblieben, sich nach ihren Befähigungen zu betätigen und oft unter großen äußeren Hemmnissen sowie schweren Entbehrungen den von ihnen als richtig erkannten Weg zu gehen. Sie bestand darin, außerhalb der offiziell anerkannten und durch bestehende Institutionen beschützten Bereiche sich neue Gebiete der Forschung zu suchen und hier, so wie im alltäglichen Leben mittelalterlichen Abgeschiedenseins oft an der Peripherie der Städte, auf Grenzgebieten eine sorgenvolle und ungewisse wissenschaftliche Existenz sich zu gründen." Neuberg (1927). On the activities of Jewish scientists in peripheral areas, cf. also Deichmann (2002) and Ringer (1983).

my case, this cannot be caused by any oppressive feeling of guilt, because the Christian staff I'm talking about acted irreproachably during the time of conflict."[781]

This is not the place to undertake an analysis of this many-faceted statement; but in judging it, Neuberg's situation in exile in America has to be taken into account, which he described to Butenandt as follows: "So as you see, my dear Butenandt, the evening of my life is, all in all, by no means a bed of roses – but perhaps it is not the purpose of life that it should be so."[782] Neuberg's unjustified generalizations nevertheless remain peculiar, and may serve to give an impression of the inner conflicts to which he was a prey in the latter phase of his life, which began with his forced retirement on March 28, 1934. The five years of humiliating experiences that Neuberg, like all German Jews, was forced to suffer, brought him to the conclusion that emigration was unavoidable if he wanted to survive.

On August 1, 1939, Neuberg terminated his registration as a resident of Dahlem, Berlin, stating that he was emigrating to Palestine.[783] According to what he himself wrote, however, it was not until two weeks later that he took the final step and left Germany: "On August 15, 1939, friends of mine in military circles in Potsdam suggested I ought to leave the country immediately."[784] These friends had obtained an identity document for him,[785] so that Neuberg was able to leave Germany for Amsterdam the following evening with two suitcases and five Dutch guilders.[786] Neuberg had lost all his money and possessions; the so-called "proper-

781 "Die juedischen Organisationen sind mit den unsere Rasse belastenden Minderwertigkeiten durchaus gezeichnet. Wenn die Sammler – hoch lebe der amerikanische Geist – von den für Israel gesammelten Geldern 50 bis 55% in die eigene Tasche fliessen lassen und Sirius fern dem Gedanken gegenüberstehen, man koenne ehrenamtlich etwas tun, so zeigt dies doch die Niederungen in der juedischen Mentalitaet. Ich bin zu alt geworden, um mich darueber zu entruesten, aber ich werde doch nachdenklich gegenueber der Tatsache, dass meine juedischen Mitarbeiter, wo in der Welt sie auch leben, nur schreiben, wenn sie etwas wollen, waehrend die christlichen schreiben, ohne jemals etwas zu verlangen. Ich glaube nicht, die juedischen Mitarbeiter schlechter behandelt zu haben als die nichtjuedischen. In meinem Fall kann nicht ein bedrueckendes Schuldgefuehl die Ursache sein, denn es handelt sich um meine christlichen Mitarbeiter, die in der Conflictzeit sich tadellos benommen haben." Neuberg to Simon, March 5, 1956, CNP, Box 15: Simon, Ernst. Simon (1902–1969), a member of Neuberg's staff since 1925, who was dismissed in 1933 and after fleeing from Germany via France, worked at the Daniel Sieff Institute in Revohot, Israel, had asked Neuberg, in connection with a claim for compensation, to make a statement to the effect that under normal circumstances he would have been able to make a career as a professor, but that this had been prevented by the Nazi reign of terror.

782 "Wie sie also sehen, lieber Herr Butenandt, zeigt sich mein Lebensabend in toto keinesfalls auf Rosen gebettet – aber vielleicht ist das auch nicht der Sinn des Lebens." Neuberg to Butenandt, March 7, 1947. CNP, Box 4: Butenandt, Adolf.

783 Police registration certificate, August 1, 1939. CNP, Box 12: Neuberg, Carl, Personal.

784 "Am 15. August 1939 legten mir meine Freunde in den Potsdamer Militärkreisen nahe, das Land sofort zu verlassen." Neuberg to Thomas, December 21, 1949. CNP, Box 15: Thomas, Karl.

785 Neuberg to Thomas, December 21, 1949. CNP, Box 15: Thomas, Karl.

786 Neuberg to Butenandt, March 7, 1947. CNP, Box 4: Butenandt, Adolf.

ty levy" imposed on Jews after the pogrom night at the end of 1938 had amounted in Neuberg's case to RM 46,400.[787]

> "In April 1939, in accordance with the Ordinance on the Utilization of Jewish Assets of 12.3.1938, I delivered up the following objects to the pawnbroker's in Berlin: 23.924 kilograms of table silver (silver bowls, baskets, jugs, mats etc.), 11.080 kg of silver spoons of various sizes, 10.338 kg of silver knives and forks, 1 gentleman's gold pocket watch, and a golden watch chain weighing 86.5 grams. I did not receive any kind of consideration for these objects."[788]

Then there was the compulsory pledging of any valuables, the so-called "*Reich* flight tax," which was assessed at RM 30,000, and the "emigration fee" of RM 4,800 payable to the Jewish community. It may be assumed that almost all of Neuberg's assets found their way into the coffers of the Nazi state.

For his part, however, Neuberg had ensured before leaving Germany that parts of the laboratory equipment that he had borrowed from his former institute[789] were properly restored to their former location. In his obituary for Neuberg, Butenandt described his recollection of this event:

> "I will never forget it: early one morning, a furniture van drew up outside the institute in Dahlem, and two of its crew, who claimed not to know who they were coming from or who had instructed them, unloaded everything that had formerly been taken off to Steglitz, from the laboratory table to a box containing simple test tubes [...]. But I found on my desk a little note in Neuberg's handwriting saying: 'Thank you for everything, I will never forget it. When you read this greeting, I will already be in Holland.'"[790]

787 This and the following details are taken from affidavits of Neuberg's dated July 16, 1952, concerning the discriminatory special levies paid by him to the state. See also: CNP, Box 7: Karpen, M.

788 "Ich habe im April des Jahres 1939, der Verordnung über den Einsatz jüdischen Vermögens vom 03.12.1938 entsprechend, folgende Gegenstände an die Pfandleihanstalt Berlin abgeliefert: 23,924kg Tafelsilber (silberne Schüsseln, Körbe, Kannen, Untersätze etc.), 11,080kg silberne Löffel unterschiedlicher Größe, 10,338kg silberne Messer und Gabeln, 1 goldene Herrentaschenuhr, 86,5g goldene Uhrkette. Irgendeinen Gegenwert habe ich für die vorbezeichneten Gegenstände nicht erhalten." Affidavit of Neuberg's, July 16, 1952. CNP, Box 7: Karpen, M.

789 Cf. Chapter 6.3.

790 "Es wird mir immer unvergessen sein: eines morgens fuhr ein Möbelwagen vors Dahlemer Institut, und es wurde von 2 Beifahrern, die angeblich nicht wußten, von wem sie kamen und wer sie beauftragt hatte, alles ausgeladen, was einst nach Steglitz gewandert war, vom Laboratoriumstisch bis zu einem Kasten mit einfachen Probierröhrchen [...]. Auf meinem Schreibtisch aber fand ich einen kleinen Zettel, auf dem mit Neubergs Schriftzügen zu lesen stand: ‚Haben Sie Dank für alles, es wird unvergessen sein. Wenn Sie diesen Gruß lesen, bin ich schon in Holland'." Butenandt (1981), 35.

6.5. THE LONG FLIGHT TO THE USA

After escaping to Holland, Neuberg worked for about five months[791] in the laboratory of Professor Johan Pieter Wibaut,[792] Director of the Institute of Organic Chemistry at the University of Amsterdam. According to Wibaut, his guest was principally concerned during this time with investigating the splitting of phosphates.[793] It was not, however, possible to find any evidence of publications of Neuberg's written during his time in Amsterdam.

In view of his precarious situation, Neuberg was forced to cash in on his scientific discoveries. He sold the rights to at least two patents, covering his processes for the "manufacture of saccharic acid from polysaccharides containing glucose" and the "manufacture of ascorbic acid from natural products," to the Deventer-based company *Koninklijke Industrieele Maatschappaj*. For the first of these patents, Neuberg received the sum of Fl. 6,000, which was to be transferred to his daughter Irene in New York. The second patent raised him a further Fl. 2,000 plus royalties, dependent on the production quantities, for a period of five years.[794] Neuberg also appears to have received personal assistance from employees of this company: They were involved in facilitating Neuberg's onward journey to Palestine, and also attempted to assist him in his efforts to rescue his sister Anna, who was still in Germany, from the Nazi terror.

Neuberg himself lived quite close to the University of Amsterdam during this time,[795] and himself stated that he received financial assistance from a member of his extended family. Neuberg mentions this fact in a letter to Thomas, remarking that this relative had also lent him money for his passage to Palestine "with no hope of ever seeing it again." This "third cousin" may have been the Amsterdam doctor Julius Neuberg, who survived the Nazi occupation in Holland and whom Neuberg continued to keep in touch with after the war.[796] In view of the progress of the war, however, it seemed impossible for him to stay in Amsterdam. In order to be able to leave the country before the threatened occupation of the Nether-

791 This is the period of time between Neuberg's departure from Germany on August 19, 1939, and his departure from the Netherlands on February 29, 1940. This latter date is found in a *Form of application for the exemption of the household and personal effects*, signed by Neuberg on May 8, 1940. CNP, Box 12: Neuberg, Carl, Personal.

792 Johan Pieter Wibaut (1886–1967) studied chemistry in Amsterdam and took his *habilitation* in 1915. In 1925 he became full Professor of Organic Chemistry at the same university. He retained this position until his retirement. Wibaut assisted a number of emigrants, and also assistants from the laboratory of Ernst Laqueur (1880–1947), who as a Jewish professor lost his job after the occupation of the Netherlands. Written communication from Dr. Knegtmans, Amsterdam, May 8, 2006.

793 Information from an article written by Wibaut in Dutch on the occasion of Neuberg's 75th birthday and published in *Chemisch Weekblatt* of July 26, 1952. CNP, Box 11: Neuberg, Carl, Birthday Congratulations.

794 Koninklijke Industrieele Maatschappaj to Neuberg, November 13, 1939 and February 9, 1940. CNP, Box 2: Ascorbic Acid.

795 Neuberg's flat was in Botticellistraat in the south of Amsterdam, as can be seen from Neuberg's correspondence with the Dutch company. CNP, Box 2: Ascorbic Acid.

796 Neuberg to Thomas, December 21, 1949. CNP, Box 15: Thomas, Karl.

lands[797] and a possible invasion of France, Neuberg, with a view to accelerating a decision on his transit visa, wrote to a number of French colleagues in January 1940 on postcards, which anyone could read. He probably chose this method with the full intention of making it clearer to the authorities monitoring the mail that he merely wished to pass through the country. The postcards were to be mailed directly in Paris by a representative of the Dutch company that Neuberg was associated with. Finally, with the assistance of French friends, he succeeded in obtaining a transit visa allowing him to travel through France.[798] In particular, his former subordinate Claude Fromageot,[799] who was Head of the Biochemical Laboratory of the University of Lyons at the time, played an important part in this, since his father, the diplomat Henri Fromageot, was a judge at the Permanent International Court of Justice in The Hague, and thus in a position to plead Neuberg's cause with the relevant people.[800] Neuberg was probably able to cross France on a German passport in late February 1940, and to sail from Marseilles for Beirut on a French troopship.[801]

As early as June 1938, there had been initial contacts between Carl Neuberg and the President of the Hebrew University in Jerusalem, Dr. Magnes.[802] These had partly to do with the latter's wish to win the illustrious scientist Neuberg as Director of the Chemical Laboratory in the Department of Cancer Research at the university in Jerusalem, and contained an offer to Neuberg to that effect. It can only be a matter of speculation through what actors and intermediaries this contact was established. It may, however, be assumed that Chaim Weizmann, a biochemist and later the first President of the State of Israel, played a not insignificant part in getting Neuberg appointed in Jerusalem. Weizmann was a great admirer of Neuberg and turned to him frequently for advice on his plans to build up chemistry and biochemistry in Palestine.[803] Andor Fedor, then Dean of the Faculty of

797 The Netherlands were attacked by the German Army Group B on May 10, 1940. The invasion of France began on June 5, and the occupation was completed with the taking of Paris on June 14, 1940 and the forced signing of the Franco-German armistice on June 22.

798 Neuberg to J. van der Lande, January 5, 1940. CNP, Box 2: Ascorbic Acid.

799 Claude Fromageot (1899–1958) had worked at Neuberg's institute in the late 1920s.

800 Claude Fromageot to Neuberg, January 15, 1940, and Henri Fromageot to Neuberg, February 17, 1940. CNP, Box 5: Fromageot, Claude.

801 Neuberg to Thomas, December 21, 1949. CNP, Box 15: Thomas, Karl.

802 Our information is based on the brief summary of the relations between the University and Neuberg made by the then Head of Administration, D. W. Senator, immediately after Neuberg's departure from Palestine. Cf. Central Archives (1940). The archives of the University of Jerusalem contain a 350-page file with correspondence relating to Neuberg's work at the Hebrew University.

803 Nachmansohn (1979), 326. The high regard was, however, mutual. On the occasion of the establishment of the State of Israel, Neuberg wrote to Ernst D. Bergmann, who was working at the Weizmann Institute in Jerusalem: "I repeat the congratulations that I expressed to you and the Institute, but in particular to the admired and honored President Weizmann. [...] The impression made upon me by his entry in the visitors' book of the KWI of which I was then the Director remains unforgettable: On the 10th anniversary of the Balfour Declaration." ("Ich wiederhole die Glückwünsche, die ich Ihnen und dem Institut, namentlich aber dem bewunderten und verehrten Präsidenten Weizmann abgestattet habe. Unvergesslich bleibt der

Medicine at the university in Jerusalem, had been in close contact with Carl Neuberg for decades in his capacity as editor of the *Biochemische Zeitschrift*. Two pupils of Neuberg's, Jesaiah Leibowitz[804] and Ernst Simon, who had been living in Palestine since 1935, probably also played a part in Neuberg's appointment to the Hebrew University. Neuberg himself had, moreover, followed with great interest the establishment of the university in Jerusalem after the First World War.[805]

Neuberg reached Jerusalem in early March 1940.[806] However, a few more months were taken up with quite protracted negotiations concerning amongst other things the budget for the department he was to head. It was not until August 1, 1940, that he received from the President of the Hebrew University, Judah Leon Magnes,[807] his certificate of appointment as Professor and Head of the Department of Cancer Research in the university's Institute of Chemistry.[808] Already at this time, initial efforts were being made by New York University to establish a research professorship for him. But Neuberg decided to accept the position offered him in Jerusalem, initially for a period of two years.[809]

The modest circumstances in which Neuberg had to work during his time in Jerusalem, the difficulties that had to be overcome in order to build up an independent academic scene in Palestine under the British mandate, and the prospect of being able to take up professorship in the United States may all have contributed to persuading Neuberg to leave the "Jewish homeland" again quite soon. In 1949, he summed up the possibilities of doing research work in Palestine as follows: "moderate working conditions, a just adequate salary, but practically no budget."[810] Neuberg made an agreement with the authorities of the Hebrew University that his move to New York should initially be for one year only, although both sides were allowed the possibility of ending their mutual

Eindruck, den mir seine Eintragung ins Gästebuch des damals von mir geleiteten KWI machte: Am 10. Jahrestage der Balfour Declaration.") Neuberg to Bergmann, January 4, 1949. CNP, Box 3: Bergmann, Ernst.

804 Jesaiah [Yeshayahu] Leibowitz (1903–1994), born in Riga, studied chemistry and philosophy, took his doctorate in 1924, then on the staff of the KWI for Biochemistry, emigrated to Jerusalem in 1935, from 1941 Professor of Biochemistry at the Hebrew University, from 1952 Professor of Organic Chemistry and Neurology. He also published articles on subjects from the fields of philosophy and the history of science.

805 Cf. Herlitz et al. (1987), Vol. IV/1, 463.

806 At this time, Neuberg was living in King George Street in Jerusalem. From a *Personnel Security Questionnaire* of the US Atomic Energy Commission, May 30, 1950. CNP, Box 16: U. S. Atomic Energy Commission.

807 Judah Leon Magnes (1877–1948), a notable rabbi in the United States, was the first Chancellor of the Hebrew University of Jerusalem after its foundation, and from 1935 until his death President of the University. Cf. Herlitz et al. (1987), Vol. 4.

808 Magnes to Neuberg, August 1, 1940. CNP, Box 6: Hebrew University of Jerusalem.

809 Neuberg to Thomas, December 21, 1949. CNP, Box 15: Thomas, Karl.

810 "Leidliche Arbeitsbedingungen, knapp ausreichendes Gehalt, aber praktisch keinen Etat." Neuberg to Thomas, December 21, 1949. CNP, Box 15: Thomas, Karl.

relationship prematurely.[811] When Neuberg left the country on November 20, 1940, however, it was a farewell for ever. Although he expressed regrets to colleagues who had remained in Palestine that his departure had turned out to be final,[812] there are also indications that accustomed as he was to the research conditions in Germany, he was not really able to adapt properly to the unfavorable conditions and the difficulties of the pioneering period in what was soon to become Israel. Andor Fodor, Professor in the Department of Biochemistry of the Hebrew University and at the same time Dean of the Faculty of Medicine, described his efforts to keep Neuberg in Jerusalem as follows:

> "If I once drew my sword from its sheath to do combat against hostile forces on behalf of your appointment in Jerusalem, it was to serve not so much you personally as rather the state of biochem. at this university, so that it might not be represented merely by young lads whom the administration can treat as it likes [...]. When you were here, you were not really able to feel at home in this milieu, and quite rightly too."[813]

In addition, it may be assumed that Neuberg's decision was influenced to a significant extent by his daughters, Irene and Marianne, who were by now both living with their families in the United States. Despite his premature departure, Neuberg remained in contact with (among others) the biochemist Leah Bloch-Frankenthal, and attempted to assist her in her further career by writing letters of recommendation.[814]

Amidst the tumult of the Second World War unleashed by Germany, it proved to be extremely difficult and dangerous to arrange a journey from Palestine to the United States. Owing to the war situation, Neuberg had to go the longer, eastward, way round to reach his destination, New York. A number of very diverse documents make it possible to reconstruct his itinerary:[815] After leaving Jerusalem, he first crossed present-day Jordan, Iraq and Iran, before embarking, probably at an Indian port, to cross the Indian and Pacific Oceans.[816] To get that far, he had had

811 It will not be possible to describe his scientific work during this period until the files in the archives of the Hebrew University have been evaluated.

812 For example, Neuberg wrote to the Director of the Physiological Institute of the University of Jerusalem, Wertheimer: "All in all, I think it was a pity that the progress of the war made it impossible for me to return. I am after all very attached to the country and some people there." ("Im Ganzen meine ich, es war doch schade, daß die Entwicklung des Krieges es mir unmöglich gemacht hat, zurückzukehren. An dem Land und einigen Leuten dort, hänge ich doch sehr stark.") Neuberg to Wertheimer, May 30, 1945. CNP, Box: Wertheimer, Ernst.

813 "Einst bin ich, gegen feindliche Landsknechte mein Schwert aus der Scheide ziehend, fuer ihre Berufung nach Jerusalem eingetreten, nicht sowohl um Ihnen persönlich zu dienen, als vielmehr dem Niveau der Biochem. an dieser Universität, auf dass hier nicht nur Juengelchen ihre Vertreter sein moegen, mit denen die Verwaltung nach Belieben umspringen kann [...]. Sie hatten sich, angekommen, in diesem Milieu nicht einfuehlen können und das mit Recht." Fodor to Neuberg, July 6, 1947. CNP, Box: Neuberg, Carl, Manuscripts – misc.

814 Neuberg to Bloch-Frankenthal, January 26, 1955. CNP, Box 3: Bloch-Frankenthal, Leah.

815 Some of these details are taken from the letter from Neuberg to Karl Thomas that has been referred to frequently already. Neuberg to Thomas, December 21, 1949. CNP, Box 15: Thomas, Karl.

816 The menus from this ship, which he must have kept as souvenirs, were among Neuberg's papers. CNP, Box 7: Java Pacific Line. The menus bear the dates November 30, 1940 to

to make use of such unorthodox means of transport as a mule and a wheelbarrow.[817] On November 30, 1940, he boarded a ship of the Java-Pacific Line, the *M. V. Bloemfontein*, and after a voyage lasting more than two months, via Ceylon, the Dutch East Indies, New Guinea and Hawaii, he finally reached Los Angeles, where he was received on January 25, 1941 by his daughter Marianne, who was living in California. He then crossed the North American continent to reach his destination, New York, having encircled two-thirds of the globe on his flight from Hitler's Germany.

The fact that Neuberg, shortly before the United States entered the war and after the American authorities had stopped admitting all the Jewish emigrants who applied for residence permits,[818] was nevertheless given permission to enter the country, was no doubt due not least – in addition to the approaches made by his daughters – to his reputation as a research scientist of international renown and editor of the *Biochemische Zeitschrift*. But his journey into exile through so many different countries was also assisted and facilitated by the commitment of well-disposed colleagues and former pupils. Although the working conditions in the various countries were not always in line with the standards he had been accustomed to from his time in Dahlem, his international scientific connections and the contacts he had built up over a period of 35 years did made it possible for him to escape, and thus also, in all probability, to survive and live on.

Although Anna Salomon, née Neuberg, was not a famous woman, it is appropriate at this point to say something about the fate of Neuberg's elder sister. Indications as to her whereabouts appear only sporadically and marginally in the documents investigated. Anna Neuberg was born in Pyrmont, her mother's home town (now Bad Pyrmont), on April 6, 1882.[819] She married a Max Salomon, about whose origins and profession no further details have been discovered.[820] There were two children of this marriage, who are mentioned by Carl Neuberg in a letter written in 1946:[821] This indicates that the son, Carl Salomon, was a doctor in Jerusalem, while the daughter, Ellen Salomon, was living after the war in Montevideo, the capital of Uruguay. After Carl Neuberg had fled to Holland in 1939, with the situation in Nazi Germany becoming more and more threatening, he attempted to help his sister to escape as well. To enablc Anna to emigrate to Palestine, he first invoked the assistance of Dutch friends who had contacts in the

January 25, 1941, indicating that Neuberg must have taken about ten days to cross the Middle Eastern desert.

817 Nachmansohn/Schmid (1988), 294.

818 Gabaccia (2002), 168–199.

819 Archives of the Federal Government/International Search Service (1986), Vol. 2, 1277.

820 The name of her husband is taken from the burial certificate of Neuberg's father dated January 26, 1916. Jewish Cemetery, Weissensee, Berlin. According to this document, the Salomon family were living at Schönhauser Allee 144, where Julius Neuberg, the father of Carl and Anna, also died.

821 Neuberg to Roth, October 9, 1946, CNP: Box 14: Roth, Walter.

Ministry of Justice there.[822] At the end of 1940, Neuberg was informed that a Jewish refugee committee was optimistic about Anna's position with regard to her leaving Germany for the Netherlands.[823] In order to arrange her escape, Neuberg paid the sum of US$ 1,000 into an account in Uruguay.[824] But as it was to turn out, it was already too late for Anna to escape:

> "Unfortunately, the news about Anna is not good. Although everything was arranged, although the ticket was paid for and all the deposits required had been put down, she did not get out of Germany with the last transport. We are trying to get her to Sweden, from where she could go and join Ellen, even though it would be difficult. It is uncertain whether we will succeed in this. And it all costs time and money."[825]

A little later, Neuberg was informed that Anna had been deported to Warsaw, where she could be contacted through a "Committee for Refugees." From time to time, Anna appears to have had the opportunity to contact relatives, who reported to her brother in the United States:

> "She writes as follows: Things are relatively good with us here, only of course one mustn't compare it with the lovely times we were able to spend together in the past. If Carl could do anything, it would of course be very welcome; indeed, his sister would be very glad of any small thing."[826]

From the United States, Neuberg went on trying, as far as was practicable, to get assistance to his sister.[827] He continued to receive news of Anna until early 1943.[828] After the war, he could only assume that his sister had failed to survive the Nazi terror. He wrote to a relative in Amsterdam, whose whole family had died in a concentration camp: "I too, as far as human knowledge goes, have to mourn the death of my sister, who was probably killed in Poland; at least, all attempts to find her have proved futile."[829]

822 Prof. van Loon of Deventer to Neuberg, January 15 and 19, 1940. CNP, Box 8: Loon, Henrik v. In these letters, van Loon reports on his efforts to obtain permission from the authorities concerned for Anna to leave for Palestine.

823 Van Loon to Neuberg, December 2, 1940, CNP, Box 8: Loon, Henrik v.

824 Neuberg to Ziffer, August 20, 1947. CNP. Box 18: Ziffer, Rudolf.

825 "Von Anna hören wir leider nichts Gutes. Obgleich alles geregelt war, obgleich die Fahrkarte bezahlt und alle geforderten Depots hinterlegt waren, ist sie mit dem letzten Transport nicht mehr aus Deutschland herausgekommen. Wir versuchen sie nach Schweden zu bringen, von wo ab sie, wenn auch schwierig zu Ellen fahren könnte. Ob das gelingt ist nicht sicher. Alles kostet Zeit und Geld." Neuberg to Nathan, April 13, 1942. CNP, Box 10: Nathan, Blanche.

826 "Sie schreibt wie folgt: Uns geht es verhältnismäßig gut nur darf man natürlich keine Vergleiche anstellen mit den schönen Zeiten, die wir zusammen verleben durften. Wenn Carl etwas tun könnte, wäre das natürlich zu begrüßen, überhaupt würde seine Schwester sich über jede Kleinigkeit sehr freuen." Kindler to Neuberg, May 9, 1942 and June 15, 1942. CNP, Box 7: Kindler, Hilde.

827 Neuberg to Kindler, July 6 and August 18, 1942. CNP, Box 7: Kindler, Hilde. In these letters he asks Hilde Kindler to send Anna whatever she could, especially clothes.

828 Neuberg to Nathan, February 28, 1943. CNP, Box 10: Nathan, Blanche.

829 "Auch ich habe nach menschlicher Berechnung den Tod meiner Schwester zu beklagen, die vermutlich in Polen umgebracht ist, jedenfalls sind alle Nachforschungen nach ihr ergebnislos gewesen." Carl Neuberg to Julius Neuberg, March 7, 1946. CNP, Box 12: Neuberg, Julius.

Anna Salomon, née Neuberg, is officially listed as having gone missing in Warsaw.[830]

830 Archives of the Federal Government/International Search Service (1986), Vol. 2, 1277.

7. EXILE IN AMERICA (1941–1956)

7.1. NEUBERG'S LIFE IN THE UNITED STATES

Emigrants displayed very diverse attitudes to Germany and the Germans after their escape into exile. In view of what they had experienced and suffered, many decided that they neither wished to have anything more to do with the country, nor to speak the German language any more. But there were a few among the emigrants who wished to return to their former home after the war.

Neuberg, having been driven out of Germany, found himself very much in two minds. He found it very difficult to get used to American culture and the American way of life. At the same time, however, the injustice he had suffered, his expulsion and the very probable murder of his sister by the Nazis had left deep wounds in Neuberg's image of Germany. In 1954, in a letter to the then Chancellor Konrad Adenauer, he summed up this dichotomy within himself with a quotation from the poet Ricarda Huch: "You can leave a country, but the country never leaves you."[831] In a letter to Otto Warburg, who had stayed at the KWG/MPG in Germany, Neuberg wrote: "Of course I think of Berlin, but Berlin doesn't think of me."[832]

During the war years, he supported Germany's adversaries to the best of his ability[833] and longed for the end of the "Third Reich."[834] When the allied air raids on German cities increased substantially in intensity from 1942 onwards, Neuberg expressed a certain satisfaction, while at the same time voicing his concern for friends who had remained in Germany: "Things must look dreadful in Germany now, and the end will, justly, be pretty gruesome for the Nazis. Let's hope that not

831 "Man kann ein Land verlassen, aber das Land verlässt einen nie." Neuberg included this quotation from Ricarda Huch in a letter written April 26, 1954 to the then German Chancellor Konrad Adenauer, thanking him for the award of the Federal Cross of Merit. CNP, Box 1: Adenauer, Konrad.

832 "Natürlich denke ich an Berlin, aber Berlin denkt nicht an mich." Neuberg to Warburg, April 15, 1949. CNP, Box 17: Warburg, Otto.

833 A passage from a letter from Milli Berlak, who was working as a secretary at the KWI for Biochemistry, to her former boss Neuberg on July 12, 1942, may serve as evidence of this: "I utterly share your wish to stand fighting on the British side." ("Ihren Wunsch auf englischer Seite kämpfend stehen zu wollen, teile ich vollkommen.") CNP, Box 3: Berlak, Millie. Later, Millie Berlak was able to escape to Great Britain.

834 "[...] I hope that we will all meet joyfully again one day and dance together as the Nazis are hanged. May God so dispose, Amen." ("[...] ich hoffe, daß wir uns alle eines Tages vergnügt wiedersehen und bei der Hinrichtung der Nazis zusammen tanzen werden. Das walte Gott, Amen." Neuberg to Berlak, June 6, 1943. CNP, Box 3: Berlak, Millie.

too many of the innocent are killed with them."[835] Neuberg was himself able to contribute to the military efforts to weaken the German war machine by making his knowledge of the locations of German facilities for producing glycerol available to the American authorities. This was of strategic significance to the extent that glycerol was used among other things as a material in the manufacture of explosives, and also for other military purposes.[836]

The flight to exile in America represented salvation for Neuberg, but at the same time expulsion from his own culture and the premature end of his scientific career. He summed up the dilemma he felt regarding his emigration in the following words:

> "Thomas, my dear colleague, you see that America is no paradise for people like us. If I had come here in 1933 I would probably still have got a job of some kind; now I'm just an old immigrant dog who arrived with only a little suitcase. I'm not really an emigrant; I was simply thrown out."[837]

Neuberg had been displaced and massively threatened in his home country because of his Jewish descent. Without a future in Berlin, he had finally decided to flee, but in exile in America he was forced to realize that this decision, which had been taken almost too late for him to be able to get out of Germany, was certainly too late for him to be able to build up a new life as a scientist in his new home. This "too late" motif recurs again and again in his letters to Germany.[838]

At the age of 64, he was forced to come to terms with the living and working conditions of a country that was foreign to him. He was profoundly put off by the style of research in the American universities, and the way scientists were treated. The principle that was customary on the general labor market in America of short-term employment according to need, followed by dismissal, was also practiced in research establishments, and this was completely incomprehensible to him.

> "No one who didn't come here aged under 20 will ever feel at home here. [...] There is no security here. You can be dismissed and paid off every Friday, and the following Monday a younger and cheaper successor will be sitting at your desk, or the shop will have been shut down completely. In this respect, the United States really is the country of unlimited possibilities."[839]

835 "Es muss jetzt scheusslich in Deutschland aussehen und das Ende wird, gerechterweise, für die Nazis ziemlich grausig sein. Hoffen wir, dass nicht zu viele von den Unschuldigen mit umkommen." Neuberg to Gaffron, February 8, 1943. CNP, Box 5: Gaffron, Hans.

836 Neuberg to the U.S. Coordinator of Information, March 26, 1943. CNP, Box 16: U.S. Coordinator of Information.

837 "Lieber Herr College Thomas, Sie sehen in Amerika ist für unsereinen nicht das Paradies. Wenn ich 1933 hergekommen waere, wuerde ich vermutlich noch irgendeine Stelle gefunden haben; jetzt bin ich ein alter immigrierter Hund, der nur mit einem Handkoefferchen angekommen ist, Emigrant bin ich eigentlich nicht, sondern nur einfach herausgeworfen." Neuberg to Thomas, December 21, 1949. CNP, Box 15: Thomas, Karl.

838 See Neuberg to Hofmann, February 1, 1949. CNP, Box 6: Hofmann, Eduard.

839 "Niemand, der nicht unter 20 Jahren hierher gekommen ist, wird hier heimisch. [...] Sicherheit gibt es hier nicht. Man kann jeden Freitag entlassen und ausgezahlt werden und am Montag sitzt ein jüngerer oder billigerer Nachfolger an ihrem Arbeitsplatz oder die Bude ist überhaupt geschlossen. In diesem Sinne ist [die] USA wahrhaft das Land der

Neuberg had been shaped by the academic milieu of Berlin University in the days of the Empire. In a speech on the occasion of a symposium at the Polytechnic Institute of Brooklyn arranged in his honor on 15 January 1955, he goes into the matter of how his style of work had been influenced and fostered by the German professorial system:

> "My only excuse for this pleasurable but unsystematic way of working is the fact that I always had a more or less tyrannical chief who made me follow up a problem to its last consequences, by light or heavy pressure. […] The older I got, the more I realized that a somewhat autocratic chief may be of incalculable benefit to the development and achievement of a young scientist."[840]

As Bruno Kisch declared, he placed particular value on conscientiousness and honesty in scientific work:

> "Neuberg had an extremely critical brain, with a sharp eye for errors in scientific work, and above all possessed an over-sensitivity bordering on allergy to any kind of dishonest or superficial way of working."[841]

His knowledge of the biochemical literature, which was unrivalled in its comprehensiveness, enabled him to recognize shoddy academic procedures very quickly, and he had no inhibitions about writing to the publishers of such publications personally to draw attention to their errors in his usual downright manner.[842] In letters to colleagues who had remained in Europe, Neuberg repeatedly complained about the way American scientists, as it seemed to him, again and again stole discoveries which were then celebrated in the United States as "new" or "groundbreaking," but which in some cases had already been described in European literature 50 or 60 years earlier. In 1947, for example, he wrote to Emil Abderhalden, who, as we know today, was himself not always the most principled of scientists:[843]

> "Here, everything that did not happen to have been done in America, or is more than ten years old, is considered to be free for the taking and has to be rediscovered. Have you never noticed that, of course with the exception of a few great American research scientists, the people here make their livelihood from pinching things?"[844]

Neuberg supposed that the reasons for this practice, which he observed repeatedly in America, lay in the fact that "only very few of their fellow countrymen have really well-grounded knowledge and the desire or ability to read work in a foreign

unbegrenzten Moeglichkeiten." Neuberg to Kobel, December 8, 1948. CNP, Box 7: Kobel, Maria.

840 CNP, Box 13: Polytechnic Institute of Brooklyn.

841 "Neuberg war ein außerordentlich kritischer Kopf, der Fehler in wissenschaftlichen Arbeiten scharfsichtig erkannte und vor allem gegen jede Art unehrlichen oder oberflächlichen Arbeitens eine ans Allergische grenzende Überempfindlichkeit besaß." Kisch (1966), 328.

842 Neuberg to Potter, June 2, 1943. CNP, Box 13: Potter, Van R.

843 Deichmann/Müller-Hill (1998), Frewer (2000), Kaasch (2000) and Kaasch/Kaasch (2003).

844 "Hier gilt als vogelfrei und muß neu entdeckt werden, was nicht gerade in Amerika geschaffen oder aelter als 10 Jahre ist. Ist Ihnen nie aufgefallen, dass selbstverstaendlich abgesehen von einigen grossen amerikanischen Forschern, die Leute hier vom Klauen leben?" Neuberg to Abderhalden, May 15, 1947, and Neuberg to Abderhalden, January 27, 1948. CNP, Box 1: Abderhalden, Emil.

language."[845] In addition, he drew attention to substantial shortcomings in the American university system. He complained that most researchers lacked independence. In particular, the American tradition of teamwork, which in his view was practiced in accordance with the rules of baseball and American football, did not in the least correspond to his understanding of scientific work. Furthermore, he felt that the individual scientist's freedom to research was restricted, since in his view, "the system of institutional popes, which came to an end in Germany with Emil Fischer, often still exists"[846] in American research establishments. The American system of teaching was also judged very critically by Neuberg. He considered most lecturers to have only very inadequate or very narrow knowledge,[847] and accused the students of being more interested in an economically profitable education than in gaining scientific insights:

> "[...] and I have always found that learners have a fine feeling for the performance of the teacher. This is the case even in America, but it takes second place to the mentality 'Which teacher will enable us to earn as many dollars as possible most quickly?'"[848]

He was utterly repelled by this marked fixation on pecuniary matters and the resulting direct dependence on economic considerations which he perceived in American society in general and in a large proportion of academics in particular. He also felt the striving to obtain grants to be a form of begging, and described a large number of American universities as being "business corporations in disguise."[849] And the fact that he was expected to pay fees to become a member of the New York Academy of Science was completely irreconcilable with his European view of the role of academic societies:

> "It is often said that universities in this country are business undertakings; but the fact that academies could be so too was new to me: to the extent that you are supposed to pay membership contributions. As a member or honorary member of the Academies in Göttingen, Prague, Leningrad, Rome, Lisbon, Lund, Copenhagen, Uppsala etc., I remember being given

845 "[…] dass nur wenige ihrer Landsleute ein fundiertes Wissen haben und Lust oder Fähigkeit besitzen, fremdsprachige Arbeiten zu lesen." Neuberg to Windaus, May 12, 1948. CNP, Box 18: Windaus, Adolf.

846 "[…] dass vielfach noch das in Deutschland mit Emil Fischer zu Ende gegangene System der Instituts-Paepste existiert." Neuberg to Windaus, July 13, 1948. CNP, Box 18: Windaus, Adolf.

847 Neuberg to Windaus, May 12, 1948. CNP, Box 18: Windaus, Adolf.

848 "[...] und ich habe immer gefunden, dass die Lernenden ein feines Gefuehl fuer die Leistungen des Lehrers haben. Dies ist selbst in Amerika so aber es tritt zurück hinter die Mentalitaet 'durch welchen Lehrer koennen wir am schnellsten moeglichst viel Dollar machen'." Neuberg to Abderhalden, January 17, 1948. CNP, Box 1: Abderhalden, Emil.

849 "[…] getarnte Aktiengesellschaften": Neuberg to Windaus, May 12, 1948. CNP, Box 18: Windaus, Adolf.

> gold medals, swords of honor and money for papers, and also being paid by these academies to travel to give lectures. I will certainly not apply for membership of that academy [...]."[850]

And Neuberg went on in his incomparably forthright manner:

> "In Berlin, in Friedrichstrasse, there was an establishment called the 'Academy of Music,' where you also paid – but for amusements and sparkling wine. It was frequented by physiologists, among others, for example by our late friend Cremer, who wanted to drag me off there once; but we drank our *sekt* somewhere else – quite unacademically."[851]

But Neuberg's criticism of the United States was not directed only against the American educational system. Rather, he saw in the whole of American society what he considered to be an excessive striving after material wellbeing. He himself, as his granddaughter Joan Armer remembers, had an aversion to everything that he considered to be personal luxury.[852] Neuberg, who had been an American citizen since the end of August 1946,[853] described his new compatriots' attitude to money with a very sharp tongue:

> "Although the 'Savior' is said to have driven the moneychangers from the temple, both Jews and Christians pray most fervently to the great god $, who not only has his altars in the home, but is worshipped in every room. It is not without interest in a Freudian way that the façades of banks and *bureaux de change* are in the form of stylized altars. There is no veil before the Holy of Holies, and it is made up not of Torah rolls, but of rolls of banknotes."[854]

He also had little time for the cultural sponsoring that is customary in the United States. Neuberg's description of the way well-off American citizens sought to purchase the admiration of their fellows through the "material" possession of works of art makes clear how little he was able to acclimatize to his new environment:

850 "Dass Universitaeten hierzulande Geschaeftsunternehmungen sein sollen wird behauptet, dass es auch Akademien sein koennen war mir neu: insofern als man Mitgliedsbeitraege zahlen soll. Ich entsinne mich als Mitglied oder Ehrenmitglied der Akademien in Goettingen, Prag, Leningrad, Rom, Lissabon, Lund, Kopenhagen, Upsala usw. goldene Medaillen, Ehrendegen und Geld für Arbeiten bekommen zu haben, auch Bezahlung für Reisen zu Vortraegen von diesen Akademien. Ich werde mich sicher nicht bei jener Akademie bewerben [...]." Neuberg to Barken, January 7, 1944. CNP, Box 2: Barken, G.

851 "In Berlin auf der Friedrichstrasse gab es ein Lokal die 'Academy of Music' man zahlte auch – aber für Amuesement und Sekt. Es wurde auch von Physiologen frequentiert; so z. B. von unserem heimgegangenen Freund Cremer, der mich einmal dorthin schleppen wollte; wir tranken dann unseren Sekt woanders – ganz unakademisch." Neuberg to Barken, Professor at the Medical School in Boston, who had suggested to Neuberg that he should join the New York Academy of Science. Barken to Neuberg, January 7, 1944. CNP, Box 2: Barken, G.

852 Written communication from Joan Armer, December 14, 1997.

853 Neuberg's US Citizen Petition is dated August 29, 1946. From a *Personnel Security Questionnaire* of the Atomic Energy Commission, March 30, 1950. CNP, Box 16: U.S. Atomic Energy Commission.

854 "Obgleich der 'Heiland' die Wechsler aus dem Tempel getrieben haben soll, beten Juden und Christen inbruenstig zum Gotte $, der nicht nur Hausaltaere hat, sondern in jedem Zimmer tief verehrt wird. Es ist im Sinne von Freud nicht ohne Interesse, dass die Front von Banken und Wechselstuben einen stilisierten Altar darstellen. Das Allerheiligste ist nicht verhuellt und besteht auch nicht aus Thora-, sondern Geldrollen." Neuberg to Fodor, July 29, 1947. CNP, Box 5: Fodor, Andor.

"You don't believe, do you, that one might be able to come to feel at home in the States at an advanced age? There is an unbridgeable cultural divide between our customs and ways of thinking. One funny example: as recently as 1940, there were labels under the works of art concerned: 'Purchased for $ x' or 'Gift of Mr. X to the value of $ x.' Sic!"[855]

Despite all this criticism, Neuberg remained in the United States. One reason for this is no doubt likely to have been his closeness to his two daughters Irene and Marianne, who had left Germany long before him and found refuge in the United States. The elder daughter, Irene, had studied chemistry in Berlin and had written her doctoral thesis at the KWI for Biochemistry under her father's supervision. In 1932 she went to the Institut Pasteur in Paris for two years, and there met her first husband, Bruno Rabinowitsch.[856] Rabinowitsch, whose parents came from Russia,[857] had grown up in Vienna and studied chemistry there. After their marriage, the couple had left France in 1934 and gone to Istanbul, where Irene had had a job as a chemical assistant. After spending some time in Russia and Palestine, they finally emigrated to the United States. In America, Irene and Bruno Rabinowitsch, who had a daughter called Joan, changed their surname to Roberts. Later, Irene married again and took the name Forrest.[858] In New York, Irene worked – mostly together with her father – as a chemical assistant at New York University, the Polytechnic Institute of Brooklyn and the Medical College of New York. In between times she was also a translator for the American Defense Department and worked in an automobile factory in Dayton, Ohio. After her father's death in 1956 she took over the direction of the Biochemical Research Laboratory of the Veterans Administration Hospital in Brockton, Massachusetts, and then from 1961 until 1980 occupied the same position in Palo Alto, California. Irene remained in contact with faculties of science, and was engaged in biochemical research both at Stanford University and at the University of San Francisco. She died at the age of 86 in 1994. Carl Neuberg had a particularly close relationship with his daughter Irene throughout his life. Apart from their scientific work, which led to a number of joint publications, Neuberg spent a lot of time with her. Father and daughter lived together in New York for a while;[859] in later years, Neuberg visited her regularly on her farm in Vermont.[860]

855 "Glauben Sie etwa, man koennte in fortgeschrittenem Alter in U.S.A. noch heimisch werden? Eine unausfuellbare kulturelle Kluft trennt Gepflogenheiten und Denkweise. Ein heiteres Beispiel: Noch 1940 stand unter den resp. Kunstwerken: 'Gekauft für x $' oder 'Geschenk von Mr. X im Werte von x $.' Sic!" Neuberg to Reinold August 17, 1948. CNP, Box 13: Reinold, Marianne.

856 Unless otherwise stated, cf. American Men & Women of Science (1995–1996), 1385; written communication from Joan Armer, March 11, 1998.

857 Bruno Rabinowitsch (1908–1968). His parents died in a concentration camp in Riga. In the US, Rabinowitsch worked as a chemist for Monsanto Chemical Corporation and was one of the inventors of the Acrilan fiber (polypropenonitrile).

858 This made it particularly difficult to trace Joan Armer, without whose assistance it would not have been possible to reconstruct these family relationships.

859 Neuberg to Abderhalden, January 6, 1948. CNP, Box 1: Abderhalden, Emil.

860 Neuberg gives this as his holiday address to F. F. Nord, for example, in July 1946. CNP, Box 12: Nord, Frans Friedrich.

Only a little information could be found about Neuberg's second daughter, Marianne.[861] She had begun to study medicine in Heidelberg before the war, but had been unable to complete her studies before fleeing from Germany. She married Fritz Lederer, with whom she had two daughters. After staying temporarily in Vienna, the Lederer family emigrated to the United States in 1938. In America, Marianne – who was also called "Fuchsi" by friends of the family[862] – initially worked at a company making measuring instruments,[863] and was later technical assistant in the Department of Nuclear Medicine at the Veterans Administration Hospital in Los Angeles. She died in 1987 at the age of 76.

Neuberg probably got to know the widow Hilde Unger during his time in Palestine. Hilde Unger, née Bacher, was born in Berlin on April 21, 1897. She appears to have been married to Walter Unger, the proprietor of the well-known Kempinski Hotel in Berlin. Her husband was almost certainly murdered in Auschwitz concentration camp.[864] After Neuberg had traveled on from Jerusalem to America, he appears to have been involved in obtaining entry visas for Hilde Unger and her daughter.[865] Hilde Unger finally arrived in New York in 1945. She kept house for Neuberg for several years[866] before they were married in 1950.[867] Hilde, who was not very well liked in Neuberg's family,[868] accompanied her husband on his first visit to Germany in 1951. But already in the course of this stay in Germany, serious differences arose between them,[869] and they separated only three years after their marriage.[870] Neuberg summed up his short second marriage in a letter as follows:

> "The rumor you have heard that I had married again is unfortunately true. This was however not a recent event, but happened five years ago. Things went awry, and, although not divorced, I have been living apart from my wife for two years now. She was as beautiful as she was cultured, and moreover quite a dish, as she was one of the proprietors of the Kempinski; but sometimes, the finest menu does not satisfy in the long run (like having plenty to eat in jail)."[871]

861 Unless otherwise stated, this information is taken from Joan Armer's written communication of December 14, 1997.
862 Durig to Neuberg, August 24, 1946. CNP, Box 4: Durig, Arnold.
863 Neuberg to Butenandt, March 7, 1947. CNP, Box 3: Butenandt, Adolf.
864 C. Neuberg to E. Neuberg, February 6, 1948. CNP, Box 12: Neuberg, Ernst.
865 Neuberg to Overberger, January 17 and February 13, 1942. CNP, Box: Overberger, G. Charles.
866 C. Neuberg to E. Neuberg, February 6, 1948. CNP, Box 12: Neuberg, Ernst.
867 Neuberg to the Administrative Headquarters of the MPG, February 14, 1953. MPG Archives, Dept. II, Rep. 1A, Neuberg's Personnel File.
868 Written communication from Joan Armer, December 14, 1997.
869 C. Neuberg to E. Neuberg, February 6, 1948. CNP, Box 12: Neuberg, Ernst.
870 Neuberg to Hofmann, March 2, 1953. CNP, Box 6: Hofmann, Eduard.
871 "Das zu Ihnen gelangte Geruecht, dass ich wieder geheiratet haette, ist leider wahr. Das habe ich zwar nicht jetzt sondern vor 5 Jahren getan. Die Angelegenheit ist schief gegangen und ich lebe seit 2 Jahren, ohne geschieden zu sein, von meiner Frau getrennt. Sie war ebenso schoen wie gebildet, an sich ueberdies schmackhaft, da sie eine der Inhaberinnen von Kempinski gewesen ist, aber manchmal befriedigt die schoenste Speisekarte nicht auf die

The time that Neuberg took off from his still very extensive research work he mainly spent in the company of his daughter Irene, whom he often visited on her farm in Vermont during vacations. By contrast, Neuberg declined an invitation from his friend and colleague Hans Aron, who was living in Chicago, to spend a summer vacation together in the Rocky Mountains, with a verse saying: "Going riding, playing golf and swimming / is certainly nothing for me; / the more so as don't know whether, in the end / the shooting and swimming girl isn't already spoken for."[872]

Neuberg's ideas of vacation and recreation were of a different nature. To him, a stay at the Marine Biological Laboratory in Woods Hole, Massachusetts, offered just as much relaxation and enjoyment. This celebrated research establishment, which continues to exist in a similar form right down to today, was established in 1888, and is one of the oldest marine biological institutions in the western world.[873] Year after year, the scientific elite of the United States and other countries assembled here, around 400 kilometers north-east of New York, in order to pursue their own research in well-equipped laboratories and libraries, cultivate personal relationships and exchange ideas with colleagues. A large number of Nobel Prize winners were frequent guests at Woods Hole.[874] While staying here, Carl Neuberg met with new colleagues as well as old acquaintances and pupils from Berlin such as Otto Meyerhof, Otto Warburg, Albert Szent-Györgi and Otto Loewi, and also David Nachmansohn and Neuberg's friend and physician Bruno Kisch. The latter reported in his memoirs that there was also a great deal of socializing at Woods Hole, and that in addition to the many high caliber scientific lectures, concerts and song recitals were arranged in the evenings.[875] Neuberg spent almost every summer there during the 1950s, generally staying for several weeks. In Woods Hole he was able to escape from the extreme heat of the New York summer, and recover from the hassle of city life.[876] The atmosphere prevailing there may perhaps also have reminded Neuberg of the spirit of Dahlem. However, he described the reasons for his annual efforts to obtain a place at

Dauer (abgegessenes Menu im Zuchthaus)." Neuberg to Heubner, August 30, 1954. CNP, Box 6: Heubner, Wolfgang.

872 "Wo man reitet, golft und schwimmt, das ist für mich nix, ganz bestimmt. / Auch weiß ich nicht, ob allerletzt, das Schieß- und Schwimm-Girl nicht besetzt." Neuberg to Aron, July 21, 1943. CNP, Box 2: Aron, Hans.

873 Unless otherwise stated, this and the following information is taken from the website of the Marine Biological Laboratory (www.mbl.edu).

874 The following scientists worked at Woods Hole, and were probably in contact with Neuberg there: August Krogh, Otto Meyerhof, Otto Warburg, Thomas Morgan, Otto Loewi, Albert Szent-Györgyi, Herbert Gasser, Isidor Rabi, Hermann Muller, John Northrop, Wendal Stanley, Selman Waksman, Fritz Lipmann, Axel Theorell, George Beadle, Joshua Lederberg, Severo Ochoa, Donald Glaser, James Watson, John Eccles, Alan Hodgkin, Andrew Huxley, Keffer Hartline, George Wald, Konrad Bloch and others.

875 Kisch (1966), 329–330.

876 Many documents among Neuberg's papers confirm this. Cf. CNP.

Woods Hole self-critically in a letter to Daniel Mazia,[877] who appears to have expressed surprise at Neuberg's diligence and zeal in his old age:

> "I think doing experimental work at my age is a testimonium paupertatis animi. That means I am not intelligent enough to have a worth-while hobby, unfortunately, I have only learned to work at the laboratory."[878]

In the course of his life, Neuberg had got to know many important scientists, and was able to repeat a wealth of anecdotes about them and their work. For this reason, and also in view of the fact that Neuberg was far from sparing with his frank and sometimes wounding criticism, his friend Bruno Kisch describes his stories as a "*chronique scandaleuse of the academic world.*"[879] Neuberg had known Kisch since the 1920s, and during his later years was linked to him by a close friendship that is reflected in their lively correspondence.[880] Although Neuberg was sometimes put off by the distinct piety of his friend, who was also Jewish,[881] his relationship with Kisch was characterized by sincere gratitude and human warmth.[882] Neuberg found another friend in the Nobel Prize winner Otto Loewi,[883] whom he frequently, according to Ines Mandl, met for lunch.

Otto Warburg was another of Neuberg's closer acquaintances during his later life, despite the fact that their relationship had at times been quite strained. It was very difficult for many emigrants to understand how Otto Warburg, although Jewish, had been able to continue his research work at the KWG relatively unmolested throughout the entire period of the Nazi dictatorship.

> "I am heretic enough to presume that Otto the Great in no way regrets having stayed in his lovely laboratory. If the Nazi gang hasn't touched him, then nobody will harm him, whoever

877 Daniel Mazia (1912–1996), American cell biologist who performed important research in the fields of mitosis and cell reproduction. After taking his doctorate in 1937, Mazia worked at Princeton University and then at the Marine Biological Laboratories in Woods Hole. From 1951 onwards he was Professor of Zoology at the University of California in Berkeley.

878 Neuberg to Mazia, May 2, 1955. CNP, Box 9: Mazia, Daniel.

879 "*Chronique Scandaleuse der gelehrten Welt.*" Kisch (1966), 328.

880 A large part of this correspondence is to be found among Bruno Kisch's papers, K2 and K3 (a–c).

881 Unlike Neuberg, Kisch observed the prescripts of his religion very exactly. Personal communication from Ines Mandl, October 25, 1996.

882 Neuberg to Kisch, July 29, 1952. SBPK, Manuscripts Department: Kisch papers, K 3 (a–c), sheet 52.

883 Otto Loewi (1873–1961) took his doctorate in Strasbourg, and was then an assistant of Harko von Noorden's in Frankfurt. As Professor of Pharmacology in Vienna and Graz, he was awarded the Nobel Prize in Medicine jointly with Sir Henry Dale (1875–1968) in 1936 for their discoveries relating to the chemical transmission of nerve stimuli. He was forced to emigrate in 1938, and took up a research professorship at the University of New York in 1940.

may occupy Berlin. Everybody will compete to make his life pleasant. As you know, he's also a great diplomat."[884]

Neuberg described him as someone who distanced himself "from both his contemporaries and his predecessors."[885] Neuberg's sarcastic comments about Warburg must doubtless be regarded in the light of his insuperable disappointment and pain at having had to leave "his" KW Institute and flee from his beloved Berlin. But in 1948 his remarks about "Otto the Great" are considerably more conciliatory in tone:

> "Warburg is here. He has grown much milder and appears to regret many things. But he holds views that differ very sharply from those of myself and others, not only about respiration but about many other things in the world and in Germany."[886]

Warburg for his part employed his best efforts in Germany to enable Neuberg to get his old institute in Dahlem back again. He visited Neuberg several times in New York, and Neuberg enjoyed his hospitality during his visits to Germany.

After his emigration, Neuberg also maintained intensive and regular correspondence with three intimate friends, even though in some cases he was critical of their behaviour:[887] the physiologist Emil Abderhalden from Halle, the Viennese physiologist Arnold Durig and the Karlsbad doctor and research scientist Paul Mayer. During the latter years of his life, however, his circle of acquaintances grew smaller and smaller, no doubt also as a result of his increasingly bad health, which greatly restricted his activities.[888] But Neuberg maintained written contact with former colleagues and subordinates until shortly before his death.

7.2. RESEARCH AND TEACHING IN EXILE

After Neuberg's flight from Nazi-ruled Germany, which lasted a year and saw him travel via Amsterdam, Palestine and the Pacific Ocean to America, a completely new period of his life began in New York. First of all he had to concern himself with finding a job, in order to earn his living. Although he had been assured of a limited term research professorship at New York University, this was funded only for two years through a donation from a relative, Walter

884 "Ich bin Ketzer genug anzunehmen, dass Otto der Grosse gar nicht bedauert in seinem schoenen Laboratorium geblieben zu sein. Wenn ihm die Nazibande nichts getan hat, wird ihm keiner ein Leid zufuegen, wer auch Berlin besetzen mag. Man wird wetteifern, ihm das Leben angenehm zu machen. Wie Sie ja wissen, ist er auch ein grosser Diplomat." Neuberg to Gaffron, February 23, 1943, CNP, Box 17: Gaffron, Hans.

885 Neuberg to Wieland, September 25, 1947, CNP, Box 17: Wieland, Heinrich O.

886 "Warburg ist hier. Er ist viel milder geworden und bereut anscheinend vieles. Er hat aber nicht nur ueber die Atmung sondern ueber manches in der Welt und in Deutschland von mir und anderen stark abweichende Ansichten." Neuberg to Wieland, December 16, 1948, CPN, Box 18: Wieland, Heinrich O.

887 See Chapter 3.1.

888 Kisch (1966), 328.

Neuberg.[889] The intention had been that Neuberg should take up his position at the beginning of September 1940;[890] but as a result of the difficulties attending his journey, as described above,[891] he was not in fact able to do so until February of the following year. He was indeed not officially registered as a member of the Faculty of the Department of Chemistry until February 1942.[892] The working conditions Neuberg found in his new laboratory at Washington Square in Manhattan[893] were, however, in no way comparable to those that he was used to from his institute in Dahlem:

> "No one who has not seen it with his own eyes can have any idea of the meagerness, indeed in most cases the squalidness of the set-up. I agree with Warburg, who happens to be here at the moment, that what passes here for a prestige laboratory would have been turned down by our cleaning women in Dahlem as a room to eat their sandwiches in. Warburg is working at the celebrated Photosynthesis Laboratory: it is no more than a damp basement hole."[894]

As is evinced by a letter to his former assistant Maria Kobel, the conditions under which he had to work had not improved even seven years later:

> "The 'laboratory' that I have at New York University is as big as Malkowski's[895] washing-up room. If I want to use a fume cupboard I have to go to an overcrowded students' laboratory, and to do polarization I have to go to a different institute. Library unimaginably scanty. Complete sets of periodicals are not to be found anywhere, and incomplete ones only after long journeys."[896]

When Neuberg took up his work at New York University, he was almost 65 years old. Despite this, his professorship, for which no further outside funds were earmarked after the first two years,[897] was extended six times in all. It was only Neuberg's persistence that enabled him to obtain these extensions. In a letter to the Head of the Faculty of Chemistry, William MacTavish, for example, he referred to other scholars he knew who also held professorships at institutions in

889 Neuberg received an annual salary of $2,500. Archives of New York University, Refugee Appointments, Record group 4.0.1, Series No. 1, Box No. 19.

890 Telegram from Voorhis to Neuberg, July 12, 1940. CNP, Box 17: Voorhis.

891 W. C. MacTavish, Chairman of the Department of Chemistry of NYU, to the Department of State, USA Visa Division, July 28, 1941. CNP, Box 12: New York University.

892 New York University to Neuberg, January 31, 1942. CNP, Box 12: New York University.

893 Archives of New York University, New York University Directory (1942–1943), 50.

894 "Von der Dürftigkeit, ja meist Schäbigkeit der Einrichtungen kann sich niemand einen Begriff machen, der sich nicht durch Augenschein überzeugt hat. Mit Warburg, der gerade hier ist, bin ich mir darüber einig, dass unsere Dahlemer Putzfrauen die hiesigen 'Prachtlabors' als Frühstücksraum ablehnen würden. Warburg arbeitet in dem berühmten Labor für Photosynthesis. Es ist nichts als ein feuchtes Kellerloch." Neuberg to Kobel, February 8, 1948. CNP, Box 7: Kobel, Maria.

895 Xaver Malkowski was a laboratory servant at the KWI for Biochemistry in Dahlem, Berlin.

896 "Das 'Labor', das ich in der New York University habe, ist so gross wie Malkowski's Spülraum. Wenn ich einen Abzug benuetzen will, muss ich in ein ueberfuelltes Studentenlaboratorium gehen, zum Polarisieren in ein anderes Institut. Bibliothek von unvorstellbarer Duerftigkeit. Zeitschriften findet man nirgendwo complett und das Uncomplette nach langen Fahrten." Neuberg to Kobel, August 17, 1948. CNP, Box 7: Kobel, Maria.

897 Neuberg to Butenandt, March 7, 1947. CNP, Box 3: Butenandt, Adolf.

New York at even more advanced ages, and unlike himself also drew regular salaries.[898] Despite this, he was only able to work at New York University until the end of 1948, when he had to give up his position "for reasons of age"[899] and was given emeritus status. The Dean of the University, in a letter of March 30, 1949 to David Price of the National Institute of Health, who had approached him on Neuberg's behalf, again pointed out that the University had an age limit for professors:

> "[...] Everything we learned about his research was excellent, and if it were not that the University administration feels it quite necessary from all-University point of view to regard the retirement age of professors at Washington Square College as between sixty-five and seventy, with seventy at the maximum, Dr. Neuberg would presumably be continuing his research in New York University. [...] Our attitude towards Dr. Neuberg is indicated by the fact that upon his retirement he was made Research Professor Emeritus. With us emeritus is not an automatic title, but given only and definitely for merit."[900]

According to Neuberg himself, however, this occurred "[...] less because of my age than because of the extreme need of my laboratory at the New York University."[901]

At the age of 71, Neuberg was compelled once again to seek other openings in order to be able to earn his living with scientific work. Initially he attempted, with the assistance of his former pupil and member of staff Friedrich Nord, to get a post at Fordham University in New York,[902] but this institution was unable to take him. Through his connections with Severo Ochoa, who later, in 1959, was awarded the Nobel Prize in Physiology and Medicine, Neuberg sought to obtain a professorship at New York Medical College, initially in vain.[903] Neuberg explained the reasons for these efforts to the director and founder of the Institute for Polymer Research at the Polytechnic Institute of Brooklyn, Hermann Mark,[904] as follows:

> "It is essential for me to be linked to a university. After 6½ years of extensions at NYU [= New York University, Authors' note], the people there don't want to leave me my little

898 Neuberg to MacTavish, February 26, 1947. CNP, Box 8: MacTavish, W. C.

899 Archives of New York University, New York University Directory (1949–1950), 21.

900 CNP, Box 17: U. S. Public Health Services.

901 In a letter of December 14, 1948 to Eugene Gisel, the Head of the Department of Chemistry of Fordham University in New York, Neuberg applied for an unpaid position as Research Professor. However, Gisel was obliged to inform Neuberg on December 23, 1948 that there was no way this could be put through. CNP, Box 5: Gisel, Eugene.

902 Nord to Gisel, November 30, 1948. CNP, Box 12: Nord, (Frans) Friedrich.

903 Exchange of letters between Neuberg and Ochoa in October 1948. CNP, Box 12: Ochoa, Severo.

904 Hermann Mark (1895–1992), born in Vienna, studied chemistry and then joined the KWI for Fiber Chemistry. After his *habilitation* (1925) he became head of the research department of the IG Farben corporation. Being Jewish, he moved to Vienna in 1933, where he held a Chair of Chemistry. After the annexation of Austria, Mark emigrated in 1938, and after first working in industry in Ontario, Canada, occupied a Chair of Chemistry at the Polytechnic Institute of Brooklyn in New York. He is considered to be one of the founders of polymer sciences. Cf. Beneke (2005).

> laboratory after my retirement. What I really need is more an academic address than a busy laboratory. I don't need money, either as a fee for myself or to run the experiments. [...] I have no ambitions to achieve a position, and would not put pressure on anyone."[905]

Mark finally succeeded in accommodating Neuberg in his institute's Department of Chemistry. From the beginning of 1949, Neuberg was listed initially as an Adjunct Professor, and was appointed Visiting Professor in October of the same year.[906] According to Mark, Adjunct Professor was the title usually given to university lecturers with the qualifications for a professorship but who did not form part of the regular faculty.[907] The title of Visiting Professor was then awarded to Neuberg "[...] in view of the worldwide reputation of Professor Neuberg."[908] Hermann Mark, who found opportunities for very many emigrant scholars from Europe to resume their scientific activity,[909] knew Neuberg from the Dahlem days, as Mark had researched at the KWI for Fiber Chemistry there.[910] Neuberg was able to conduct experimental work at the Polytechnic Institute of Brooklyn for a total of six years, working conditions there being, from Neuberg's point of view, no better than before:

> "I'm now a Visiting Professor at the Polytechnic Institute, with my salary doubled (2 x 0 = 0). I have a workplace in a students' laboratory and permission to buy chemicals and also to employ ancillary staff with my own money. I've not yet succeeded in finding anyone stupid enough to give me money for this."[911]

After a symposium with the title *"Current Peptide and Protein Research"* had been held in honor of Carl Neuberg in January 1955,[912] Neuberg gave up his laboratory at the Polytechnic Institute in April of the same year, and from then until his death worked only in the Biochemical Department of New York Medical College.[913] Neuberg had been working there with a variety of staff since 1951, so that at times he had two labs available, both directly connected to university

905 "Der Anschluß an eine Hochschule ist für mich von wesentlicher Bedeutung. Nachdem, ich 6½ Jahre lang in der NYU verlängert worden bin, wollen die Leute bei dem dort eingetretenen Retirement mir mein kleines Laboratorium nicht belassen. Was ich eigentlich brauche ist mehr eine akademische Adresse, als ein betriebsames Laboratorium. Ich brauche kein Geld, weder als Honorar für mich noch für die laufenden Untersuchungen. [...] Ich ambiere auch keine Stellung und würde niemanden bedrängen." Neuberg to Mark, December 29, 1948. CNP, Box 8: Mark, Hermann F.

906 Personnel form of May 31, 1950. CNP, Box 13: Polytechnic Institute.

907 Mark to Neuberg, January 10, 1949. CNP, Box 8: Mark, Hermann F.

908 Mark to Dean Burk, April 1, 1949. CNP, Box 8: Mark, Hermann F.

909 Personal communication from Ines Mandl, October 25, 1996.

910 Neuberg to Mark, May 3, 1950. CNP, Box 8: Mark, Hermann F.

911 "Mit doppeltem Gehalt (2 x 0 = 0) bin ich jetzt Visiting Professor am Polytechnic Institute. Dort habe ich einen Arbeitsplatz in einem Studentenlaboratorium und die Erlaubnis, mir für mein Geld Chemikalien zu kaufen und auch Hilfskraefte zu besorgen. Bisher ist es mir noch nicht gelungen, Dumme zu finden, die mir dieses Geld geben." Neuberg to Thomas, December 21, 1949. CNP, Box 15: Thomas, Karl.

912 This symposium took place on January 15, 1955. Those who spoke in addition to Neuberg were Jesse Greenstein, Lyman Craig, Barbara Low and Heinrich Waelsch. CNP, Box 13: Polytechnic Institute.

913 Neuberg to Oster, September 21, 1954. CNP, Box 13: Polytechnic Institute.

faculties.[914] After his flight from Germany, where he had been expelled from his university positions in 1934, Neuberg never again occupied a teaching post at a university.

As Neuberg only held research or honorary professorships at these various institutions, he had no regular budget available for his scientific work. He had to raise funds himself, in the form of so-called grants, in order to carry out his experiments and pay his assistants. These grants were awarded to scientists by public or private institutions and foundations on application in the form of scholarships or assistance, if the responsible advisory body considered the research project for which the application was made to be worthy of support. The funds were, however, not disbursed directly to the scientists themselves, but to the institutions at which the research projects were to be carried out. This may have been another important reason for Neuberg's efforts to always remain attached to an academic institution. In view of the conditions Neuberg had worked under previously, however, it is understandable that he was not very happy about this type of research funding:

> "You get your chemicals and a starvation wage for an assistant, generally $100–200 a month, from grants, which, unlike the situation in the *Notgemeinschaft [der Deutschen Wissenschaft* – the Emergency Association of German Science: Authors' note] you have to beg for in a most undignified manner. Never for more than a year, and not to be spent as you see fit, but through the universities, which often, as smart business enterprises, deduct a few percent for 'administrative expenses.'"[915]

Neuberg's work at this time focused first and foremost on the area of hydrotropy – the investigation of sparingly soluble substances. Neuberg was awarded assistance from numerous sources for his work on various aspects of this field over a period of several years. A large proportion of the papers he left behind, which are in the keeping of the American Philosophical Society in Philadelphia, consist of correspondence with various organizations that granted funds to Neuberg or rejected his applications. The bodies that furnished Neuberg with funds were in some cases corporations[916] or private foundations such as the Ella Sachs Plotz

914 The fact that Neuberg had already been working at the Medical College since 1951 is shown on the one hand by his corresponce with the management of this institution, and on the other hand by the fact that publications of his in these years originate from the various laboratories. Cf. for example Benjamin to Kleiner, February 23, 1951. CNP, Box 12: New York Medical College.

915 "Chemikalien und Hungergehalt für einen Assistenten, meist 100–200 $ monatlich, erhält man aus Grants, um die man – anders als bei der Notgemeinschaft – hoechst unwuerdig betteln muss. Nie fuer laenger als ein Jahr und nicht etwa zu freier Disposition, sondern ueber die Universitaeten, die vielfach als smarte Geschaeftsunternehmungen sich fuer 'Verwaltung' etliche Perzente abziehen." Neuberg to Kobel, August 17, 1948. CNP, Box 7: Kobel, Maria.

916 Neuberg received, for example, $3,600 from Swift & Company for a paper on "The Biochemical Formation and Significance of the Ozones." Swift & Company to Neuberg, May 2, 1944. CNP, Box 15: Swift & Company.

Foundation,[917] but in other cases public sector organizations such as the United States Public Health Service, which provided Neuberg with funds over a period totaling at least eight years.[918] Neuberg also received a one-off research grant from the MPG in 1952.[919] A further foundation, the Madison Foundation for Biochemical Research, assisted Neuberg by providing a laboratory. However, when this foundation went bankrupt Neuberg was compelled to move into the Medical College with his entire laboratory.[920] Neuberg described his dependence on American industry to Abderhalden, who in a previous letter had complained bitterly about the very modest conditions he himself was having to work in in Europe, as follows:

> "I don't get a cent from the university, and have to earn my living laboriously by working in industry in fields that are quite remote from my own. As you know, I have never been one to work in the area of technical applications."[921]

Neuberg had arrived in the USA almost penniless. The research professorship set up for him at New York University thanks to a donation from a relative was not well-endowed, and, as has already been mentioned, was limited to two years. Neuberg had to seek out other ways of earning his livelihood and consequently worked first and foremost as a consultant to industrial companies.

Initially, he was only able to earn relatively small fees for writing expert opinions[922] or for advisory discussions.[923] Later, however, he also succeeded in obtaining proper consulting contracts with a variety of companies. Initially, he was a consultant to the chemical concern Merck & Co. in Rahway, New Jersey, and later also advised the New York company Interchemical Corporation.[924]

917 Ella Sachs Plotz Foundation to Neuberg, June 16, 1947. CNP, Box 4: Ella Sachs Plotz Foundation.

918 CNP, Box 16: U. S. Public Health Services. From these documents, some of which are difficult to follow and incomplete, it can be determined that Neuberg received further grants from the following organizations: American Cancer Society, United States Atomic Energy Commission, United States Department of the Navy – Office of Naval Research, Society Sigma XI of Columbia University, Sugar Research Foundation, National Science Foundation and Dazian Foundation of Medical Research, NYC.

919 Notes on a meeting held on August 12, 1952. MPG Archives, Dept. II, Rep. 1A, Neuberg's Personnel File.

920 Neuberg to Butenandt, June 17, 1954. CNP, Box 3: Butenandt, Adolf. It is unclear how long the facilities at the Foundation remained available to Neuberg. The correspondence with the Madison Foundation shows that Neuberg had been a member of the Foundation since November 1950, and since February 1951 also "Scientific Director." CNP, Box 8: Madison Foundation for Biochemical Research.

921 "Von der Universitaet bekomme ich keinen Cent und muss mir meinen Unterhalt durch Taetigkeit in der Industrie auf mir gaenzlich fernliegenden Gebieten muehselig verdienen. Wie sie wissen, hat mir die Arbeit für die Technik nie gelegen. " Neuberg to Abderhalden, January 3, 1946. CNP, Box 1: Abderhalden, Emil.

922 Neuberg to Kobel, August 17, 1948. CNP, Box 7: Kobel, Maria.

923 Hamburger to Neuberg, December 30, 1941. CNP, Box 6: Hamburger, Adolf.

924 Neuberg worked for Merck from 1942 to 1946 and for Interchemical Corporation from 1943 to 1949. Personnel Security Questionnaire of the Atomic Energy Commission, March 30, 1950. CNP, Box 16: U. S. Atomic Energy Commission.

Furthermore, Neuberg had the opportunity to do research work at Merck's pharmaceutical factory.[925] This, however, entailed a degree of inconvenience:

> "Things have got somewhat better for me as a result of the fact that a number of industries are assisting me; in particular, one very big factory has made a laboratory available to me. It would be ideal, only getting there takes 1¾ hours in each direction! That's a lot for an old man, and at any rate makes considerable demands on one's flexibility and sleep economy."[926]

Interchemical Corporation was one of the biggest American chemical companies, and in the early 1940s was attempting to extend its range of products into the field of pharmaceuticals.[927] This corporation too, which was located in central Manhattan, made a laboratory available to Neuberg, where according to Ines Mandl he was able, at least from 1945 to 1949, to work on experiments every day, with complete freedom to do whatever work he felt inclined to.[928] The results of his research work ultimately led to a number of publications.[929] Neuberg's relations with these companies seem to have gone beyond the purely businesslike. The Vice President of Interchemical Corporation, Albert Gessler,[930] for example, gave a speech in Neuberg's honor before the American Society of European Chemists and Pharmacists (ASEC) on the occasion of his 70th birthday in December 1947. Neuberg was grateful to Interchemical for making the company's penthouse available to European scientists who were staying in New York at the ASEC's invitation.[931]

Warwick Chemical Company[932] was another business where Neuberg was employed for a while. In addition, he also received royalties under contracts with

925 "I will not get the inquiry from Merck that you mention, since I have already been in contact with the company for a long time, and for some time have had a well equipped laboratory available there [...]." ("Die angekündigte Anfrage von Merck werde ich nicht bekommen, denn ich stehe seit langem mit der Firma in Verbindung und habe seit länger[er] Zeit ein gut eingerichtetes Labor daselbst zur Verfügung [...].") Neuberg to Aron, November 14, 1942. CNP, Box 2: Aron, Hans.

926 "Es ist etwas besser für mich dadurch geworden, dass mich manche Industrien unterstützen, speciell habe ich in einer ganz grossen Fabrik ein Labor zur Verfügung gestellt bekommen. Es wäre ideal, aber die Reise dauert je 1¾ Stunde[n]! Das ist für einen alten Mann reichlich und stellt immerhin einige Anforderungen an die Elastizität und die Schlafökonomie." Neuberg to Aron, September 30, 1942. CNP, Box 2: Aron, Hans.

927 Fritzsching to Neuberg, April 14 and 15, 1943. CNP, Box 5: Fritzsching, Erwin.

928 Thus Ines Mandl was able to perform the experiments for her Master's thesis in this laboratory under Neuberg's direction. Written communication from Ines Mandl, November 6, 2006.

929 See for example Neuberg/Mandl (1948).

930 Albrecht Gessler had stayed in Berlin as a young scientist and had attended Neuberg's lectures at the Agricultural College. Later, he was Vice President and Director of the Research Department of Interchemical.

931 For example, the Nobel Prize winner Otto Warburg, who spent time in New York at Neuberg's invitation in 1948, stayed there. Neuberg to Warburg, September 19, 1948. CNP, Box 17: Warburg, Otto.

932 Neuberg worked for this company at least from October 1942 to March 1943. Warwick Chemical Company to Neuberg, June 1, 1943. CNP, Box 17: Warwick Chemical Company.

major industrial companies to whom he had assigned patent rights.[933] Neuberg declined offers to work for further companies, since he did not value such consulting work and regarded it merely as a means to an end. Links with a university had absolute priority for him. Neuberg wrote to Hans Aron in Chicago, who had attempted in vain to get Neuberg a position at the university there, as follows:

> "My gratitude towards you and Farmer is none the less for that. If the latter's love is not purely platonic, he should get me a good academic position, in which I can direct Simplicius Simplicissimus's flatuses at the whole of industry not in simplex, but in quadruplex."[934]

And in a 1947 letter he wrote:

> "Apart from suffering from diabetes, I am not in bad shape; but unfortunately, instead of pursuing pleasanter activities, I am forced to keep on slogging away in industry in order to earn my living for as long as I can manage."[935]

Neuberg had no longer been a university lecturer in the classic sense since having to leave Germany. At New York University, Neuberg, whose attitude to American students has been mentioned above, merely had the status of a research professor, to which no teaching duties were attached.[936] Later, his deteriorating health no longer allowed him to cope with the stress that would have been associated with regular lecturing or supervision,[937] though this was envisaged for him at the Polytechnic Institute of Brooklyn.[938] There are, however, indications that Neuberg did still work with students, at least in the laboratory.[939] This was something that had been very important to him even in his youth, as he sought to arouse young people's enthusiasm for his subject; furthermore, it was in this field that he believed his strengths as a teacher lay.

933 CNP, Box 12: Neuberg, Carl, Patents.

934 "Meine dankbare Gesinnung Ihnen und Farmer gegenüber ist darum nicht geringer. Wenn die Liebe des Letzteren nicht nur platonisch ist, soll er mir eine gute academische Stelle beschaffen, in der ich der gesamten Industrie die Flatusse des Simplizius Simplizissimus nicht simplex, sondern quadruplex entbieten kann." Neuberg to Aron, November 14, 1942. CNP, Box 2: Aron, Hans.

935 "Abgesehen von einem Diabetes, bin ich auch leidlich 'im Stande', leider aber, statt angenehmere Tätigkeiten auszuüben, dazu gezwungen, durch Arbeiten in der Industrie mich ausschleimend, meinen Lebensunterhalt zu verdienen solange es noch geht." Neuberg to Plesch, August 11, 1947. CNP, Box 13: Plesch, Johannes.

936 Cf. Archives of New York University, New York University Directory (1941–1942) and (1949–1950).

937 Medical certificate for Carl Neuberg, issued by Bruno Kisch on September 29, 1950. CNP, Box 7: Kisch, Bruno.

938 Neuberg was to give a course on "The Chemistry of Carbohydrates." Mark to Dean Raymond E. Kirk, April 1, 1949. CNP, Box 8: Mark, Hermann F. Whether this course actually took place is unclear, but improbable in view of the medical certificate just mentioned.

939 Evidence of this can be found in two letters sent by the Head of the Department of Chemistry at the Polytechnic Institute of Brooklyn to students named Salvesen and London, informing them that they would be working on a research project directed by Neuberg. Kirk to Salvesen and London, June 15, 1951. CNP, Box 13: Polytechnic Institute.

7.3. NEUBERG'S AMBIVALENT ATTITUDE TO GERMANY AFTER 1939

Whether Neuberg maintained any more or less intensive contact with acquaintances from his home town after his flight is not known. In earlier days, however, he frequently refered to his emotional ties to Germany and the Hanover region. In a letter of thanks on the occasion of his appointment as a Corresponding Member of the Academy of Sciences in Göttingen in 1921, for example, he wrote: "As a native of Hanover, I appreciate this high honor all the more for its having been accorded me by the learned body of my home province."[940] But even after his flight from Germany, Neuberg liked to think back to the days of his youth. This can be seen from a postcard found among Kisch's papers, which Neuberg sent his friend and doctor from vacation in 1947.[941] He also mentions this "attachment" in a 1949 letter to Fritz Micheel:[942]

> "From Pyrmont, the home town of my grandparents, I once made [a trip] with my late wife through that district, which she did not know, and she immediately understood my imperishable attachment to the places of my youthful reminiscences."[943]

Despite all this, Neuberg seems to have spoken very little about his youth in Hanover after his flight to the United States. His close colleague Ines Hochmuth Mandl could not remember him ever having talked about it to her.[944] His feelings for the Hanover region, and also his feeling of belonging to Germany, are at least to some extent explainable in terms of the close relationships he had with his mother Alma and his wife, who had died so early:

> "I too have known the feeling of really belonging, for I have been really close to only two people in my life: my mother, who died of blood poisoning (after cutting her finger) at the age of 54, in a time when there was still no sulfonamide or penicillin, and secondly my wife, who was carried away by leukemia at the age 44."[945]

In the early post-war years, Neuberg's feelings for Germany were severely put to the test, since in many cases former adversaries or opportunistic scientists who

940 "Die hohe Ehre empfinde ich als Hannoveraner umso schätzenswerter, da sie mir von der gelehrten Körperschaft meiner Heimatprovinz zuteil geworden ist." Letter to the Presiding Secretary of the Academy, Runger, December 2, 1921. Archives of the Academy of Sciences, Göttingen, Pers. 20.

941 SBPK Manuscript Department, Kisch papers K3 (a–c), sheet 45.

942 Fritz Micheel (1900–1982) worked at the KWI for Chemistry from 1924 to 1928 and was Professor of Organic Chemistry at the University of Münster from 1946 onwards.

943 "Von Pyrmont, der Heimat meiner Grosseltern machte ich einmal mit meiner verstorbenen Frau [eine Reise] durch die erwaehnte, ihr unbekannte Gegend und sie begriff sofort meine unvergaengliche Anhaenglichkeit zu den Staetten meiner Jugenderinnerungen." Neuberg to Micheel, April 19, 1949. CNP, Box 9: Micheel, Fritz.

944 Personal communication from Ines Mandl, October 25, 1996.

945 "Das Gefühl der ganz großen Zugehörigkeit habe auch ich kennengelernt, denn wirklich verbunden bin ich nur zwei Menschen in meinem Leben gewesen: meiner Mutter, die mit 54 Jahren an einer Blutvergiftung (Schnitt in den Finger) starb, zu einer Zeit, wo es noch keine Sulfonamide oder Penicillin gab, und zweitens meiner Frau, die mit 44 Jahren von einer Leukämie dahin gerafft wurde." Neuberg to Anschütz, July 12, 1953. CNP, Box 2: Anschütz, Ludwig.

had profited from the expulsion of Jewish academics did not suffer any just punishment or appropriate consequences. The following quotation demonstrates that he was able to judge Butenandt's attitude very clearly as that of one who had exploited the Nazi system and that of the young Federal Republic to his own advantage:

> "The information I have does not confirm that colleagues in the medical or scientific faculties in Germany have lost their positions, are starving or lacking a livelihood. Of those I know, all of those whom I would wish it and above all those whom I do not wish it are living under tolerable, in some cases even excellent conditions and have managed to obtain secure positions, even if they were declared members of the Party, like my successor Butenandt for instance."[946]

Especially where the murder of his sister Anna is concerned, his understandable and justifiable embitterment becomes clear:

> "My only sister was murdered by the Nazis, too. It pains me all the more that as I had always predicted, and despite all the military defeats, the German mentality has remained victorious. The big amnesty that the Nazis have just been granted by the Allies confirms this to the full."[947]

Neuberg sympathized with the American position that not only those who had demonstrably committed crimes should be punished, but that those people also had to be called to account who had profited personally from the crimes of the Nazi regime, or had at least tacitly assented to them. Thus he wrote to Windaus in connection with the Nuremberg Trials in 1947:

> "I don't feel very happy about the trial of the I.G. people either. But don't overlook the differences of mentality. They are not seeking to make a distinction between active Nazis and fellow travelers who supported the regime for whatever reasons."[948]

Neuberg, himself engaged in a struggle to get the sums paid out that he were due to him from his pension entitlements, also registered with disappointment the lack of willingness on the German side to pay compensation:

946 "Meine Informationen lauten auch anders, dass die Kollegen der medizinischen und naturwissenschaftlichen Fächer in Deutschland ihre Stellung verloren haetten, hungern und erwerbslos waeren. Von meinen Bekannten leben alle, denen ich es wuensche und vor allem die, denen ich es nicht wuensche unter ertraeglichen, zum Teil sogar ausgezeichneten Bedingungen und haben es verstanden, selbst wenn sie declarierte Mitglieder der Partei gewesen sind, wie beispielsweise mein Nachfolger Butenandt, sich gesicherte Positionen zu verschaffen." Neuberg to Durig, October 2, 1946. CNP, Box 4: Durig, Arnold.

947 "Meine einzige Schwester ist auch von den Nazis umgebracht [worden]. Umso seltsamer berührt es mich, dass, wie ich es immer vorausgesagt habe, trotz aller militärischen Niederlagen, die deutsche Mentalität siegreich geblieben ist. Die grosse Amnestie, die die Nazis eben von den Alliierten erfahren haben, bestätigt dies vollinhaltlich." Neuberg to Frank, December 27, 1946. CNP, Box 5: Frank, Alfred.

948 "Den Prozess gegen die I.G. Leute halte auch ich nicht fuer gluecklich. Aber uebersehen Sie nicht die Verschiedenheit der Mentalitaet. Hier will man nicht den Unterschied machen zwischen aktiven Nazis und den Mitlaeufern, die das Regime, aus welchen Gruenden auch immer, unterstuetzt haben." Neuberg to Windaus, December 23, 1947. CNP, Box 18: Windaus, Adolf.

> "What annoys us all, and also Americans in important positions, is the complete lack of willingness to make restitution. The lack of the most basic feeling for injustice among the German authorities makes it difficult to always contradict the protagonists of *realpolitik*. The argument is: 'What are we to think of people who do not pay compensation to those who were unjustly thrown out, but do pay salaries to those who took over positions that became free under suspicious circumstances or who were newly appointed to such positions, without just compensation for those who were expelled?'"[949]

Of Neuberg's former colleagues and staff in Germany, many did not renew contact with him until long after the end of the war, and in his view much too late. After everything he had suffered in Germany, he always refused on principle to take the first step towards restoring relations with his former home country.[950] Furthermore, he seemed to feel himself to be forgotten even among the academic community back home:

> "But I do not suffer from any illusions with regard to my influence in Germany. I heard from former French and English colleagues that they had received with mixed feelings invitations to a conference of German biochemists taking place in the summer. They were amazed – though I was not – that I knew nothing about the whole event."[951]

Despite everything, Neuberg assisted many of his colleagues who had remained in Germany by sending them letters, at their request, confirming their "irreproachable conduct" during the Nazi period.[952]

As his personal and academic contacts to Germany gradually revived, Neuberg seems to have recovered a more positive attitude towards his country of origin. This change was no doubt also favored by Neuberg's critical attitude towards the American mentality that has already been described. A letter to the biochemist Fritz Micheels, who was living in Münster, makes clear how closely he was following developments in the young Federal Republic:

> "It is my most sincere wish to experience the rise of Germany and I completely subscribe to your views with regard to the great qualities and regrettable political naivety of the German people. Only yesterday I was able to convince myself of the upturn that is taking place and of

949 "Was uns alle, aber auch massgebliche Amerikaner verstimmt, ist der voellige Mangel an Wiedergutmachungswillen. Das Fehlen jedes primitiven Gefuehls fuer Unrecht bei deutschen Behoerden macht es schwer, immer den Realpolitikern zu widersprechen. Man argumentiert: 'Was soll man von Leuten halten, die zu Unrecht Davongejagte nicht entschaedigen, die Gehaelter an die unter verdaechtigen Umstaenden in ein freigemachtes Amt Eingesetzte zahlen oder auf solche Stellen neu berufen, ohne gerechte Entschädigung der Verjagten.'" Neuberg to Anschütz, February 1, 1949. CNP, Box 2: Anschütz, Ludwig.

950 Neuberg declared this in a letter to Agnes Linke, in which he also expressed his disappointment at not having had any signs of life from his former subordinates at the Institute. Neuberg to Linke, January 26, 1948. CNP, Box 8: Linke, Agnes.

951 "Aber ich gebe mich keinen Illusionen ueber meinen Einfluss in Deutschland hin. Von ehemaligen franzoesischen und englischen Mitarbeitern hoerte ich, dass sie mit gemischten Gefuehlen aufgenommene Einladungen zu einer im Sommer stattfindenden Tagung deutscher Biochemiker erhalten haben. Sie waren erstaunt – ich war es nicht – dass ich von der ganzen Veranstaltung nichts wusste." Neuberg to Anschütz, March 28, 1948. CNP, Box 2: Anschütz, Ludwig.

952 Examples can be found in the correspondence with Ludwig Anschütz, Eduard Hofmann and Friedrich Glum. Cf. CNP.

> Germany's capabilities as I spent several hours wandering around a German industrial exhibit. In some objects – e.g. luxury glass and painted porcelain – I noticed some regrettable concessions to American taste."[953]

With regard to the development of discoveries and innovations, Neuberg saw Europe, and Germany in particular, as enjoying superiority over the United States for historical reasons. He believed that this applied particularly to biochemistry, whose outstanding representatives at that time either came from Germany or were trained there:

> "The progress of biochemistry is the progress of chemistry. Incidentally, the European, and above all the German, influence stands out a mile: Cori, Wald, Ochoa, Lipmann, Waelsch, Schoenheimer, Nachmansohn, Clark, Du Vigneaud etc. are all either of German origin or received postgraduate training in Germany."[954]

When he was offered the chairmanship of an international Enzyme Nomenclature Conference planned to take place in Germany in 1955,[955] Neuberg turned it down in view of his advanced age, but did not hesitate to recommend that another German should be chosen. Neuberg proposed his former pupil Friedrich Nord:

> "You are well enough acquainted with my attitude to know that it would be my most heartfelt wish for a German to chair this commission, since it is in Germany that there are the most children who have lain in the cradle of enzyme chemistry."[956]

Except during the immediate post-war years, Neuberg always felt himself very closely attached to Germany, and never lost his deep admiration for the country's cultural and scientific achievements.[957] This feeling took on sharper contours again in the later post-war years, and Neuberg followed German efforts to rebuild the country very sympathetically. He expressed this quite frequently in letters to colleagues and former members of his staff in Germany:

> "I share with you, my dear Herr Hofmann, the wish that Germany should recover quickly. As I recently wrote to Windaus, I am of the same opinion as Fichte: 'You should believe in

953 "Es ist mein inniger Wunsch, noch den Aufstieg Deutschlands zu erleben und ich unterschreibe voll ihre Ansichten ueber die grossen Qualitaeten und die bedauerliche politische Naivitaet des deutschen Volkes. Von dem Aufschwung und der Leistungsfaehigkeit Deutschlands habe ich mich gerade gestern durch ein mehrstuendiges Herumwandern in der deutschen Industrieausstellung ueberzeugt. Bei manchen Dingen – wie z. B. Luxusglas und bemaltes Porzellan – habe ich bedauerliche Concessionen an den amerikanischen Geschmack bemerkt." Neuberg to Micheel, April 19, 1949. CNP, Box 9: Micheel, Fritz.

954 "Die Fortschritte der Biochemie sind die Fortschritte der Chemie. Uebrigens ist der europaeische und darunter in erster Linie der deutsche Einfluss leicht zu erkennen. Cori, Wald, Ochoa, Lipmann, Waelsch, Schoenheimer, Nachmansohn, Clark, Du Vigneaud u. a. sind deutschen Ursprungs oder postgraduated in Deutschland geschult." Neuberg to Lynen, June 6, 1951. CNP, Box 8: Lynen, Fedor.

955 Ammon to Neuberg, February 7, 1955. CNP, Box 1: Ammon, R.

956 "Sie kennen meine Einstellung hinreichend um zu wissen, dass ich auf das allerwaermste wuensche, dass der Vorsitz dieser Commission von einem Deutschen gefuehrt wird, weil ja in den Wiegen der Enzymchemie die meisten Kinder in Deutschland gelegen haben." Neuberg to Ammon, November 25, 1955. CNP, Box 1: Ammon, R.

957 Nachmansohn (1979), 323.

Germany's future, on the rebirth of its people; do not allow yourself to be robbed of this faith, despite all that has happened'."[958]

Neuberg could not understand the decision of some emigrants to completely cut connections with their German roots after the Holocaust, by anglicizing their names or even refusing to use the German language.[959] For him, by contrast, identification with his origins was so much part of himself that he even gave clear signals of setting himself apart. Bruno Kisch reports that Neuberg put up a notice on his office door bearing the warning "Europe begins here!"[960] It is not surprising that this attitude was not always appreciated: "Thanks to my views, I am considered in certain circles here to be a 'Nazi'. A strange situation: Not tolerated in Germany, and here a hardly tolerable German."[961]

During the post-war years, Neuberg was asked from various quarters in Germany how he viewed the question of returning home. In this respect, however, Neuberg repeatedly emphasized that he had not received any offers from relevant institutions that would have opened up appropriate prospects for him to live and work in Germany.[962] Neuberg found a particular advocate in Berlin in his colleague Warburg, who had been tolerated in Germany from 1933 to 1945[963] and renewed contact with Neuberg after the war during a number of visits to the United States. Warburg attempted to get Neuberg's former institute restored to him:

"In the meantime, however, I have after all written to Berlin with regard to your being given back your institute, and I would like to think there is a chance of it happening. I pointed out that this would be the first case of an attempt to atone for injustice done to scientists not only

958 "Mit Ihnen, lieber Herr Hofmann, teile ich den Wunsch, dass Deutschland sich schnell erhole. Ich halte es mit Fichte, wie ich juengst an Windaus schrieb: 'Du sollst an Deutschlands Zukunft glauben, an seines Volkes Auferstehen, lass diesen Glauben Dir nicht rauben, trotz allem was geschehen'." Neuberg to Hofmann, February 26, 1948. CNP, Box 6: Hofmann, Eduard.

959 "He [Max Bergmann, Authors' note] was an excellent scholar, but of a character quite foreign to me, with a stupid hostility to everything German. He never spoke German again." ("Er war ein vortrefflicher Gelehrter, aber ein mir wesensfremder Character, albern in seiner Deutschfeindlichkeit. Er sprach nie mehr deutsch.") Neuberg to Kobel, August 17, 1948. CNP, Box 7: Kobel, Maria.

960 Kisch (1966), 329.

961 "Wegen meiner Einstellung gelte ich in manchen hiesigen Kreisen für eine [sic] ‚Nazi'. Seltsame Situation: In Deutschland untragbar und hier ein nur schwer zu ertragender Deutscher." Neuberg to Glum, March 2, 1948. CNP, Box 5: Glum, Friedrich.

962 Neuberg describes this, for example, in a letter to Adolf Butenandt of March 7, 1947, in which he states that Heinrich Wieland had asked him about his ambitions to live in Germany again. CNP, Box 2: Butenandt, Adolf.

963 Warburg was probably able to keep his position as Director of the KWI for Cell Physiology because the Institute was completely financed by the Rockefeller Foundation – and to some extent also because he had been awarded the Nobel Prize in 1931. He was, nevertheless, not immune to attacks from the *Reich* Ministry for Science, the Arts and National Education. Cf. Moser (2006), 203–204.

> in empty phrases, but through actions. And that I could not imagine what could do more to restore the prestige of Dahlem. Now we must wait and see how successful I have been."[964]

Adolf Windaus too approached the MPG with regard to an opportunity for Neuberg to work there.[965] Neuberg, however, remained skeptical about these efforts. In a letter to the former Secretary General of the KWG, he referred in particular to his advanced age, which would certainly have made it considerably more difficult or even impossible for him to manage a large institute again.[966] On the other hand, he certainly enjoyed imagining what official reactions might be expected to any demands he might make:

> "Inquiries from German colleagues as to whether I would like to return to Germany have caused me to wonder what the situation would be like if I were suddenly to lay claim to my old institute? Of course, there is no serious question of it, but it would be an amusing situation."[967]

A former member of Neuberg's staff in Dahlem, Eduard Hofmann, showed remarkable commitment in his attempts to achieve justice for his former Director. At that time, as Neuberg himself remarked, Hofmann was the only one of his pupils who had obtained an independent professorship in Germany.[968] In a series of letters to Otto Hahn, President of the MPG, Hofmann pleaded for reparation to be made to Neuberg,[969] even though he was well aware that there was no question of Neuberg again taking up a leading position in a research establishment.[970] Even

964 "Ich habe inzwischen aber doch nach Berlin geschrieben wegen der Rückgabe ihres Instituts und ich möchte glauben, dass diese Sache [eine] Chance hat. Ich habe darauf hingewiesen, dass dies der erste Fall wäre, in dem man nicht durch Phrasen, sondern durch Taten versuchen würde, das gegen Wissenschaftler begangene Unrecht wieder gut zu machen. Und dass ich mir nichts denken könnte, was das Prestige von Dahlem mehr wiederherstellen könnte. Nun wollen wir den Erfolg abwarten." Warburg to Neuberg, December 1948. CNP, Box 17: Warburg, Otto.

965 This can be seen from a memorandum on a conversation between Windaus and Telschow on November 23, 1948. MPG Archives, Dept. II, Rep. 1A, Neuberg's Personnel File.

966 Neuberg to Glum, December 4, 1947. CNP, Box 5: Glum, Friedrich.

967 "Anfragen deutscher Kollegen, ob ich nach Deutschland zurueckkehren wolle, haben mich dazu veranlasst, einmal nachzudenken, wie die Situation denn waere, wenn ich ploetzlich mein altes Institut beanspruchte? Im Ernst kommt die Sache natuerlich nicht in Frage, aber amuesant waere die Situation schon." Neuberg to Linke, January 26, 1948. CNP, Box 8: Linke, Agnes.

968 Neuberg to Hofmann, August 5, 1948. CNP, Box 6: Hofmann, Eduard.

969 "It must surely be the aim of all of us to get a man to come back to Germany who always felt himself to be a German and who left Germany only at the last minute, in 1939." ("Es muß doch unser aller Bestreben sein, den Mann wieder nach Deutschland zu bringen, der sich stets als Deutscher fühlte und der erst in letzter Minute, im Jahre 1939, Deutschland verließ.") Hofmann to Hahn, February 11, 1949. MPG Archives, Dept. II, Rep. 1A, Neuberg's Personnel File.

970 "It would of course be presumptuous of me to have any ambitions to direct an institute today." ("Selbstverstaendlich waere es eine Vermessenheit von mir, heute noch die Leitung eines Instituts zu ambieren.") Neuberg to Hofmann, April 18, 1949. And: "It should not be forgotten that I have grown 10 years older and no longer have the constitution of a horse and the inexhaustible energy that I used to have." ("Vergessen darf man nicht, dass ich 10 Jahre

though Neuberg, who was appointed an External Member of the MPG in 1948,[971] did not return to the organization's employment, it may be assumed that the personal efforts of many scientists, some of them of great renown, were at least helpful to Neuberg in pursuing his pension claims.

As has already been shown, the post-war years brought Carl Neuberg back into closer contact with his home country. His understandably ambivalent relationship with Germany improved thanks to the efforts of his former colleagues and staff to obtain a certain degree of reparation for him. His feeling of belonging to German culture and the German way of life was unbroken. Conditions in the United States were not easy for Neuberg, including and especially from a material point of view, and the prospect of being able to receive his pension after all was no doubt a tempting one. All these things may have strengthened the exile in his resolve to revisit Germany. He had entertained initial ideas in the respect as early as 1949:

> "To my profound regret, I shall have to postpone my plans to travel to Europe for the following reason: I have just been awarded a government fellowship, limited to one year, and have been informed that I cannot start the new task associated with the fellowship by going abroad on vacation."[972]

Two years later the journey proved possible. Neuberg reached Bremerhaven with his second wife Hilde on board the *M. S. Gripsholm* on September 14, 1951.[973] During his time in Germany he visited Dahlem, staying with Otto Warburg.[974] The two research scientists became quite attached to each other:

> "To use an American expression, 'we miss you terribly here'; it was so nice being able to chat with a clever man between photosynthesis and cancer experiments, just like in the old days. Probably, you will find you like Europe again, despite all its sins, so I don't exclude the possibility that you will turn up here every year in future on your way to Gastein. This is not just an empty courtesy but is meant quite seriously, that you should stay with us every year."[975]

älter geworden bin u. nicht mehr auf der Höhe meiner früheren Pferde-Gesundheit u. damals unverwüstlichen Arbeitskraft stehe.") Neuberg to Hofmann, February 20, 1949. CNP, Box 6: Hofmann, Eduard.

971 Hahn to Neuberg, October 16, 1948. CNP, Box 9: Max-Planck-Gesellschaft.

972 "Die Plaene für meine Reise nach Europa muss ich zu meinem lebhaften Bedauern aus folgendem Grunde zurueckstellen. Ich habe gerade ein für ein Jahr befristetes Fellowship des Governement erhalten und bin dahin belehrt worden, dass ich die mit dem Fellowship verbundene neue Taetigkeit nicht mit einem Urlaub in Uebersee beginnen koenne." Neuberg to Schlitter (Basler Chemische Gesellschaft), September 24, 1949. CNP, Box 14: Schlitter, E.

973 Neuberg to Telschow, June 31, 1951. MPG Archives, Dept. II, Rep. 1A, Neuberg's personnel file. Neuberg's journey, on which he was accompanied by Mrs. Hilde Neuberg, lasted from September 4 to October 31, 1951. Undated letter to the U. S. Atomic Energy Commission. CNP, Box 16: U. S. Atomic Energy Commission.

974 Neuberg to Jander, February 1, 1952. CNP, Box 7: Jander, G.

975 "Um einen amerikanischen Ausdruck zu gebrauchen: 'we miss you terribly here'; es war so nett zwischen photosynthesis und cancer experiments wie in alten Zeiten mit einem klugen Mann zu plaudern. Wahrscheinlich wird Ihnen Europa trotz all seiner Sünden doch wieder gut gefallen und so halte ich es nicht für ausgeschlossen, dass Sie nun jedes Jahr auf der Durchreise nach Gastein bei uns auftauchen. Es soll keine leere Redensart sein, sondern ist

Neuberg was intending to carry out experiments on the hydrotropy of sparingly soluble substances at the Gastein research institute near Innsbruck,[976] at the invitation of Ferdinand Scheminzky.[977] It may be assumed that Neuberg, who had earlier been a passionate climber, did not spend his entire vacation in the laboratory, but also enjoyed the world of the mountains, which he loved so much.[978]

On this first visit to Germany, Neuberg also met some of his former staff who had always taken his part.[979] At the same time, he used his time in the country to step up pressure on the Springer Verlag with regard to his demands for compensation in respect of his dismissal as editor of the *Biochemische Zeitschrift,*[980] and to pursue his pension claims with the Administrative Headquarters of the MPG in Göttingen.[981]

Neuberg had made an initial effort to assert his claim for the payment of a pension in 1947. Among other things, he made use of his revived contact with Adolf Butenandt, to whom he described his economic circumstances in a letter:

> "Perhaps I may still be able to look forward in my old age to what used to be described as the well-earned rights of a civil servant. I could certainly do with it, as I am practically living from hand to mouth."[982]

Butenandt appears to have advised Neuberg to approach Otto Hahn in his capacity as President of the KWG.[983] Neuberg submitted his claims to Hahn, and in addition to listing his earlier emoluments, also expressly referred to the period totaling more than 37 years that he had spent in the service.[984] In addition to Adolf

ganz ernst gemeint, dass Sie jedes Jahr bei uns wohnen sollen." Warburg to Neuberg, October 10, 1951. CNP, Box 17: Warburg, Otto.

976 Scheminzky to Neuberg, May 31, 1951, and Neuberg to Scheminzky, June 21, 1951. CNP, Box 14: Scheminzky, F.

977 Ferdinand Scheminzky (1899–1973), Professor of Balneology and Electrophysiology at the Gastein research institute of the Austrian Academy of Sciences.

978 Thus Scheminzky, on November 28, 1951, requested him to compile a list of the experiments he had performed in Gastein after all, so that could be included in the annual report. "If you didn't carry out any investigations of your own in Gastein, I must nevertheless ask you to act 'as if'." ("Haben Sie in Gastein selbst keine Untersuchungen durchgeführt, so muß ich Sie doch bitten, so zu tun 'als ob'.") CNP, Box 14: Scheminzky, F.

979 For example, he visited Herbert Collatz in Berlin and met Maria Kobel, Ernst Auhagen and Otto von Schönebeck in Frankfurt. Collatz to Neuberg, December 17, 1951, and Neuberg to Collatz, January 12, 1952. CNP, Box 4: Collatz, Herbert.

980 Neuberg visited Julius Springer on September 23, 1951. Julius Springer to Ferdinand Springer, September 23, 1951. ASV, E. 72.1, Neuberg, Carl.

981 Neuberg to Jander, February 1, 1952. CNP, Box 7: Jander, G.

982 "Vielleicht schaut für mich auf meine alten Tage doch noch etwas heraus, was man früher als wohlerworbene Rechte der Beamten bezeichnete. Brauchen könnte ich es, da ich sozusagen von der Hand in den Mund lebe." Neuberg to Butenandt, September 25, 1947. CNP, Box 3: Butenandt, Adolf.

983 Neuberg to Butenandt, December 4, 1947. CNP, Box 3: Butenandt, Adolf.

984 Neuberg to Hahn, October 1, 1947. CNP, Box 6: Hahn, Otto.

Butenandt and Otto Hahn, Heinrich Wieland, Adolf Windaus and even Richard Kuhn[985] offered to assist Neuberg in asserting his claims for compensation.[986]

As the newly established MPG did not consider itself to be the legal successor to the KWG, and the latter's pension fund had apparently been confiscated by the Russian occupation authorities,[987] the pension claims of former employees of the Society were referred to the responsible state governments.[988] Due to bureaucratic obstacles, the payment of the amounts due to Neuberg, although approved, was delayed until June 1951.[989] Despite his forced resignation under pressure from the Nazi government, Neuberg was awarded only a pension by the responsible authorities and not remuneration as a Professor Emeritus, which reduced his emoluments by 25%.[990] It may assumed that Neuberg's situation, nevertheless, improved as a result of the regular payments he now received;[991] this remark is not, however, intended to play down in any way the injustice that he suffered.

As it was highly unlikely that Neuberg, now 75 years old, would be able to work under the MPG umbrella again, his former assistant Hofmann attempted to procure an academic position for him elsewhere in Germany. Hofmann was a member of the faculty of the *Technische Hochschule* (College of Advanced Technology) in Munich, which enabled him to organize at least a one-year guest professorship at the college's Faculty of Agriculture for Neuberg.[992] It was to take up this position that Neuberg set out on his second trip to Germany in May 1952, this time together with his daughter Irene.[993] What concrete tasks and

985 Richard Kuhn (1900–1967) studied chemistry in Munich, and took his doctorate under Richard Willstätter. In 1929 he became Director of the KWI for Medical Research in Heidelberg (MPI from 1950 onwards). After the war he was also Professor of Biochemistry at the University of Heidelberg. He was awarded the 1938 Nobel Prize in Medicine for his work on carotenoids and vitamins, but was unable to receive it until 1945. As recent research has shown, Kuhn was implicated in human experiments during the Third Reich. Cf. Schmaltz (2006), 152–155.

986 Memorandum of November 23, 1948. MPG Archives, Dept. II, Rep. 1A, Neuberg's Personnel File. Also: Kuhn to Neuberg, June 18, 1948. CNP, Box 7: Kuhn, Richard.

987 Neuberg to Anschütz, June 1, 1948. CNP, Box 2: Anschütz, Ludwig.

988 Memo initialled Sg./Ri of the MPG Göttingen, December 23, 1949. CNP, Box 9: Max-Planck-Gesellschaft.

989 Neuberg's pension payments, retrospective for a year, were paid partly to accounts in New York and partly to Frankfurt am Main. Seeliger to Neuberg, May 28 and June 15, 1951. MPG Archives, Dept. II, Rep. 1A, Neuberg's Personnel File.

990 Neuberg to Hamburger, November 26, 1953. CNP, Box 6: Hamburger, Adolf.

991 It may be mentioned at this point that there is an entry in Neuberg's Personnel File at the MPG stating that his pre-war Personnel File was no longer extant: "[...] most of the newer files were either lost due to the repeated and sometimes very overhasty moves, or else were confiscated or got out of the way by a wide variety of interested parties." ("[...] das Gros der neueren Akten ging entweder bei den wiederholten, manchmal recht überstürzten Umzügen verloren oder wurde von den verschiedensten Interessenten beschlagnahmt oder beiseite gebracht.") Gertrud Krumm to Glum, November 24, 1947. MPG Archives, Dept. II, Rep. 1A, Neuberg's Personnel File.

992 Rheinfelder (Bavarian State Ministry of Education) to Reindel (Munich College of Advanced Technology), June 27, 1952. CNP, Box 13: Reindel, F.

993 Neuberg to Fromageot, April 23, 1952. CNP, Box 5: Fromageot, Claude.

opportunities the position of Guest Professor entailed for Neuberg must remain open. It may be assumed that it was rather a kind of honorary professorship, although Neuberg certainly also considered remaining in Bavaria for some substantial period of time.[994] What is certain is that he celebrated his 75th birthday at Weihenstephan, near Munich, and on this occasion received a large number of congratulations and honors,[995] which provoked the following comment to his former secretary:

> "On my 75th birthday I received many demonstrations of affection from former members of staff. For example, I received a new honorary doctorate, a variety of honorary memberships, a medal and other bric-a-brac, only no money."[996]

Although the guest professorship had been established for a period of one year, Neuberg did not spend the whole time in Munich, since despite his advanced age he was continuing to invest a lot of time in his scientific activity in New York.[997] But already in the following year, he embarked once more, again with his daughter Irene, for his third and final visit.[998]

During his stays in Germany, Neuberg was invited to give lectures at a variety of institutions, in which he generally spoke about the results of his latest research into the field of hydrotropy.[999] In 1953, for example, he gave the opening speech at the convention of the Deutsche Physiologisch-Chemische Gesellschaft (German Physiological Chemistry Society) in Homburg an der Saar, on which occasion he was also appointed an honorary member of that society.[1000] Neuberg foresaw that this would be his last journey to Germany:

> "At my age, I do not make any real travel plans, although I do dream of coming to Munich in 1956 for the inauguration of the new Institute of Biochemistry, and seeing you and my other old friends there again. You will no doubt think it would make more sense for me to book a

994 "It is by no means out of the question that I might decide to spend a few years in Bavaria." ("Es ist nicht ausgeschlossen, dass ich mich dazu entschliesse, einige Jahre in Bayern zu verleben.") Neuberg to Reiter, November 20, 1951. CNP, Box 13: Reiter, Melchior.

995 He received congratulations from some 200 different people. Neuberg to Kisch, August 13, 1952. SBPK, Manuscript Department: Kisch papers, K3 (a–c).

996 "Zu meinem 75. Geburtstag habe ich viele Beweise der Anhänglichkeit früherer Mitarbeiter erhalten. Zum Beispiel erhielt ich ein neues Ehrendoktorat, verschiedene Ehrenmitgliedschaften, eine Medaille und anderen Tinev, nur kein Geld." Neuberg to Berlak, August 20, 1952. CNP, Box 3: Berlak, Millie.

997 Neuberg returned to New York in October 1952, as a number of different letters show. MPG Archives, Dept. II, Rep. 1A, Neuberg's Personnel File.

998 This is shown by a letter of Neuberg's to Bruno Kisch, August 4, 1953. SBPK, Manuscript Department: Kisch papers, K3 (a–c). Neuberg, incidentally, always traveled to Germany by ship, since he was forbidden to fly for medical reasons. This is indicated in a letter from Neuberg to W. Lautsch at the Free University of Berlin, June 7, 1953. Archives of the Free University of Berlin, HSA FUB: Rektorat/Aussenamt, file "Aussenkommission: Gastprofessoren. Buchstaben Na–Nh," 1950s and 1960s, Neuberg correspondence.

999 Neuberg held lectures at the Universities in Munich, Göttingen and Basle, and at the Bayer corporation in Elberfeld and BASF in Ludwigshafen. From an overview of Neuberg's lectures for New York Medical College. CNP, Box 12: New York Medical College.

1000 The convention took place in Homburg an der Saar on October 1, 1953. CNP, Box 4: Deutsche Physiologisch-Chemische Gesellschaft.

ticket at the Styx Travel Agency for a voyage to the Asphodel Fields, on a ship under the command of the great Captain Charon."[1001]

1001 "Eigentliche Reiseplaene mache ich in meinen Jahren nicht, obgleich ich doch davon traeume, im Jahre 1956 zur Einweihung des neuen Institutes für Biochemie nach München zu kommen und dann Sie und die anderen alten Freunde wiederzusehen. Sie werden meinen[,] es sei angebrachter, beim Reisebuero Styx ein Billet vormerken zu lassen zur Fahrt nach der Asphodelos-Wiese auf einem Schiff das faehrt unter dem Commando des grossen Capitaen Charon." Neuberg to Kobel, October 7, 1955. CNP, Box 7: Kobel, Maria.

8. NEUBERG'S SCIENTIFIC REPUTATION

8.1. HONORS AND AWARDS

Carl Neuberg was a scientist of international renown who achieved major advances, particularly through his outstanding work in the field of the chemistry of fermentation.[1002] He rendered important services to the development of the then still young discipline of biochemistry, not only through the results of his own research, but also and in particular by his editing of the *Biochemische Zeitschrift*. The introductory text on the website of the American Philosophical Society, for example, points out these merits with the words: "To say that Carl Neuberg [...] was a pioneer in biochemistry is to understate the case: he coined the term."[1003] In the course of his life, Neuberg received a multiplicity of honors in recognition of his research. On his 70th and 75th birthdays in particular, he was awarded numerous distinctions by the international scientific community.

On the occasion of his 75th birthday, Neuberg was finally also to have been honored by the Academy of Sciences of the German Democratic Republic (East Germany) and the Humboldt University (HU) in Berlin. An Institute for Medicine and Biology had been set up within the newly-established German Academy of Sciences,[1004] and Fritz Windisch, a former pupil of Neuberg's, had been appointed to its Chair of Biochemistry.[1005] It was probably thanks to Windisch's initiatives that Neuberg, now living in the United States, was to be honored both by the GDR-based Academy of Sciences and by the Humboldt University, with whose scientists Neuberg was linked by a lively correspondence.[1006] It was planned amongst other things to award Neuberg an honorary doctorate, which, however, he felt unable to accept, probably in view of the "McCarthyism" rampant in the United States at that time:

> "Perhaps you would be interested in having a copy for the personnel files of a letter to me from the German Academy of Sciences in Berlin. I have refused all honors, including the intended award of an honorary doctorate by the Humboldt University and also election as a member of the Academy, about which the Secretary of the Academy had addressed an inquiry to me. As things are, it is only natural that I should act in this way."[1007]

1002 For an account of the development of Neuberg's scheme of fermentation, see Engel (1996).

1003 Cf. www.amphilsoc.org/library/mole/n/neuberg.htm, 02/21/2006.

1004 Sauer et al. (1960), 141.

1005 Windisch emphasized in a letter to Neuberg that the chair he had been appointed to was the one formerly held by Neuberg. He himself had been awarded it for work including research into fermentation. Windisch to Neuberg, March 2, 1948. CNP, Box 18: Windisch, Fritz.

1006 For example with Erich Thilo, the director of the Institute of Inorganic Chemistry, who invited Neuberg to an event in commemoration of Emil Fischer; Neuberg, however, declined to attend. CNP, Box 15: Thilo, Erich.

1007 "Vielleicht interessiert Sie für die Personalakten die Kopie eines mir von der deutschen Akademie der Wissenschaften in Berlin [gesandten] Schreibens. Ich habe alle Ehrungen

He then explained to Fritz Windisch the reasons for his declining, first and foremost among which was the desire not to offend any potential sources of funding:

> "Certainly the matter has nothing in the least to do with politics, and in this respect too I am non-political, as I have been all my life. In America there prevails a mentality which is not immediately predictable, and which I am basically totally indifferent to. But I would like to avoid anything that might cause the bodies that approve money, from whom I personally do not receive any salary but my staff do, to be subjected to reprisals of any kind."[1008]

But none of this could deter the Faculty of Agriculture and Horticulture of the Humboldt University[1009] from awarding Neuberg, whose "whole life was spent at or within the ambit of the University of Berlin and the Agricultural College that was always associated with it,"[1010] a posthumous honorary doctorate in agricultural sciences.[1011]

But in his country of exile too, tribute was paid to Neuberg's achievements. In 1955, the last year of his life, the symposium *"Current Peptide and Protein Research"* was organized in his honor at the Polytechnic Institute of Brooklyn.[1012] To indicate the extent of Neuberg's international reputation, there follows a summary in tabular form of the awards he received and of his memberships and honorary memberships of various scientific societies.

inclusive des beabsichtigten Ehrendoktorates der Humboldt-Universität abgelehnt, ebenso die Wahl zum Mitglied der Akademie, nachdem der Sekretär der Akademie eine Anfrage an mich gerichtet hatte. Nach Lage der Dinge habe ich natürlich so gehandelt." Neuberg to Mark, July 9, 1952. CNP, Box 8: Mark, Hermann F.

1008 "Die Angelegenheit hat gewiss auch nicht das geringste mit Politik zu tun, ich bin auch in dieser Beziehung, wie mein ganzes Leben lang unpolitisch gewesen. In Amerika herrscht eine nicht direkt vorauszusehende Mentalität, die mir an sich hoechst gleichgueltig ist. Aber ich moechte vermeiden, dass die Geld bewilligenden Koerperschaften, von denen ich persoenlich kein Gehalt beziehe, jedoch meine Mitarbeiter, irgendwelchen Ressentiments ausgesetzt werden." Neuberg to Windisch, October 12, 1955. CNP, Box 18: Windisch, Fritz.

1009 The new name of the Berlin Agricultural College, which Neuberg had long worked at, after its incorporation into the Humboldt University.

1010 Neuberg to Windisch, January 9, 1956. CNP, Box 18: Windisch, Fritz.

1011 The honorary doctorate was conferred upon Neuberg two weeks after his death, in recognition in particular of his work in the fields of fermentation and enzymes and the chemical mechanisms involved in the precipitation of casein when milk is coagulated with rennet. Deed of conferment of June 14, 1956. CNP, Box 10: Neuberg, Carl, Awards and Diplomas.

1012 CNP, Box 10: Neuberg, Carl, Awards and Diplomas.

HONORARY DOCTORATES, MEDALS AND DECORATIONS I [1013]

1913	Knight's Cross, 1st Class, of the Domestic Order of the Ernestinian House of Saxony	Carl Eduard Duke of Saxony
1913	Order of the Red Eagle, 4th Class	King of Prussia
1913	Cross of Merit, 3rd Class	Prince of Waldeck und Pyrmont
1914	Medjidi Order, 3rd Class	HIM Sultan Mehmed V, Turkey
1915	Commander's Cross of the Order of the Oak Crown	Grand Duchess of Luxembourg
1915	Medal of Honor of the Red Cross, 2nd Class	Prince Imperial Franz Salvator, Austria-Hungary
1918	Iron Cross 2nd Class (white ribbon, black edging)	German Reich
1918	Iron Cross 2nd Class (black ribbon, white edging)	German Reich
1918	Cross of Honor, 1st Class	Red Cross Society, Sofia
1922	Emil Fischer Medal	Deutsche Chemische Gesellschaft (now Gesellschaft Deutscher Chemiker – Society of German Chemists)
1923	Honorary Doctorate in Medicine	University of Breslau[1014]
1925	Honorary Doctorate in Biology	University of St. Petersburg[1015]
1929	Honorary Doctorate in Engineering Science	University of Danzig
1930	Delbrück Commemorative Medal	Experimental and Teaching Institute of the Brewing Industry, Berlin[1016]
1930	Medal for persons of particular merit	Notgemeinschaft der deutschen Wissenschaft
1930	Scheele Medal	Society of Swedish Chemists[1017]
1930	Delbrück Medal	Berlin Academy of Agriculture[1018]
1931	Pasteur Medal	Société de Chimie Biologique, Paris[1019]

1013 Where not otherwise stated, the following information is taken from CNP, Box 10: Neuberg, Carl, Awards and Diplomas.

1014 MPG Archives, Dept. I, Rep. 1A, 1448, 25.

1015 Habel (1951), 455.

1016 Experimental Institute to MPG Administrative Headquarters, October 2, 1930. MPG Archives, Dept. I, Rep. 1A, 2032, 67.

1017 KWG (1931).

1018 Ibid.

1019 KWG (1932).

HONORARY DOCTORATES, MEDALS AND DECORATIONS II

1931	Leblanc Medal	Société Chimique de France[1020]
1931	Golden Berzelius Medal	Swedish Society of Physicians[1021]
1932	Honorary Doctorate in Medicine und Surgery	University of Palermo
1932	Honorary Professor of Natural Sciences	State Institute for Further Medical Training, Leningrad
1933	Honorary Doctorate in Law	University of St. Andrews, Scotland[1022]
1934	Palmas academias 1st Class	Portuguese Academy of Sciences
1934	Commander's Cross of the Order pour l'Instruction publique	Portuguese Ministry of Education[1023]
1935	Cross of Honor for Front Combatants	German Reich (represented by Adolf Hitler)
1937	Austrian War Commemoration Medal	Republic of Austria
1947	Carl Neuberg Medal	American Society of European Chemists
1948	Honorary Doctorate in Agriculture	University of Milan[1024]
1952	Honorary Doctorate in Medicine	Free University of Berlin[1025]
1952	Medaille de la Société de Chimie Biologique	Société de Chimie Biologique, France
1956	Federal Cross of Merit	Federal Republic of Germany[1026]
1956	Honorary Doctorate in Agricultural Sciences	Humboldt University, Berlin

1020 KWG (1932).
1021 Swedish Academy of Sciences. Svenska Läkaresällskapets förhandlingar. Stockholm (1931), 271–274.
1022 University of St. Andrews. University Senatus Minutes (1933), 36.
1023 KWG (1935).
1024 CNP, Box 2: Antoniani, Claudio.
1025 Archives of the Free University of Berlin [abbreviated below as AFU]. HAS FUB: Med. Fak./Dekanat, Minutes of Faculty Meetings (meeting of June 16, 1952).
1026 Neuberg was informed of the award of this medal by the German Ambassador in New York, Heinz L. Krekeler, on March 22, 1954. CNP, Box: Krekeler, Heinz L. The initiative for this honor came from Adolf Butenandt, who emphasized in a letter to Otto Hahn how good his relationship to his predecessor had always been. Butenandt to Hahn, November 26, 1953. MPG Archives, Dept. II, Rep. 1A, Neuberg's Personnel File.

SCIENTIFIC SOCIETIES I

1921	Corresponding Member of the Academy of Sciences, Göttingen[1027]
1922	Member of the German Academy of Sciences Leopoldina, Halle[1028]
1922	Member of the Royal Academy of Sciences, Lund
1925	Corresponding Member of the Academy of Sciences of the USSR, Leningrad
1926	Corresponding Member of the Royal Society of Physicians, Budapest
1927	Honorary Member of the Stefan Tisza Scientific Society, Debrecen, Hungary
1927	Honorary Member of the Ukrainian Medical Society, Charkov
1927	Honorary Member of the Biological Society, Vienna
1927	Honorary Member of the Academia Medica, Rome
1927	Scientific Member of the Advisory Board of the Institut International d'Agriculture, Rome[1029]
1928	Honorary Member of the Czech Academy of Agricultural Sciences, Prague
1929	Member of the Royal Danish Academy of Sciences and Letters, Copenhagen[1030]
1930	Vice President of the German Chemistry Society[1031]
1932	Honorary Member of the Academy of Sciences, Barcelona
1932	Corresponding Member of the Academy of Sciences, Lisbon

1027 Archives of the Academy of Sciences, Göttingen. Pers. 20, Neuberg, Carl (1052), letter of thanks. See also CNP, Box 10: Neuberg, Carl, Awards and Diplomas.
1028 Archives of the German Academy of Sciences Leopoldina. M No. 3476.
1029 Neuberg was "German Representative" here until 1935. Who was Who in America? (1966).
1030 Archives of the Kongelige Danske Videnskabernes Selskab. Prot. No. 213/1929.
1031 KWG (1931).

SCIENTIFIC SOCIETIES II

1932	Honorary Member of the Finnish Academy of Sciences, Helsinki[1032]
1932	Honorary Member of the Biologiske Selskap, Copenhagen
1932	Honorary Member of the Dutch Society for Biochemistry[1033]
1932	Honorary Member of the Dutch Society for Pharmacology[1034]
1932	Honorary Member of the Berlin Physiological Society[1035]
1934	Honorary Member of the Academy of Sciences, Lisbon
1936	Honorary Member of the Royal Scientific Society, Uppsala
1947	Honorary Member of the Rudolf Virchow Medical Society, New York
1947	Member of the New York Academy of Science
1948	External Scientific Member of the MPG
1949	President und later Honorary President of the American Society of European Chemists
1952	Honorary Member of the Imperial Society of Agriculture, Japan[1036]
1952	Honorary Member of the Japanese Biochemical Society
1953	Honorary Member of the German Physiological Chemistry Society[1037]

1032 Proceedings of the Finnish Academy of Sciences (1932).
1033 KWG (1933).
1034 Ibid.
1035 Ibid.
1036 CNP, Box 7: Kaiser-Wilhelm-Gesellschaft.
1037 Ammon/Dirscherl (1956).

8.2. THE NON-AWARD OF THE NOBEL PRIZE

The Nobel Prize, which is doubtless the most important and highest recognition that can be awarded to a scientist, was never awarded to Neuberg, as it was to so many of his colleagues at the KWG.[1038] Although he was nominated for the Nobel Prize no less than 25 times – seven times for the Prize in Medicine and 18 times for that in Chemistry – between 1919 and 1935, it was never awarded to him, though he was even short-listed five times. Neuberg is thus one of those scientists who were most frequently regarded by their colleagues as worthy of receiving the Nobel Prize[1039] without ever actually having been awarded it.[1040] The Norwegian scientific historian Ragnar Björks, in a paper analyzing the award of the Nobel Prizes in Physiology and Medicine during the first 50 years, spoke of "The Fate of Carl Neuberg," remarking ironically: "[...] this is a story about how *not* to get a Prize."[1041]

In his paper, Ragnar Björk investigated the Nobel Committee's criteria for the award of the prize and came to the conclusion that the three criteria derived from Alfred Nobel's will, "it is a discovery and/or invention," "it has benefited mankind" and "it is recent"[1042] were again and again interpreted to Neuberg's disadvantage. In particular, his studies on alcoholic fermentation or his first description of carboxylase[1043] in 1911 did not appear to the Committee to be of clear medical relevance.[1044] In addition, Björk maintains that a further principle, not based on Nobel's will, namely the "principle of rotation," was of importance in conjunction with the third criterion, "it is recent," in the non-award of the Nobel Prize to Neuberg.[1045] In view of this fourth criterion, the Nobel Committee avoided awarding the Prize several times in succession to such a specialized discipline as biochemistry. And the biochemical "slot" had already been filled by the award to Euler-Chelpin and Harden in 1929, and did not come up again until 1937, when Neuberg's career had been prematurely ended by the Nazi regime, so that now Neuberg was tripped up by the "recent" criterion.[1046] Ironically it was Neuberg, himself short-listed for the Nobel Prize in Chemistry in 1929, who had personally proposed Harden.

1038 On this point cf. Kazemi (2002).

1039 Björk (2001), 401. Cf. also the table on the following page.

1040 Neuberg's former boss, August von Wassermann, was for example proposed for the Nobel Prize in Medicine 28 times. Also never chosen were Friedrich Paschen and Theodor Curtius. Crawford/Heilbron (1990), 851–852, cf. also Crawford et al. (1987) and Björk (2001).

1041 Björk (2001).

1042 Ibid., 396.

1043 Nowadays called pyruvate decarboxylase. Karlson (1990), 311.

1044 Björk (2001), 407.

1045 Ibid., 406–407.

1046 Ibid., 407.

NOMINATIONS OF NEUBERG FOR THE NOBEL PRIZE IN MEDICINE OR CHEMISTRY[1047]

	Medicine	Chemistry	Nominated for Chemistry by
1919	1	–	
1920	2	–	
1921	–	1	
1922	–	–	
1923	–	–	
1924	–	–	
1925	1	–	
1926	–	–	
1927	–	–	
1928	–	–	
1929	–	2	Arthur Binz# and Ernst Zerner#
1930	1	2	Richard Pfeiffer# and Richard Willstätter#
1931	–	3	Max Bergmann,# Heinrich Wieland* and Richard Willstätter#
1932	1	6	Georg Barke,# Hans von Euler-Chelpin+, Oskar Loew,# Wilhelm Prantl,# Heinrich Wieland* and Richard Willstätter#
1933	–	1	Heinrich Wieland*
1934	1	2	Erwin Bauer,# Heinrich Wieland*
1935	–	1	Felix Ehrlich#

1047 After Crawford et al. (1987) and Björk (2001).
Neuberg proposed as sole laureate.
* Wieland always proposed Neuberg jointly with O. Diels.
+ Euler-Chelpin had proposed Neuberg jointly with G. Embden.

After the war, Neuberg wrote to the subsequent winner of the Nobel Prize in Medicine, Hans A. Krebs,[1048] as follows:

> "It was I who got Harden the Nobel Prize, as he was not proposed by anyone else but me. He was, as my friend Paul Mayer told me, perfectly well informed about this. When Mayer asked him whether, in view of the threatening clouds that were beginning to lower over Germany, he could not take me with him to England, he said that there was no position free that was worthy of me, but that he would see if he could get me made an FRS [Fellow of the Royal Society, Authors' note]. That would have meant a lot. But Harden took this intention with him to the grave."[1049]

Whether those who did not rate Neuberg's scientific achievements so highly also felt themselves strengthened in their reservations by Neuberg's personal manner must remain speculation. Neuberg's forthright, sometimes sarcastic way of exercising criticism no doubt made him unpopular with some of his colleagues. But equally, critical voices expressing their displeasure at Neuberg's being passed over were to be heard again and again from among his colleagues. Neuberg's scientific merits remain undoubted and his claim to a Nobel Prize is impressively evinced by Björk's analysis; but that said, comments made after the war such as the following by Emil Abderhalden, and in particular the intentions behind them, are to be regarded with considerable skepticism:

> "I was enraged to hear how the Nobel Prize for Medicine has been award this year. Of course I have a very high regard for the Cori family, but it was your turn. If an honor is to be given in the field of carbohydrates, then you are way out in front [...]"[1050]

1048 Hans A. Krebs (1900–1981), born in Hildesheim, took his doctorate in Hamburg in 1925 and worked with Otto Warburg at the KWI for Biology. In 1932 Krebs took his *habilitation* in internal medicine in Freiburg. As a German-Jewish scientist he emigrated in 1933 to England, where he did research and taught in Cambridge and Oxford. In 1937 Krebs described the citrate cycle which is also named after him (the citric acid cycle or Krebs cycle). For the description of coenzyme A he received the Nobel Prize in Medicine jointly with Fritz A. Lipmann (1899–1986) in 1953. Sauer (2001).

1049 "Harden, der von keiner anderen Seite als von mir vorgeschlagen war, verschaffte ich den Nobelpreis. Er war, wie mir mein Freund Paul Mayer erzaehlte, darueber genau orientiert. Auf dessen Frage, ob man mich bei der heraufziehenden Ungemuetlichkeit in Deutschland nach England nehmen koennte, meinte er, man habe keine meiner wuerdigen Stelle frei, aber er wolle sehen ob er mich zum FRS machen koennte. Dies haette viel bedeutet. Aber Harden hat diese Absicht mit ins Grab genommen." Neuberg to Krebs, January 17, 1955. CNP, Box 7, Krebs, Hans A.

1050 "Zornig habe ich vernommen, wie in diesem Jahr der medizinische Nobelpreis verteilt worden ist. Ich schätze natürlich die Familie Cori sehr, jedoch warst Du dran. Wenn schon auf dem Kohlenhydratgebiet geehrt wird, dann stehst Du mit grossem Abstand an erster Stelle [...]." Abderhalden to Neuberg, November 6, 1947. CNP, Box 1: Abderhalden, Emil.

8.3. THE "CARL NEUBERG MEDAL"

Carl Neuberg received a particular honor from the New York based *American Society of European Chemists and Pharmacists* (ASEC). This society, established in 1940 as an organization to assist fellow scientists newly arrived in America,[1051] drew its members – Neuberg was made an honorary member in 1942[1052] – primarily from the ranks of scientists expelled from Europe. As was reported by Neuberg's close colleague, Ines Mandl, the purpose of the ASEC was first and foremost to cultivate mutual relationships which were used to exchange information on available jobs, scholarships and grants, and on the qualifications of members of staff. These networks were of particular benefit to young scientists and to other European emigrants in the "New World."[1053] Regular meetings were organized to cultivate the areas of science that members had worked on together, and to obtain information on other fields of work.[1054]

On the occasion of Carl Neuberg's 70th birthday, the ASEC created the *Carl Neuberg Medal,* to be awarded annually, and awarded it initially in 1947 to the person whose name it bore. According to the statutes, the medal was to be awarded every year to an outstanding scientist who had earned merit through particular achievements in the border area between biochemistry and medicine. The decision as to who was to receive the medal was taken by a committee specially elected for the purpose.[1055] The initial award of the medal to Carl Neuberg took place during a ceremonious assembly. Severo Ochoa spoke on the scientific achievements of the guest of honor, and Albert Gessler, together with Erwin Schwenk, reported on the time when both knew Neuberg as a young professor and director of the institute in Berlin. Neuberg himself spoke about the cultural divide that he experienced between Europe and the United States. He emphasized the differences in the leisure activities of the various generations, contrasting the passive enthusiasm of the non-participating public at a baseball match in America at that time with the thoroughly adventurous activities of various scientists of his acquaintance in his generation:

> "And to remain with the members of my own discipline, I will only mention that Wieland and Otto Hahn were reckless skiers, that Hans von Euler daringly dropped incendiary bombs on Paris – regardless of which, or in a deeper sense with a high regard for which, he has recently been made a Member of the Académie des Sciences in Paris – that Hermann von Schroeter, 40 years ago, was an enthusiastic balloonist, that Hans Fischer would get onto his motorbike in Munich on a Saturday afternoon and dash at breakneck speed over the notorious Zirler Berg to the Brenner and back in order to not miss his Saturday round of *skat* with Hoenigschmidt and Wieland, that Warburg was a daredevil horse rider, who, having once

1051 The organization's original name was "Mutual Aid Society of European Chemists and Pharmacists"; this was changed in 1943. CNP, Box 6: Goldschmiedt, Henry.

1052 Goldschmiedt to Neuberg, December 1, 1942. CNP, Box 6: Goldschmiedt, Henry.

1053 Personal communication from Ines Mandl, October 25, 1996.

1054 For example, Neuberg arranged for Otto Warburg to give a lecture to the ASEC when he was staying in New York in 1948. Warburg to Neuberg, September 12, 1948. CNP, Box 17: Warburg, Otto.

1055 American Society of European Chemists and Pharmacists (1947).

> injured his spine in a fall, was hardly cured before he was spurring his horse on again in order to test experimentally, as he always said, whether the beast would shy at his mother-in-law's ghost a second time."[1056]

There are indications that in later years the ASEC was renamed *Carl Neuberg Society for International Scientific Relations.*[1057] No precise date for this renaming could however be found.[1058] In the 1960s the Society's members increasingly lost interest in it, which was perhaps also due to the declining need for what was originally a welfare organization; in consequence of which, multiplying personal differences gradually led to the Society's disintegration.[1059]

On the occasion of the 100th anniversary of Carl Neuberg's birth in 1977, the Society was revived for a short time. In that year, the *Carl Neuberg Medal – 100th Anniversary Medal* was awarded to Ines Mandl.

No further information could be discovered as to what became of the ASEC or the Carl Neuberg Society for International Scientific Relations in the time after 1980. It can only be assumed that the attractiveness of the Society continued to dwindle as the generation of scientists who had personally experienced the fate of expulsion and emigration departed the scientific scene.

1056 "Um bei den Fachgenossen zu bleiben, so erwaehne ich nur, dass Wieland und Otto Hahn wilde Skilaeufer waren, dass Hans von Euler tollkuehn Brandbomben auf Paris warf – dessen ungeachtet oder in tieferem Sinne dessen hochgeachtet, ist er juengst Mitglied der Académie de Sciences in Paris geworden – dass Hermann von Schroeter vor 40 Jahren begeisterter Freiballonfahrer war, dass Hans Fischer Sonnabend mittags in München aufs Motorrad stieg und in rasender Fahrt ueber den verrufenen Zirler Berg zum Brenner und zurueck eilte, um ja den samstaegigen Skat mit Hoenigschmidt und Wieland nicht zu versaeumen, dass Warburg ein verwegener Reiter war und als er sich bei einem Sturz das Rueckrat verletzt hatte, kaum genesen, wieder sein Pferd jagte, um wie er immer sagte, experimentell zu pruefen, ob der Gaul ein zweites Mal vor dem Gespenst seiner Schwiegermutter scheuen wuerde." American Society of European Chemists and Pharmacists (1947), 6.

1057 Düwell (1990), 765.

1058 The change of name may have taken place in 1964, since a file among Neuberg's papers that is marked with the new name dates from that year. CNP, Box 3: Carl Neuberg Society for International Scientific Relations.

1059 This appears from a note that Neuberg's daughter Irene added to the documents on the Carl Neuberg Society. CNP, Box 3: Carl Neuberg Society for International Scientific Relations.

HOLDERS OF THE CARL NEUBERG MEDAL[1060]

1947	Carl Neuberg
1948	Fritz A. Lipmann
1949	Jesse Greenstein
1950	Friedrich F. Nord
1951	Severo Ochoa
1952	Harlan Wood
1953	Heinrich Waelsch
1954	David Nachmansohn
1955	Feodor Lynen
1956	Henry Lardy
1957	Moses Kunitz
1958	Erwin Chargaff
1959	Horace Barker
1960	Sarah Ratner
1961	Erich Baer
1962–1976	*Medal was not awarded during this period*
1977	Ines Mandl
1978	Louis E. Sitzbach
1979	*Medal not awarded*
1980	Hugh Fudenberg

1060 Ines Mandl remembers that the Carl Neuberg Medal was also awarded to Bernhard Honecker. Written communication from Ines Mandl, November 6, 2006. It has not been possible to determine the year.

9. EPILOGUE

We started this book with two quotations from Erwin Chargaff, which allow us to gain some impression of the kind of fracture must have taken place in the biography of a scientist who, like Carl Neuberg, was forced into emigration. Only in exceptional cases were such people able to take up the earlier threads of their life again. Often they were obliged, in order to "survive" both physically and mentally, to set out in completely new directions. When we concern ourselves with the lives of emigrants today, we are not really able to understand what it means to be torn so brutally out of the continuity of the life one had planned for oneself.

In encyclopedias published up to the 1990s, we can trace how the name Neuberg was initially relegated to being no more than a marginal note in history. But if you enter the name "Carl Neuberg" as a search argument in the internet today, more than 300,000 entries appear within seconds. Most of these are mentions of the street address of Hanover Medical School (*Medizinische Hochschule Hannover* – MHH) as the source of scientific publications, as since January 1, 1997, the School's official address has been "Carl-Neuberg-Strasse 1."

The story of this naming is closely linked to the process of coming to terms with the Nazi past of German universities. Although the MHH was not established until 1965, it chose Konstanty Gutschow[1061] as its architect and subsequently honored him for his achievements: From 1978 onwards, immediately after Gutschow's death, there were internal discussions within the School as to whether the road in which it stood should not bear his name. Finally, on July 14, 1981, it was announced that the official street address of the MHH was to be "Konstanty-Gutschow-Strasse 8."[1062] However, after the publication in 1988 of *Träume in Trümmern. Planung zum Wiederaufbau zerstörter Städte im Westen Deutschlands 1940–1950*[1063] ("Dreams in Ruins. The Planning of the Reconstruction of Devastated Cities in Western Germany, 1940-1945"), a group of academics and students at the MHH began to investigate Gutschow's role during the Nazi period. This inevitably threw up the question of to whether the School ought to remain linked to its architect by the street name conferred in his honor. After lengthy

1061 Konstanty Gutschow (1902–1978), became a self-employed architect after completing his studies. In 1939 he was appointed "Architect of the Elbe Shore" for Hamburg, and was incorporated into Albrecht Speer's staff from 1943 onwards. In the denazification procedure he was classified as an "Offender" ("belastet") in 1945 and as a "Follower" ("Mitläufer") in 1949. From 1947 onwards he was involved in the reconstruction of towns in northern Germany; in 1949 he became advisor to Hanover City Council and thus was involved in the planning of Hanover Medical School in 1961.

1062 Cf. Archives of Hanover Medical School. File 1/History of the Establishment and Curare special edition (1994).

1063 Durth/Gutschow (1988), 508.

internal debates, the members of the Senate of the MHH finally resolved on November 11, 1994, that the use of the name should be discontinued.

In a subsequent public debate within the School it was then decided that the institution's address should be linked to the name of Carl Neuberg. Since January 1, 1997, the official address of Hanover Medical School has been *Carl-Neuberg-Strasse 1*. This renaming in his honor is a symbolic indication that the MHH is a medical research establishment from the time following the Second World War and the Holocaust. The recognition of Carl Neuberg emphasizes on the one hand that biochemistry was a major factor in medical progress during the 20th century, and on the other hand that a German scientist of Jewish origin who was driven out of his country played an essential role in this development.

In the course of the preparations for and the implementation of the renaming of the street, however, it very quickly became clear that there was only very scanty knowledge of Neuberg's life, so that it was perceived as an obligation, and that not only in view of the renaming of the street, to restore an awareness of the life and scientific achievements of this notable research scientist to the consciousness of the German scientific and historical communities. An exploration of the materials in the archives of the Max Planck Society, the Springer publishing house, the American Philosophical Society and many other sources made it possible to draw a picture of the life of the impassioned research scientist Carl Neuberg in all its diverse facets, after which it was even more apparent what barbaric consequences Neuberg, and all other known and unknown refugees from Nazi Germany, had suffered through their being driven into exile, and from the resulting fractures in both their private lives and their professional careers.

Neuberg's scientific achievements are undisputed. He earned himself a place of honor in the field of biochemistry, one of the most important scientific disciplines in the development of medicine in the 20th century. He certainly did not invent the word "biochemistry," or its German equivalent "Biochemie": This honor is due to the Viennese chemist Vinzenz von Kletzinsky,[1064] who intended to use the word to denote a further subsidiary discipline within biology. Nor was Neuberg the first to use the term in the field of scientific periodicals: Franz Hofmeister, who edited the periodical *"Beiträge zur Chemischen Physiologie und Pathologie"* from 1902 onwards, used the term "Biochemie" in the journal's subtitle.[1065] Despite all this, it is clear that the definition and demarcation of the content of the subject "biochemistry" were determined by the periodical founded by Neuberg in 1906 and edited by him, and that it was this that enabled the term to prevail in the general nomenclature of science. Neuberg himself went into this matter in a letter to the later Nobel Prize winner Fedor Lynen:

> "Does biochemistry really exist as a separate subject? Are Emil Fischer, Wieland, Willstaetter, Euler chemists or biochemists? The fact that the problems may be approached from two sides, i.e. also from the medical side, does not say much. Not all that much anatomical, morphological or purely physiological knowledge is required to work in the field

1064 Vinzenz Kletzinsky (1826–1881). Cf. Völckel (2003), 150.

1065 On the history of the term "biochemistry," see also Jaenicke (2002) 492–494 and Lieben (1970).

> successfully. The name biochemistry, which I popularized with the title 'Biochemische Zeitschrift' (Hopkins followed a few months later with his 'Biochemical Journal') and which was then transferred to the Institute in Dahlem, means nothing more than the wish to avoid 'competition' with the legitimate Institutes of Chemistry. Advances in biochemistry are advances in chemistry."[1066]

Neuberg saw himself as a chemist. This was a result of his training and his scientific socialization. It is also reflected in his passion for chemical experiments. Again and again he experienced an almost childish pleasure when he was able, for example, to observe the crystallization of a protein in the test tube.[1067]

Neuberg's outstanding ability also arose out of his enormous diligence and his highly developed talent for organization. This was shown, for example, in the way he coped with the double burden of managing an extraordinarily active KWG research institute while at the same time editing with great commitment a highly reputed scientific journal, which he succeeded in getting established at the Springer Verlag within just a few years. Thanks to his proofreading and editing of the *Biochemische Zeitschrift*, Neuberg possessed an immense knowledge of scientific literature. This in conjunction with his excellent memory made him a kind of "database" for many of his colleagues, and it was an invaluable aid to many research scientists to be able to consult him. Ines Mandl remembered a whole series of telephone calls in which colleagues asked Neuberg about remote sources in the literature relating to a specific scientific problem. Neuberg, who was always generous in dispensing information and passing on his knowledge, doubtless always had an answer to hand. This saved many a colleague an enormous amount of time and trouble.[1068] Nord describes his admirable ability to bring order into a wealth of detailed information:

> "Neuberg was both a man of science and of art, keenly aware of the mysteries surrounding us. He, the scientist and the artist, constantly strove to reconcile and harmonize the new with the familiar, struggling to establish partial order in total chaos."[1069]

1066 "Gibt es ueberhaupt eine spezielle Biochemie? Sind Emil Fischer, Wieland, Willstaetter, Euler Chemiker oder Biochemiker? Dass die Probleme von 2 Seiten, d. h. auch von der Medicin her in Angriff genommen werden koennen, besagt nicht viel. Was an anatomischen, morphologischen oder rein physiologischen Kenntnissen fuer die erfolgreiche Betaetigung benoetigt wird, ist nicht allzuviel. Der Name Biochemie, den ich mit dem Titel 'Biochemische Zeitschrift' populaer machte (Hopkins folgte einige Monate spaeter mit dem 'Biochemical Journal') und der dann auf das Dahlemer Institut ueberging, bedeutete nichts anderes, als dass eine 'Concurrenz' mit den legitimen Chemischen Instituten vermieden werden sollte. Die Fortschritte der Biochemie sind die Fortschritte der Chemie." Neuberg to Lynen, June 6, 1951. CNP, Box 8: Lynen, Fedor.

1067 Personal communication from Ines Mandl, October 25, 1996.

1068 Mandl (1978), 41. Among the papers left by Neuberg are a large number of inquiries addressed to him with regard to sources in the scientific literature. How valuable Neuberg's almost encyclopedic knowledge of his subject was to his colleagues is also described by David Nachmansohn. Cf. Nachmansohn (1979), 323–325.

1069 Nord (1958) 7.

Furthermore, Neuberg was particularly distinguished by his unconditional recognition of the scientific achievements of other research scientists.[1070] He himself formulated his attitude as follows:

> "For myself, under the term competitors I have always understood not rivals, but fellow strivers, and have always had the deepest respect for the achievements of others, and never any feeling of grudging them their success."[1071]

In addition, he possessed a sound knowledge of history and literature. For example, he corresponded with his former assistant Maria Kobel about Thomas Mann, whom he greatly admired, and his works.[1072] Bruno Kisch mentions that Neuberg, who had been robbed of his own library through his flight from Germany, always borrowed one or two volumes from his library when he visited him.[1073] When Neuberg, as a result of his declining health, began to suffer from sleeplessness, this did at least have the advantage of allowing him more time to read.[1074] His wide-ranging and diverse academic training, in conjunction with his creativity and his inimitable humor, again and again found expression in short verses or plays on words:

> "[...] I am a mercenary and apart from the Axis will serve whoever offers the most. The (hal)-Bardic odes and the Morgensterns (2 volumes of poems) I will bring with me! The Busy Lizzy and the Iron Maiden still have to be provided."[1075]

1070 Neuberg admired unreservedly, for example, the research done by Nobel Prize winner Otto Warburg. Kisch (1966), 328.

1071 "Fuer mich habe ich unter dem Wort Konkurrent nicht den Wettbewerber, sondern den Mitstrebenden verstanden und habe stets den tiefsten Respekt vor der Leistung anderer empfunden, aber nie das Gefuehl der Missgunst gehabt." Neuberg to Thomas, February 11, 1952. CNP, Box 15: Thomas, Karl.

1072 "With your attitude to things artistic, I can well understand your enjoyment of Thomas Mann's Joseph novel. Personally, I have always utterly admired the convincing manner in which Mann produced this psychologically magnificent novel out of just of a few lines of the Bible. From him I learned the name of Potiphar's wife, who is not named in the Old Testament: the name Eni has come down to us from Egyptian sources." ("Dass Sie Freude am Josephsroman von Thomas Mann haben, begreife ich bei Ihrer Einstellung zu künstlerischen Dingen. Persönlich habe ich immer aufs Höchste bewundert, mit welcher Anschaulichkeit Mann aus den wenigen Zeilen in der Bibel den psychologisch großartigen Roman gemacht hat. Aus ihm habe ich auch den Namen der Frau Potifar, die im alten Testament nicht genannt wird, kennengelernt. Der Name Eni ist uns [aus] ägyptischen Dokumenten überliefert.") Neuberg to Kobel, October 7, 1955. CNP, Box 7: Kobel, Maria.

1073 Kisch (1966), 328.

1074 Lipmann et al. (1956).

1075 "[...] ich bin ein Landsknecht und diene ausser der Axis jedem Meistbietenden. Helle-Bardengesaenge und Morgensterne (2 Gedichtbaende) bringe ich mit! Das fleissige Lieschen und die Eiserne Jungfrau muessen gestellt werden!" This passage contains a number of puns, in particular on "Morgenstern," which, as well as being the name of the humorous poet Christian Morgenstern, also denotes the iron flail that was one of the weapons borne by the medieval "Landsknecht" mercenaries. Neuberg to Aron, September 30, 1942, with regard to the possibility that Neuberg might obtain an academic position in Chicago. CNP, Box 2: Aron, Hans.

Neuberg detested everything that in his eyes represented a lack of intellectual occupation.[1076] His granddaughter remembers that he invested the television set with the label "*Volksverdummungsapparat*" – "device for dumbing down the people." Furthermore, she declared that Neuberg was unable to make anything of children until they were at least 13 or 14 years old and able to converse "intelligently."[1077] Neuberg was particularly unsympathetic to "American music," which once caused his daughter Irene, when inviting him to visit her, to comment mockingly: "We have a guest-room and it's very quiet, too. Radio only working for News, so you need not be afraid of jazz."[1078]

It is also clear that Neuberg, who in his youth had been an enthusiastic and active mountaineer, thought little of American sports such as boxing or baseball, or of the type of passive enthusiasm demonstrated by the public at such events: "There is another world apart from that of the screaming spectators of a football game."[1079] He was similarly disinclined to spend money on things that in his eyes represented personal luxury. He continued, for example, to use the New York subway, even when his economic situation would long have permitted him to let himself be driven in a taxi again.[1080] Furthermore, in the papers he left behind, repeated indications of Neuberg's generosity and readiness to help can be found, for example the many parcels and gifts of clothing that he sent through the *Care* organization to relieve distress in post-war Germany.

Obituaries and appreciations of Neuberg refer again and again not only to his scientific achievements, but also to his critical spirit, which frequently also expressed itself in sarcasm and his often very caustic humor.[1081] Ines Mandl mentioned that he would rather risk jeopardizing a friendship than renounce a witty punch line.[1082] This mirrors Neuberg's desire in his professional field always

1076 In fact, however, his interests were much wider. In a letter to Hans Gaffron, for example, who had worked with Otto Warburg at the KWI for Biology, he expresses himself on the subject of environmental protection, bewailing the terrible destruction of the forests: "The conscience of the world and the timber dealers will not be offended by the dreadful fact that more than 15,600 trees are felled for every Sunday edition of the Times. Where will that lead?" Neuberg to Gaffron, November 16, 1954. CNP, Box 5: Gaffron, Hans.

1077 His strict demands with regard to table manners were always felt to be a great burden by his grandchildren when he visited the family. On the day after the end of such a visit, the family always celebrated "pigs' day," when the children were allowed to eat with their fingers or put their feet on the table. Written communication from Joan Armer, December 14, 1997.

1078 Irene Roberts to Neuberg, March 20, 1943. CNP, Box 13: Roberts, Irene.

1079 "Es gibt eine andere Welt als die der beifallkreischenden Zuschauer eines Footballgames." Neuberg to Gaffron, November 26, 1946. CNP, Box 5: Gaffron, Hans.

1080 Written communication from Joan Armer, December 14, 1997.

1081 Nachmansohn (1979), 322.

1082 Personal communication from Ines Mandl, New York City, October 25, 1996. To some extent, as his granddaughter Joan Armer reports, the content of his humor was fairly unrepeatable. Neuberg could tell dubious jokes for hours on end, doing battle with his son-in-law and the latter's brother. Written communication from Joan Armer, December 14, 1997. One harmless sample may be mentioned, which Neuberg sent to his colleague Otto Loewi when the latter expressed surprise at Neuberg's unbroken zeal for work: "You have not often made mistakes, but you are wrong to assume that my creative power is

to present matters with clarity and precision, and the great gift he had for doing so; in which, however, he seldom sought to govern his tongue, which some of his colleagues resented very strongly.[1083] Maybe this quality, together with the uncompromising and unbending manner in which he pursued his convictions and his objectives, also explains his extravagantly bizarre idea of a worthy death:

> "It was emotion we were seeking, the emotion for which and from which we all ultimately live. Personally, I have always infinitely regretted that the events of 1933 deprived me of the opportunity to spend the winter at Advent Bay on Spitzbergen. The German Navy had invited me to go on that trip, and I had been thrilled to accept the invitation. Perhaps then the dream which has accompanied me all my life, and which seems to many to be perverse, would have been fulfilled, that of going down on a warship with its guns blazing, to finish like the Russian painter Vereshchagin at the Battle of Tsushima[1084] or like Nelson at Trafalgar. Whether there was ever a Lady Hamilton behind such wishes must remain an unanswered question."[1085]

At the end of his life, however, his tone became quieter. In a letter to his former colleague Maria Kobel, in which he told her among other things about the imminent completion of his work on hydrotropy, which he considered to be a worthy conclusion to his scientific career, he summed up:

> "With advancing age, I have increasingly learned modesty, i.e. the recognition of what one is capable of oneself, and what greater ones are able to achieve on great problems. That must also be why, apart from the cat, the donkey has always been my favorite animal. He is distinguished not only by the elegance of his movements, but by the humility which is characteristic of him throughout his life. You know me well enough to know that I have never been vain, and I am so least of all in old age."[1086]

undiminished. I always think of the wise remark made by a urologist in Vienna. An old gentleman came to him complaining that he couldn't urinate properly any more. The doctor asked, 'How old are you?' When the visitor said 98, the doctor replied: 'Then you've pished enough.'" Neuberg to Loewi, March 16, 1954. CNP, Box 8: Loewi, Otto.

1083 Nachmansohn (1979), 322.

1084 Neuberg is mistaken here. The celebrated war painter Vasili Vereshchagin, who had been invited by Admiral Stepan Makarov, Commander of the Russian Pacific Fleet, to observe the Russo-Japanese War, died with the Admiral when the Russian flagship "Petropavlovsk" hit a mine and sank on April 13, 1904. The Battle of Tsushima did not take place until May 1905. Similarly, Nelson did not go down with his ship at Trafalgar; he died in the battle, but "HMS Victory" survived and returned (with his body) to England.

1085 "Es war die Emotion, die wir suchten, die Emotion für die und von der wir ja letzten Endes alle leben. Persoenlich habe ich es unendlich bedauert, durch die Ereignisse von 1933 um die Ueberwinterung auf Spitzbergen in der Advent-Bay herumgekommen zu sein. Zu dieser Reise hatte mich die Reichsmarine eingeladen und ich hatte die Einladung mit Begeisterung angenommen. Vielleicht waere dann auch mein vielen pervers erscheinender, mein ganzes Leben lebendiger Traum erfuellt, auf einem feuernden Kriegsschiff unterzugehen, wie der russische Maler Wereschtschagin in der Seeschlacht von Tschuschima oder wie Nelson vor Trafalgar zu enden. Ob jemals eine Lady Hamilton hinter solchen Wuenschen stand, bleibe dahingestellt." American Society of European Chemists and Pharmacists (1947), 6.

1086 "Gelernt habe ich mit zunehmendem Alter in verstaerktem Masse die Bescheidenheit, d. h. die Selbsterkenntnis was man selber vermag und was an grossen Problemen groessere vollbringen. Darum wohl ist ausser der Katze immer der Esel mein Lieblingstier gewesen. Nicht nur die ihm eigene Eleganz der Bewegungen zeichnet ihn aus, sondern seine Demut, die

The strength of personality that characterized Carl Neuberg becomes apparent once again at the end of his life[1087] through a remark by his friend and physician[1088] Bruno Kisch:

> "His character [was demonstrated] also by the fact that he never, like so many other emigrants, wasted his time grumbling. He never complained about the loss of his position or of so much property, and as the doctor treating him I admired the fortitude with which he accepted his lot."[1089]

In March 1956, Neuberg went down with severe viral pneumonia, from which he never recovered. Carl Neuberg died in New York City on May 30, 1956, at the age of 78, and was buried at the Jewish Cemetery in Hoboken, New Jersey.

characteristisch ist fuer ihn waehrend seines ganzen Lebens. Sie kennen mich genau genug, um zu wissen, dass ich nie ueberheblich gewesen bin und das bin ich am wenigsten im Alter." Neuberg to Kobel, March 8, 1956. CNP, Box 7: Kobel Maria.

1087 Neuberg suffered for many years from an illness causing dizziness, the origin of which is unknown. At times, the resulting symptoms were so pronounced that Neuberg was not able to go out without assistance. Neuberg to Mayer, September 6, 1943. CNP, Box 9: Mayer, Paul.

1088 Neuberg had developed geriatric diabetes, for which by this time he had to take insulin (Neuberg to Durig, July 3, 1946. CNP, Box 4: Durig, Arnold) and a cardiac insufficiency that increasingly restricted his activities. He had been a heavy smoker for many years, of which he said: "I am not surrounded by compliant maidens, and I only drink cognac in small glassfulls, but I still smoke 35 cigarettes a day." ("Von Wunschmaedchen bin ich nicht umgeben, Cognac trinke ich nur aus kleinen Glaesern, aber 35 Cigaretten rauche ich immer noch am Tage." C. Neuberg to E. Neuberg, November 20, 1947. CNP, Box 12: Neuberg, Ernst.

1089 "Sein Charakter [zeigte sich] auch darin, dass er nie wie so viele andere Emigranten sich in Klagen erging. Er hat sich nicht über den Verlust seiner Stellung oder großer Vermögenswerte beklagt und als sein behandelnder Arzt habe ich die Geduld bewundert, mit der er sein Geschick hinnahm." Kisch (1966), 328.

10. LIST OF ABBREVIATIONS

AFU	Archives of the Free University, Berlin
AHU	Archives of the Humboldt University, Berlin
ANYU	Archives of New York University, New York
APS	American Philosophical Society, Philadelphia
ARRACB	Annual Reports of the Royal Agricultural College (Königliche Landwirtschaftliche Hochschule) Berlin
ASEC	American Society of European Chemists and Pharmacists
ASV	Archives of the Springer Verlag
CNP	Carl Neuberg Papers
KWG	Kaiser-Wilhelm-Gesellschaft (Society)
KWI	Kaiser Wilhelm Institute
LBIL	Leo Baeck Institute Library, New York
LSCSA	Lower Saxony Central State Archives, Hanover
MHH	Medizinische Hochschule Hannover (Hanover Medical School)
MPG	Max Planck Gesellschaft (Society)
MPI	Max Planck Institute
NSDAP	Nationalsozialistische Deutsche Arbeiterpartei (National Socialist German Workers' Party – Nazi Party)
RM	reichsmark
SBPK	Staatsbibliothek zu Berlin, Preussischer Kulturbesitz (State Library of the Prussian Cultural Heritage, Berlin)
ZLC	Zehlendorf Local Court, Zehlendorf District Registry, Berlin
ZTH	Zehlendorf Town Hall, Zehlendorf District Registry, Berlin

11. BIBLIOGRAPHY OF WORKS BY CARL NEUBERG AND HIS ASSOCIATES

11.1. PUBLICATIONS BY CARL NEUBERG

1899

1. Wohl, Alfred/Neuberg, Carl: Über die Darstellung des Acroleïns. Berichte der Deutschen Chemischen Gesellschaft 32 (1899), 1352–1354.
2. Neuberg, Carl: Zur Erkennung und Bestimmung des Formaldehyds. Berichte der Deutschen Chemischen Gesellschaft 32 (1899), 1961–1964.
3. Neuberg, Carl: Über eine Verbindung der Glucuronsäure mit p-Bromphenylhydrazin. Berichte der Deutschen Chemischen Gesellschaft 32 (1899), 2395–2398.
4. Neuberg, Carl: Über die Reinigung der Osazone und zur Bestimmung ihrer optischen Drehungsrichtung. Berichte der Deutschen Chemischen Gesellschaft 32 (1899), 3384–3388.
5. Wohl, Alfred/Neuberg, Carl: Über saure Ester der Borsäure. Berichte der Deutschen Chemischen Gesellschaft 32 (1899), 3488–3492.
6. Neuberg, Carl: Über die quantitative Bestimmung des Phenols im Harn. Hoppe-Seylers Zeitschrift für physiologische Chemie 27 (1899), 123–134.

1900

7. Neuberg, Carl: Zur Kenntnis des Acroleïns und Glycerinaldehyds. Über saure Ester der Borsäure. Über die Reinigung der Osazone und zur Bestimmung ihrer optischen Drehungsrichtung. Diss. phil. Berlin 1900.
8. Neuberg, Carl: Über die Harnpentose, ein optisch inactives, natürlich vorkommendes Kohlehydrat. Berichte der Deutschen Chemischen Gesellschaft 33 (1900), 2243–2254.
9. Wohl, Alfred/Neuberg, Carl: Zur Kenntnis des Glycerinaldehyds. Berichte der Deutschen Chemischen Gesellschaft 33 (1900), 3095–3110.
10. Neuberg, Carl: Zur Kenntnis der Glucuronsäure. Berichte der Deutschen Chemischen Gesellschaft 33 (1900), 3315–3323.
11. Mayer, Paul/Neuberg, Carl: Über den Nachweis gepaarter Glucuronsäuren im normalen Harn. Hoppe-Seylers Zeitschrift für physiologische Chemie 29 (1900), 256–273.
12. Neuberg, Carl: Über Löslichkeitsverhältnisse von Osazonen. Hoppe-Seylers Zeitschrift für physiologische Chemie 29 (1900), 274–280.
13. Neuberg, Carl: Über die Farbenreactionen von Zuckern. Hoppe-Seylers Zeitschrift für physiologische Chemie 31 (1900/01), 564–573.

14. Neuberg, Carl: Über den Nachweis der Bernsteinsäure. Hoppe-Seylers Zeitschrift für physiologische Chemie 31 (1900/01), 574–578.

1901

15. Neuberg, Carl: <autoreview> Ueber Farbenreactionen von Zuckern. Hoppe-Seylers Zeitschrift für physiologische Chemie 31 (1900/01), 564–573; in: Centralblatt für die medizinischen Wissenschaften 39 (1901), 482–483.
16. Neuberg, Carl: <autoreview> Über den Nachweis der Bernsteinsäure. Hoppe-Seylers Zeitschrift für physiologische Chemie 31 (1900/01), 574–578; in: Centralblatt für die medizinischen Wissenschaften 39 (1901), 469.
17. Neuberg, Carl/Wohlgemuth, Julius: Über das Verhalten der drei Arabinosen im Thierkörper. Berichte der Deutschen Chemischen Gesellschaft 34 (1901), 1745–1749.
18. Neuberg, Carl/Wolff, Hans[-Georg]: Über den Nachweis von Chitosamin. Berichte der Deutschen Chemischen Gesellschaft 34 (1901), 3840–3846.
19. Neuberg, Carl: Über Kohlehydratgruppen im Albumin aus Eigelb. Berichte der Deutschen Chemischen Gesellschaft 34 (1901), 3963–3967.
20. Blumenthal, Ferdinand/Neuberg, Carl: Über Entstehung von Aceton aus Eiweiß. Deutsche Medizinische Wochenschrift 27 (1901), 6–7.
21. Wohlgemuth, Julius/Neuberg, Carl: Über das physiologische Verhalten der stereoisomeren Arabinosen. Verhandlungen des Congresses für Innere Medicin 19 (1901), 408–412.
22. Neuberg, Carl: Über die wichtigsten Fortschritte auf dem Gebiete der Chemie und Physiologie der Kohlehydrate. Zeitschrift für klinische Medizin 42 (1901), 483–494.
23. Neuberg, Carl: Über die sogenannten Furfurolproben auf Kohlehydrate. Zeitschrift des Vereins der deutschen Zuckerindustrie 51 (1901), 270–279.
24. Neuberg, Carl: <review> M. Krüger/J. Schmidt, Bestimmung des Aminosäurestickstoffs im Harn. Zeitschrift für physiologische Chemie 31, 556; in: Centralblatt für die medicinischen Wissenschaften 39 (1901), 452–453.
25. Neuberg, Carl: <review> W. Huiskamp, Über der Eiweißkörper der Thymusdrüse, Zeitschrift für physiologische Chemie 32, 145–191; in: Centralblatt für die medicinischen Wissenschaften 39 (1901), 500–502.
26. Neuberg, Carl: <review> D. Kurajeff, Über das Protamin an den Spermatozoen der Accipenser stellatus (und über Silurin), Zeitschrift für physiologische Chemie 32, 197–200; in: Centralblatt für die medicinischen Wissenschaften 39 (1901), 531.
27. Neuberg, Carl: <review> W. Pröscher, Zur Kenntnis der Ehrlich'schen Dimethylamidobenzaldehydreaktion, Zeitschrift für physiologische Che-

mie 31, 520; in: Centralblatt für die medicinischen Wissenschaften 39 (1901), 547–548.
28. Neuberg, Carl: <review> F. A. Levene/C. Alsberg, Zur Chemie der Paranucleinsäuren, Zeitschrift für physiologische Chemie 31, 543–558; in: Centralblatt für die medicinischen Wissenschaften 39 (1901), 563–564.
29. Neuberg, Carl: <review> D. Kurajeff, Über das Jodprodukt des Oxyhämoglobins, Zeitschrift für physiologische Chemie 31, 527; in: Centralblatt für die medicinischen Wissenschaften 39 (1901), 579–580.
30. Neuberg, Carl: <review> D. Lawrow, Über die Ausscheidung des Antipyrins aus dem tierischen Organismus, Zeitschrift für physiologische Chemie 32, 111–120; in: Centralblatt für die medicinischen Wissenschaften 39 (1901), 595.
31. Neuberg, Carl: <review> M. Krüger/J. Schmidt, D., Einfluss des Coffeins und Theobromins an der Ausscheidung des Purinkörper im Harn, Zeitschrift für physiologische Chemie 32, 104–110; in: Centralblatt für die medicinischen Wissenschaften 39 (1901), 611.
32. Neuberg, Carl: <review> A. Oswald, Zur Kenntnis [des] Thyreoglobins, Zeitschrift für physiologische Chemie 32, 121–144; in: Centralblatt für die medicinischen Wissenschaften 39 (1901), 627–628.
33. Neuberg, Carl: <review> M. Nencki/J. Zalewski, Über die Reduktionsprodukte des Hämins und über die Constitution des Hämins und seine Derivate, Zeitschrift für physiologische Chemie 34, 997–1010;
34. Neuberg, Carl: <review> Zur Chemie des Chlorophylls: Abbau des Phyllocyanins zum Hämopyrol, Zeitschrift für physiologische Chemie 34, 1687; in: Centralblatt für die medicinischen Wissenschaften 39 (1901), 642–643.
35. Neuberg, Carl: <review> E. Fischer, Synthese der γ-γ-Diaminovaleriansäure, Berichte der Deutschen Chemischen Gesellschaft 34 (1901), 454; in: Centralblatt für die medicinischen Wissenschaften 39 (1901), 676.
36. Neuberg, Carl: <review> Fr. Kutscher, Überführung der rechtsdrehenden Arginin in der optisch inaktive Modifikation, Zeitschrift für physiologische Chemie 32, 476; in: Centralblatt für die medicinischen Wissenschaften 39 (1901), 692.
37. Neuberg, Carl: <review> J. Habermann/R. Ehrenfeld, Einwirkungen des nascirenden Chlors auf Casein, Zeitschrift für physiologische Chemie 32, 467; in: Centralblatt für die medicinischen Wissenschaften 39 (1901), 709.
38. Neuberg, Carl: <review> A. Hammarsten, Über die Galle des Eisbären, Zeitschrift für physiologische Chemie 32, 435; in: Centralblatt für die medicinischen Wissenschaften 39 (1901), 724.
39. Neuberg, Carl: <review>, A. Salaskin, Über die Bildung des Leucinimids bei peptischer und tryptischer Verdauung des Oxyhämoglobins resp. Globulins, Zeitschrift für physiologische Chemie 32, 592; in: Centralblatt für die medicinischen Wissenschaften 39 (1901), 790.

40. Neuberg, Carl: <review> O. Folin/Ph. Shaffer, Über der quantitative Bestimmung der Harnsäure im Harn, Zeitschrift für physiologische Chemie 32, 552; in: Centralblatt für die medicinischen Wissenschaften 39 (1901), 819.
41. Neuberg, Carl: <review> K. Glaessner, a) Über die Vorstufen der Magenfermente, b) Über die örtliche Verbreitung der Profermente in der Magenschleimhaut, Beiträge zur chemischen Physiolologie und Pathologie 1, 1–33; in: Centralblatt für die medicinischen Wissenschaften 39 (1901), 834–835.
42. Neuberg, Carl: <review> K. Glaessner/L. Langstein, Zur Kenntnis der Entstehung der Kynurensäure in Organismen, Beiträge zur chemischen Physiolologie und Pathololologie 1, 34–43; in: Centralblatt für die medicinischen Wissenschaften 39 (1901), 851.
43. Neuberg, Carl: <review> T. Mochizuki, Zur Kenntnis der tryptischen Eiweißspaltung, Beiträge zur chemischen Physiolologie und Pathololologie 1, 44–50; in: Centralblatt für die medicinischen Wissenschaften 39 (1901), 851.

1902

44. Neuberg, Carl: <autoreview> Über den Nachweis von Chitosamin. Berichte der Deutschen Chemischen Gesellschaft 34 (1901), 3840–3846; in: Centralblatt für die medizinischen Wissenschaften 40 (1902), 547.
45. Neuberg, Carl: <autoreview> Über Kohlehydratgruppen im Albumin aus Eigelb. Berichte der Deutschen Chemischen Gesellschaft 34 (1901), 3963–3967; in: Centralblatt für die medizinischen Wissenschaften 40 (1902), 547.
46. Neuberg, Carl: Über die Pentosen des Thierkörpers. Archiv für Physiologie 1901/02 (1902), 544.
47. Neuberg, Carl/Heymann, Felix: Zur Kenntnis des Pseudomucins. Beiträge zur chemischen Physiologie und Pathologie 2 (1902), 201–213.
48. Neuberg, Carl: <autoreview> Zur Kenntnis des Pseudomucins. Beiträge zur chemischen Physiologie und Pathologie 2 (1902); 201–213, in: Centralblatt für die medizinischen Wissenschaften 40 (1902), 611.
49. Neuberg, Carl: Zur Methodik der Kjeldahlbestimmung. Beiträge zur chemischen Physiologie und Pathologie 2 (1902), 214–215.
50. Neuberg, Carl: <autoreview> Zur Methodik der Kjeldahlbestimmung. Beiträge zur chemischen Physiologie und Pathologie 2 (1902), 214–215; in: Centralblatt für die medizinischen Wissenschaften 40 (1902), 563.
51. Neuberg, Carl/Blumenthal, Ferdinand: Über die Bildung von Isovaleraldehyd und Aceton aus Gelatine. Beiträge zur chemischen Physiologie und Pathologie 2 (1902), 238–250.
52. Neuberg, Carl: <autoreview> Über die Bildung von Isovaleraldehyd und Aceton aus Gelatine. Beiträge zur chemischen Physiologie und Pathologie 2 (1902), 238–250; in: Centralblatt für die medizinischen Wissenschaften 40 (1902), 789.

53. Neuberg, Carl: Über die Isolierung von Ketosen I. Berichte der Deutschen Chemischen Gesellschaft 35 (1902), 959–966.

54. Neuberg, Carl: <autoreview> Über die Isolierung von Ketosen. I. Berichte der Deutschen Chemischen Gesellschaft 35 (1902), 959–966; in: Centralblatt für die medizinischen Wissenschaften 40 (1902), 646.

55. Neuberg, Carl: Über die Constitution der Pankreasproteid-Pentose. Berichte der Deutschen Chemischen Gesellschaft 35 (1902), 1467–1473.

56. Neuberg, Carl: <autoreview> Über die Constitution der Pankreasproteid-Pentose. Berichte der Deutschen Chemischen Gesellschaft 35 (1902), 1467–1473; in: Centralblatt für die medizinischen Wissenschaften 40 (1902), 659.

57. Neuberg, Carl: Über l-Xylonsäure. Berichte der Deutschen Chemischen Gesellschaft 35 (1902), 1473–1475.

58. Neuberg, Carl/Neimann, Wilhelm: Eine Methode zur Isolierung von Aldehyden und Ketonen. Berichte der Deutschen Chemischen Gesellschaft 35 (1902), 2049–2056.

59. Neuberg, Carl: Über die Isolierung von Ketosen II. Berichte der Deutschen Chemischen Gesellschaft 35 (1902), 2626–2633.

60. Neuberg, Carl: Über Cystein I. Berichte der Deutschen Chemischen Gesellschaft 35 (1902), 3161–3164.

61. Neuberg, Carl/Wolff, H[ans Georg]/Neimann, W[ilhelm]: Über d-Glucosamin. Berichte der Deutschen Chemischen Gesellschaft 35 (1902), 4009–4023.

62. Neuberg, Carl/Wohlgemuth, Julius: Über d-Arabinose, d-Arabonsäure und die quantitative Bestimmung von Arabinose. Hoppe-Seylers Zeitschrift für physiologische Chcmie 35 (1902), 31–40.

63. Neuberg, Carl/Wohlgemuth, Julius: Über das Verhalten stereo-isomerer Substanzen im Thierkörper I. Über das Schicksal der 3 Arabinosen im Kaninchenleibe. Hoppe-Seylers Zeitschrift für physiologische Chemie 35 (1902), 41–69.

64. Neuberg, Carl/Wohlgemuth, Julius: Über die Darstellung der r- und l-Galactose. Hoppe-Seylers Zeitschrift für physiologische Chemie 36 (1902), 219–226.

65. Neuberg, Carl/Strauss, H.: Über Vorkommen und Nachweis von Fruchtzucker in den menschlichen Körpersäften. Hoppe-Seylers Zeitschrift für physiologische Chemie 36 (1902), 227–238.

66. Salkowski, E[rnst Leopold]/Neuberg, Carl: Die Verwandlung von d-Glucuronsäure in l-Xylose. Hoppe-Seylers Zeitschrift für physiologische Chemie 36 (1902), 261–267.

67. Neuberg, Carl/Neimann, Wilhelm: Synthese der Euxanthinsäure. Centralblatt für die medizinischen Wissenschaften 40, 32 (1902), 529–530.

68. Neuberg, Carl/Wolff, Hans [Georg]: Über eine neue Oxyaminosäure, in: Centralblatt für die medizinischen Wissenschaften 40 (1902), 530.

69. Neuberg, Carl: Über Pentosen des Thierkörpers. Archiv für Anatomie und Physiologie. Physiologische Abteilung (1902), 544.

70. Neuberg, Carl: Über die sogenannten Furfurolproben auf Kohlehydrate. Zeitschrift des Vereins der deutschen Zuckerindustrie 52 (1902), 245–247.

71. Neuberg, Carl: Eine Reaction auf Ketozucker und deren Trennung von den Aldosen. Zeitschrift des Vereins der deutschen Zuckerindustrie 52 (1902), 237–247.

72. Neuberg, Carl: <review> R. Camps, Über Liebig's Kynurensäure und das Kynurin, Zeitschrift für physiologische Chemie 33, 390–411; in: Centralblatt für die medizinischen Wissenschaften 40 (1902), 35–36.

1903

73. Neuberg, Carl/Langstein, Leo: Ein Fall von Desamidierung im Tierkörper; zugleich ein Beitrag zur Frage nach der Herkunft des Glykogens. Archiv für Anatomie und Physiologie; Physiologische Abteilung 1902/03 (1903), 514–516.

74. Neuberg, Carl/Wolff, Hans [Georg]: Über γ- und ß-2-Amino-d-glucoheptonsäure. Berichte der Deutschen Chemischen Gesellschaft 36 (1903), 618–620.

75. Neuberg, Carl: Über die Spaltung von racemischen Aldehyden und Ketonen. Berichte der Deutschen Chemischen Gesellschaft 36 (1903), 1192–1194.

76. Orgler, A[rnold]/Neuberg, Carl: Über Chondroitinschwefelsäure und das Vorkommen einer Oxyaminosäure im Knorpel I. Hoppe-Seylers Zeitschrift für physiologische Chemie 37 (1902/03), 407–426.

77. Mosse, Max/Neuberg, Carl: Über den physiologischen Abbau von Jodalbumin. Hoppe-Seylers Zeitschrift für physiologische Chemie 37 (1902/03), 427–441.

78. Salkowski, Ernst [Leopold]/Neuberg, Carl: Zur Frage der biochemischen Verwandlung von Kohlehydraten der d-Reihe in solche der l-Reihe. Hoppe-Seylers Zeitschrift für physiologische Chemie 37 (1902/03), 464–466.

79. Neuberg, Carl/Mayer, Paul: Über das Verhalten stereoisomerer Substanzen im Thierkörper II: Über das Schicksal der drei Mannosen im Kaninchenleibe. Hoppe-Seylers Zeitschrift für physiologische Chemie 37 (1902/03), 530–544.

80. Neuberg, Carl/Mayer, Paul: Über krystallisierte i-Mannose. Hoppe-Seylers Zeitschrift für physiologische Chemie 37 (1902/03), 545–547.

81. Neuberg, Carl: <review> A. Oswald, Weiteres über das Thyreoglobulin, Beiträge zur chemischen Physiolologie und Pathololologie 2, 545–556; in: Centralblatt für die medicinischen Wissenschaften 41 (1903), 3–4.

82. Neuberg, Carl: <review> P. L Sörensen, Studien über Aminosäuren. Chemisches Zentralblatt 74 (1903) II, 33–35.

83. Neuberg, Carl: <review> H. Sertz, Über die Veränderungen des sogenannten bleischwärzenden Schwefels im Verhältnis zum Gesamtschwefel

bei der Keimung von Lupinen (Lupinus angustifolius). Chemisches Zentralblatt 74 (1903) II, 126.

84. Neuberg, Carl: <review> E. Schulze, Beiträge zur Kenntnis der Zusammensetzung und des Stoffwechsels der Keimpflanzen. Chemisches Zentralblatt 74 (1903) II, 126.

85. Neuberg, Carl: <review> L. Michaelis, Über Hemmungen der Präzipitinreaktion. Chemisches Zentralblatt 74 (1903) II, 127.

86. Neuberg, Carl: <review> Eduard Spiegler, Über das Haarpigment. Chemisches Zentralblatt 74 (1903) II, 128–129.

87. Neuberg, Carl: <review> Karl Glaessner, Über die antipyretische Wirkung des Blutes. Chemisches Zentralblatt 74 (1903) II, 129.

88. Neuberg, Carl: <review> Henri Anten, Über den Verlauf der Ausscheidung des Jodkaliums im menschlichen Harn. Chemisches Zentralblatt 74 (1903) II, 129.

89. Neuberg, Carl: <review> Otto Loewi, Untersuchungen zur Physiologie und Pharmakologie der Nierenfunktionen. Chemisches Zentralblatt 74 (1903) II, 129–130.

90. Neuberg, Carl: <review> Th. Pfeiffer, Über die Resorption wässriger Salzlösungen aus dem menschlichen Magen. Chemisches Zentralblatt 74 (1903) II, 130.

91. Neuberg, Carl: <review> Julius Pohl, Über Allantoinausscheidung bei Intoxikationen. Chemisches Zentralblatt 74 (1903), 130–131.

92. Neuberg, Carl: <review> B. Slowtzoff, Beiträge zur vergleichenden Physiologie des Hungerstoffwechsels. 1. Der Hungerstoffwechsel der Insekten. Chemisches Zentralblatt 74 (1903) II, 131.

93. Neuberg, Carl: <review> Otto Loewi, Über Eiweißsynthese im Tierkörper. Chemisches Zentralblatt 74 (1903) II, 131.

94. Neuberg, Carl: <review> Wilhelm Wiechowski, Über den Einfluß der Analgetica auf die intracranielle Blutzirkulation. Chemisches Zentralblatt 74 (1903) II, 131–132.

95. Neuberg, Carl: <review> Hermann Hildebrandt, Zur Pharmakologie der Kampfergruppe. Chemisches Zentralblatt 74 (1903) II, 132.

96. Neuberg, Carl: <review> H. Hayashi/K. Muto, Über die Ursache der Atemlähmung bei der Coniin- und Blausäurevergiftung. Chemisches Zentralblatt 74 (1903) II, 132.

1904

97. Neuberg, Carl/Salomon, Georg: Über Vorkommen von Heteroxanthin im normalen Hundeharn. Ein Beitrag zur Lehre von der Methylierung im Tierkörper; in: Beiträge zur wissenschaftlichen Medizin und Chemie. Festschrift zu Ehren des 60. Geburtstages von Ernst Salkowski. Berlin: Hirschwald 1904, 37–44.

98. Neuberg, Carl: Zur Kenntnis der Pyrrolreaktion; in: Beiträge zur wissenschaftlichen Medizin und Chemie. Festschrift zu Ehren des 60. Geburtstages von Ernst Salkowski. Berlin: Hirschwald 1904, 271–277.

99. Neuberg, Carl/Rauchwerger, Dora: Über eine neue Reaktion auf Cholesterin; in: Beiträge zur wissenschaftlichen Medizin und Chemie. Festschrift zu Ehren des 60. Geburtstages von Ernst Salkowski. Berlin: Hirschwald 1904, 279–284.
100. Neuberg, Carl: Beitrag zur Frage nach der Zuckerbildung aus Fett im Organismus nach gemeinsam mit Ferdinand Blumenthal angestellten Versuchen. Archiv für Anatomie und Physiologie/Physiologische Abteilung 1903/04 (1904), 571–572.
101. Neuberg, Carl/Silbermann, Martin: Untersuchungen in der Glycerinsäurereihe I. Über d- und l-Glycerinsäure. Berichte der Deutschen Chemischen Gesellschaft 37 (1904), 339–341.
102. Neuberg, Carl/Silbermann, Martin: Untersuchungen in der Glycerinsäurereihe II. Die Verwandlung von Diaminopropionsäure in Isoserin. Berichte der Deutschen Chemischen Gesellschaft 37 (1904), 341–345.
103. Neuberg, Carl: Die Methylphenylhydrazinreaction der Fructose. Berichte der Deutschen Chemischen Gesellschaft 37 (1904), 4616–4618.
104. Neuberg, Carl/Milchner, R.: Über das Verhalten der Kohlehydrate bei der Autolyse und zur Frage nach der Bindung der Kohlehydratgruppe in den Eiweisskörpern. Berliner Klinische Wochenschrift 41 (1904), 1081–1084.
105. Neuberg, Carl/Richter, P[aul] F[riedrich]: Über das Vorkommen von freien Aminosäuren (Leucin, Tyrosin, Lysin) im Blute bei akuter Leberatrophie. Deutsche Medizinische Wochenschrift 30 (1904), 499–501.
106. Neuberg, Carl: Die Physiologie der Pentosen und der Glukuronsäure. Ergebnisse der Physiologie 1. Abt.: Biochemie 3 (1904), 373–452.
107. Loewy, [A.]/Neuberg, Carl: Über Cystinurie. I. Hoppe-Seylers Zeitschrift für physiologische Chemie 43 (1904/05), 338–354.
108. Loewy, [A.]/Neuberg, Carl: Zur Kenntnis der Diamine I. Hoppe-Seylers Zeitschrift für physiologische Chemie 43 (1904/05), 355–357.
109. Neuberg, Carl: Chemisches zur Carcinomfrage I. Über die Wirkungswiese des Radiums bei Karzinom. Zeitschrift für Krebsforschung 2 (1904), 171–182.
110. Neuberg, Carl: Über das Amyloid. Archiv für Anatomie. Verhandlungen der Deutschen Pathologischen Gesellschaft 1904, 19–32.

1905

111. Neuberg, Carl/Federer, Max: Über d-Amyl-phenyl-hydrazin. Berichte der Deutschen Chemischen Gesellschaft 38 (1905), 866–868.
112. Neuberg, Carl/Federer, Max: Über die Spaltung von Racemkörpern I. Berichte der Deutschen Chemischen Gesellschaft 38 (1905), 868–874.
113. Neuberg, Carl/Manasse, A[rmand]: Die Isolirung der Aminosäuren. Berichte der Deutschen Chemischen Gesellschaft 38 (1905), 2359–2366.
114. Neuberg, Carl: Chemisches zur Carcinomfrage II. Über anomale fermentative Vorgänge beim Krebs. Berliner Klinische Wochenschrift 42 (1905), 118–119.

115. Neuberg, Carl: Über kolloidale Erdalkalisalze. Chemiker-Zeitung 29 (1905), 1044.
116. Neuberg, Carl/Neimann, Wilhelm: Neue Reaktionen und Derivate der Glukuronsäure VII. Mitteilung über Glukuronsäure. Hoppe-Seylers Zeitschrift für physiologische Chemie 44 (1905), 97–113.
117. Neuberg, Carl/Neimann, Wilhelm: Synthese 'gepaarter' Glukuronsäuren VIII. Mitteilung über Glukuronsäure. Hoppe-Seylers Zeitschrift für physiologische Chemie 44 (1905), 114–126.
118. Neuberg, Carl/Neimann, Wilhelm: Quantitative Bestimmung 'gepaarter' Glukuronsäuren IX. Mitteilung über Glukuronsäure. Hoppe-Seylers Zeitschrift für physiologische Chemie 44 (1905), 127–133.
119. Neuberg, Carl/Silbermann, Martin: Untersuchungen in der Glycerinsäurereihe III. Die Konfiguration der Glycerinsäure. Ein Beitrag zur Frage der Beziehung zwischen Zuckern und Aminosäuren. Hoppe-Seylers Zeitschrift für physiologische Chemie 44 (1905), 134–146.
120. Neuberg, Carl/Silbermann, Martin: Synthese der Oxyaminobernsteinsäure. [= Synthese von Oxy- und Diaminosäuren I]. Hoppe-Seylers Zeitschrift für physiologische Chemie 44 (1905), 147–156.
121. Neuberg, Carl/Mayer, Paul: Über Cystein II. Hoppe-Seylers Zeitschrift für physiologische Chemie 44 (1905), 472–497.
122. Neuberg, Carl/Mayer, Paul: Über d-, l- und r-Proteincystin. Hoppe-Seylers Zeitschrift für physiologische Chemie 44 (1905), 498–510.
123. Neuberg, Carl: Synthese von Oxy- und Diaminosäuren II. Über Diaminokorksäure und Diaminosebacinsäure. Hoppe-Seylers Zeitschrift für physiologische Chemie 45 (1905), 92–109.
124. Neuberg, Carl: Zur Kenntnis der Diamine II. Eine neue Synthese der Diamine. Hoppe-Seylers Zeitschrift für physiologische Chemie 45 (1905), 110–120.
125. Neuberg, Carl: Zur Bestimmung der Glukuronsäure. Hoppe-Seylers Zeitschrift für physiologische Chemie 45 (1905), 183–184.
126. Neuberg, Carl: Notiz über den Nachweis von Fructose neben Glucosamin. Hoppe-Seylers Zeitschrift für physiologische Chemie 45 (1905), 500.
127. Neuberg, Carl: Zur Frage der Bildung des Erdöls. Zeitschrift für angewandte Chemie 18 (1905), 1606.
128. Neuberg, Carl/Grosser, Karl: Eine neue schwefelhaltige Substanz aus dem Hundeharn. Centralblatt für Physiologie 19 (1905), 316.

1906

129. Albu, Albert/Neuberg, Carl: Physiologie und Pathologie des Mineralstoffwechsels nebst Tabellen über die Mineralstoffzusammensetzung der menschlichen Nahrungs- und Genussmittel, sowie der Mineralbrunnen und -bäder. Berlin: Springer 1906.
130. Neuberg, Carl: Chemisches zur Karzinomfrage. 3. Mitteilung; in: Zur Feier der Vollendung der Instituts-Neubauten, hrsg. von Johannes Orth.

Arbeiten aus dem Pathologischen Institut zu Berlin. Berlin: Hirschwald 1906, 593–607.

131. Neuberg, Carl: Alkoholische Gärung; in: Practicum der Physiologischen und Pathologischen Chemie nebst einer Anleitung zur anorganischen Analyse für Mediciner, hrsg. von Ernst Salkowski. 3., verm. Aufl. Berlin: Hirschwald 1906, 240–242.

132. Neuberg, Carl: Die Hydrierung des Cholesterins. Berichte der Deutschen Chemischen Gesellschaft 39 (1906), 1155–1158.

133. Neuberg, Carl/Strauss, H.: Zur Frage der Zusammensetzung des Reststickstoffs im Blut und in serösen Flüssigkeiten. Berliner Klinische Wochenschrift 43 (1906), 258–260.

134. Neuberg, Carl/Manasse, A[rmand]: Zur Bestimmung der Aminosäuren im Harn. Berliner Klinische Wochenschrift 43 (1906), 734.

135. Neuberg, Carl/Neimann, Ernst: Über gelatinöse Erdalkalisalze I. Biochemische Zeitschrift 1 (1906), 168–176.

136. Neuberg, Carl: Synthese von Oxy- und Diaminosäuren III. Mitt. über γ-γ-1-Diaminoazelaïnsäure, γ-ß-Diaminobuttersäure und γ-Oxy-ß-aminobuttersäure. Biochemische Zeitschrift 1 (1906), 282–298.

137. Neuberg, Carl: Über die Entstehung optisch aktiver Fettsäuren in der Natur. Biochemische Zeitschrift 1 (1906), 368–379.

138. Neuberg, Carl/Ascher, Erich: Über optisch-aktive γ-ß-Diaminopropionsäure und ß-Thioglyzerinsäure. Biochemische Zeitschrift 1 (1906), 380–382.

139. Albu, Albert/Neuberg, Carl: Chemisches zur Karzinomfrage IV. Über das Vorkommen von Indol im Mageninhalt bei Karzinomen. Biochemische Zeitschrift 1 (1906), 541–544.

140. Neuberg, Carl: Zur Kenntnis des Tryptophans. Charité-Annalen 30 (1906), 424–426.

141. Neuberg, Carl: Notiz über die Reaktion von Cholesterin mit γ-Methylfurfurol. Hoppe-Seylers Zeitschrift für physiologische Chemie 47 (1906), 335.

142. Beitzke, H[ermann]/Neuberg, Carl: Zur Kenntnis der Antifermente. Virchows Archiv für pathologische Anatomie und Physiologie und für klinische Medizin 183 (1906), 169–179.

1907

143. Neuberg, Carl: Die selteneren Störungen des Kohlehydratstoffwechsels; in: Handbuch der Pathologie des Stoffwechsels, hrsg. von Carl Harko von Noorden, Bd. 2, Berlin: Hirschwald 1907, 212–245.

144. Neuberg, Carl: Die selteneren Störungen des Eiweißstoffwechsels; in: Handbuch der Pathologie des Stoffwechsels, hrsg. von Carl Harko von Noorden Bd. 2, Berlin: Hirschwald 1907, 464–488.

145. Neuberg, Carl: Die Oxalurie; in: Handbuch der Pathologie des Stoffwechsels, hrsg. von Carl Harko von Noorden, Bd. 2, Berlin: Hirschwald 1907, 490–497.

146. Neuberg, Carl: Die Phosphaturie; in: Handbuch der Pathologie des Stoffwechsels, hrsg. von Carl Harko von Noorden, Bd. 2, Berlin: Hirschwald 1907, 500–505.
147. Neuberg, Carl: Rarer Dearrangements of Carbohydrate Metabolism. In: Metabolism and Practical Medicine (Handbuch der Pathologie des Stoffwechsels, hrsg. von Carl Harko von Noorden) ed. by I. Walker Hall. London: Heinemann 1907, 716–744.
148. Neuberg, Carl: The Rarer Disturbances of Protein Metabolism. In: Metabolism and Practical Medicine (Handbuch der Pathologie des Stoffwechsels, hrsg. von Carl Harko von Noorden) ed. by I. Walker Hall. London: Heinemann 1907, 1021–1045.
149. Neuberg, Carl: Oxaluria. In: Metabolism and Practical Medicine (Handbuch der Pathologie des Stoffwechsels, hrsg. von Carl Harko von Noorden) ed. by I. Walker Hall. London: Heinemann 1907, 1046–1054.
150. Neuberg, Carl: Phosphaturia. In: Metabolism and Practical Medicine (Handbuch der Pathologie des Stoffwechsels, hrsg. von Carl Harko von Noorden) ed. by I. Walker Hall. London: Heinemann 1907, 1055–1060.
151. Neuberg, Carl: Zur synthetischen Darstellung von optisch aktivem Petroleum [Eine Entgegnung an J. Lewkowitsch]. Berichte der Deutschen Chemischen Gesellschaft 40 (1907), 4477–4478.
152. Neuberg, Carl/Rosenberg, E[gon]: Lipolyse, Agglutination und Hämolyse I. Berliner Klinische Wochenschrift 44 (1907), 54–56.
153. Salkowski, Ernst [Leopold]/Neuberg, Carl: Zur Kenntnis der Phenolglukuronsäure. Biochemische Zeitschrift 2 (1907), 307–311.
154. Neuberg, Carl/Popowsky, Nikolaus: Über Indolaminopropionsäure und ihre Halogenverbindungen Tryptophanreaktion. Biochemische Zeitschrift 2 (1907), 357–382.
155. Loewy, [A.]/Neuberg, Carl: Über Cystinurie. Biochemische Zeitschrift 2 (1907), 438–454.
156. Neuberg, Carl/Mayer, Paul: Zwei neue Bildungsweisen des Isoserins. Biochemische Zeitschrift 2 (1907), 116–120.
157. Neuberg, Carl: Zur Kenntnis der Raffinose. Abbau der Raffinose zu Rohrzucker und d-Galaktose. Biochemische Zeitschrift 3 (1907), 519–534.
158. Neuberg, Carl/Marx, Fritz: Über den Nachweis kleiner Mengen von Raffinose. Biochemische Zeitschrift 3 (1907), 535–538.
159. Neuberg, Carl, Marx, Fritz: Reduktionen in der Zuckerreihe mittels metallischen Calciums. Biochemische Zeitschrift 3 (1907), 539–545.
160. Neuberg, Carl/Reicher, C. [K]: Lipolyse, Agglutination und Hämolyse II. Biochemische Zeitschrift 4 (1907), 281–291.
161. Langstein, Leo/Neuberg, Carl: Zur Kenntnis der Beschaffenheit des Harns von Kälbern in den ersten Lebenstagen. Biochemische Zeitschrift 4 (1907), 292–298.
162. Neuberg, Carl/Brahn, B[enno]: Über die Inosinsäure. Biochemische Zeitschrift 5 (1907), 438–450.

163. Neuberg, Carl/Ascher, Erich: Notiz über Desaminocystin und Aminoäthandisulfid. Biochemische Zeitschrift 5 (1907), 451–455.
164. Neuberg, Carl/Rosenberg, Egon: Über die γ-Naphthylisocyanatverbindungen einiger Aminosäuren. Biochemische Zeitschrift 5 (1907), 456–460.
165. Neuberg, Carl: Verschiedenes über Tryptophan. (Jodtryptophan. Tryptophansilber. Optisch-inaktives Tryptophan). Biochemische Zeitschrift 6 (1907), 276–282.
166. Neuberg, Carl/Ascher, Erich: Bildung von Isoserin aus γ-ß-Dibrompropionsäure. Biochemische Zeitschrift 6 (1907), 559–562.
167. Neuberg, Carl/Reicher, K.: Lipolyse, Agglutination und Hämolyse III. Münchener klinische Wochenschrift 49 (1907), 1725–1727.
168. Neuberg, Carl: Die Entstehung des Erdöls (Künstliche Darstellung von optisch aktivem Petroleum). Sitzungsberichte der Königlich Preußischen Akademie der Wissenschaften zu Berlin 1907, 449–455.
169. Neuberg, Carl: Über kolloidale und gelatinöse Kalzium- und Magnesiumverbindungen. Sitzungsberichte der Königlichen Preußischen Akademie der Wissenschaften zu Berlin (1907), 820–822.
170. Neuberg, Carl: Spaltung von Raffinose und Saccharose und d-Galactose. Zeitschrift des Vereins der deutschen Zuckerindustrie 57 (1907), 440–453.
171. Neuberg, Carl: <review> Hermann Kunz-Kause, Über den Anteil der Chemie an der Entwickelung der medicinischen Wissenschaften. Leipzig: Tauchnitz 1907; in: Berliner Klinische Wochenschrift 44 (1907), 1392–1393.

1908

172. Neuberg, Carl: Reduktion von Aminosäuren zu Aminoaldehyden. Berichte der Deutschen Chemischen Gesellschaft 41 (1908), 956–963.
173. Neuberg, Carl/Brahn, B[enno]: Über Inosinsäure. Berichte der Deutschen Chemischen Gesellschaft 41 (1908), 3376–3381.
174. Neuberg, Carl/Rosenberg, Egon: Über die bei der Eiweißfäulnis auftretenden Fettsäuren sowie über die optisch-aktive Valeriansäure und Capronsäure. Biochemische Zeitschrift 7 (1908), 178–190.
175. Neuberg, Carl/Rosenberg, Egon: Verwandlung von optisch-inaktivem Triolein in ein optisch-aktives Glycerid und eine optisch-aktive Säure. Biochemische Zeitschrift 7 (1908), 191–198.
176. Neuberg, Carl: Die Entstehung des Erdöls. Bildung von optisch-aktiver Naphtha aus drehenden Fettsäuren. Biochemische Zeitschrift 7 (1908), 199–212.
177. Neuberg, Carl: Elektrolysen in der Zuckerreihe. Biochemische Zeitschrift 7 (1908), 527–528.
178. Neuberg, Carl: Enzymatische Umwandlung von Adrenalin. Biochemische Zeitschrift 8 (1908), 383–386.

179. Neuberg, Carl/Rewald, Bruno: Über l- und d-Capronsäure. Spaltung von 3-Methylpentansäure in die optisch aktiven Komponenten. Biochemische Zeitschrift 9 (1908), 403–413.
180. Neuberg, Carl/Rewald, Bruno: Über kolloidale und gelatinöse Erdalkalisalze II. (Verbindungen des Calciums, Magnesiums und Strontiums). Biochemische Zeitschrift 9 (1908), 537–550.
181. Neuberg, Carl: Beziehung des cyclischen Inosits zu den aliphatischen Zuckern. Biochemische Zeitschrift 9 (1908), 551–556.
182. Neuberg, Carl: Zur Frage der Konstitution des „Phytins". Biochemische Zeitschrift 9 (1908), 557–560.
183. Neuberg, Carl: Lipolyse, Agglutination und Hämolyse IV. Biochemische Zeitschrift 11 (1908), 400–403.
184. Neuberg, Carl: Depolymerisation der Zuckerarten. Biochemische Zeitschrift 11 (1908), 337–341.
185. Mandel, John A./Neuberg, Carl: Zur Kenntnis der „Glucothionsäure". Biochemische Zeitschrift 13 (1908), 142–147.
186. Mandel, John A./Neuberg, Carl: Naphtoresorcin als Reagens auf einige Aldehyd- und Ketosäuren. Biochemische Zeitschrift 13 (1908), 148–151.
187. Brasch, W./Neuberg, Carl: Biochemische Umwandlung der Glutaminsäure in n-Buttersäure. Biochemische Zeitschrift 13 (1908), 299–304.
188. Neuberg, Carl: Chemische Umwandlungen durch Strahlenarten I. Biochemische Zeitschrift 13 (1908), 305–320.
189. Neuberg, Carl: Über die Reaktion der Gallensäuren mit Rhamnose bzw. γ-Methylfurfurol. Biochemische Zeitschrift 14 (1908), 349–350.
190. Neuberg, Carl: Reduktion von Aminosäuren zu Aminoaldehyden. Hoppe-Seylers Zeitschrift für physiologische Chemie 58 (1908), 84–91.
191. Neuberg, Carl/Rosenberg, E[gon]: Bildung und Zersetzung von Fetten in der Natur. Petroleum 3 (1908), 729–734, 831–840.
192. Neuberg, Carl: Verhalten von an Eiweiß gebundenem Quecksilber. Therapeutische Monatshefte 22 (1908), 580–581.
193. Neuberg, Carl: in: Verhandlungen der 79. Versammlung deutscher Naturforscher und Ärzte 1907 (1908) 1. Hälfte: Naturwiss., 120.
194. Neuberg, Carl: Zur chemischen Kenntnis der Melanome. Virchows Archiv für pathologische Anatomie und Physiologie 192 (1908), 514–512.
195. Neuberg, Carl/Rewald, Bruno: Über kolloidale und gelatinöse Verbindungen der Erdalkalien. Zeitschrift für Chemie und Industrie der Kolloide 2 (1907/08), 321–324, 354–357.

1909

196. Neuberg, Carl: Kohlehydrate; in: Handbuch der Biochemie des Menschen und der Tiere, hrsg. von Carl Oppenheimer. Bd. 1. Jena: Fischer 1909, 159–225.

197. Neuberg, Carl: Chemie der Neubildungen; in: Handbuch der Biochemie des Menschen und der Tiere, hrsg. von Carl Oppenheimer. Bd. 2,2. Jena: Fischer 1909, 377–385.
198. Neuberg, Carl: Über die Pentose der Inosinsäure und des Pankreas. Berichte der Deutschen Chemischen Gesellschaft 42 (1909), 2806–2809.
199. Neuberg, Carl: Dimitri Ivanovitsch Kurajeff †. Biochemische Zeitschrift 16 (1909), 1–3.
200. Neuberg, Carl: Bemerkung über die „Glucothionsäuren". Biochemische Zeitschrift 16 (1909), 250–253.
201. Neuberg, Carl: Notiz über Phytin. Biochemische Zeitschrift 16 (1909), 406–410.
202. Neuberg, Carl: Chemische Umwandlungen durch Strahlenarten II. Wirkungen des elektrischen Gleichstromes. Biochemische Zeitschrift 17 (1909), 270–291.
203. Neuberg, Carl/Brahn, B[enno]: Notiz über Inosinsäure. Biochemische Zeitschrift 17 (1909), 293–296.
204. Neuberg, Carl/Cappezzuoli, Cesare: Biochemische Umwandlung von Asparagin und Asparaginsäure in Propionsäure und Bernsteinsäure. Biochemische Zeitschrift 18 (1909), 424–430.
205. Neuberg, Carl: Verhalten von racemischer Glutaminsäure bei der Fäulnis. Biochemische Zeitschrift 18 (1909), 431–434.
206. Neuberg, Carl/Karczag, Lázló: Verhalten von d, l-γ-Aminoisovaleriansäure (d, l-Valin) bei der Fäulnis. Biochemische Zeitschrift 18 (1909), 435–439.
207. Neuberg, Carl/Kansky, Eugen: Über die Isolierung aliphatischer Alkohole. Biochemische Zeitschrift 18 (1909), 445–449.
208. Neuberg, Carl/Kansky, Eugen: Über die Reduktion von Aminosäuren zu Aminoaldehyden. Biochemische Zeitschrift 20 (1909), 450–462.
209. Kikkoji, T./Neuberg, Carl: Über das Verhalten von Aminoacetaldehyd im tierischen Organismus. Biochemische Zeitschrift 20 (1909), 463–467.
210. Kikkoji, T./Neuberg, Carl: Notiz über die Anwendung des Wasserstoffsuperoxyds bei Oxydasenversuchen. Biochemische Zeitschrift 20 (1909), 523–525.
211. Neuberg, Carl: Über eine Beziehung des Pyridins zu den Zuckerarten. Biochemische Zeitschrift 20 (1909), 526–530.
212. Neuberg, Carl: Abbau einiger Di- und Oxyaminosäuren. Biochemische Zeitschrift 20 (1909), 531–536.
213. Blumenthal, Ferdinand/Jacoby, E./Neuberg, Carl: Zur Frage der autolytischen Vorgänge in Tumoren. Chemisches zur Carcinomfrage VI. Medizinische Klinik 5 (1909), 1595–1597.
214. Neuberg, Carl: Über chemische Wirkungen von Strahlenarten. Kritische Bemerkungen zur Balneotherapie der Gicht. Zeitschrift für Balneologie, Klimatotherapie und Kurort-Hygiene 2 (1909/1910), 41–43.

215. Beitzke, H[ermann]/Neuberg, Carl: Zur Frage der synthetischen Wirkung der Antifermente. Zeitschrift für Immunitätsforschung und experimentelle Therapie I: Originale (1909), 645–650.

1910

216. Neuberg, Carl: Anomalien des Eiweißabbaues; in: Handbuch der Biochemie des Menschen und der Tiere, hrsg. von Carl Oppenheimer. Bd. 4,2. Jena: Fischer 1910, 334–371.

217. Neuberg, Carl/Pollak, Hermann: Über Kohlenhydrat-phosphorsäureester II. Über Saccharoseschwefelsäure und die Phosphorylierung von Eiweiß. Berichte der Deutschen Chemischen Gesellschaft 43 (1910), 2060–2068.

218. Neuberg, Carl: Über die Pankreas-Pentose. Bemerkung zur gleichbetitelten Mitteilung von P. A. Levene und W. A. Jacobs. Berichte der Deutschen Chemischen Gesellschaft 43 (1910), 3501–3502.

219. Neuberg, Carl/Pollak, Hermann: Über Kohlenhydratphosphorsäureester I. Über Saccharosephosphorsäure. Biochemische Zeitschrift 23 (1910) 515–517.

220. Neuberg, Carl/Scott, Leonard/Lachmann, Samuel: Elektrolytischer Abbau von Mono- und Disaccharidsäuren sowie von Oxyaminosäuren. Biochemische Zeitschrift 24 (1910), 152–165.

221. Neuberg, Carl: Über Oxydationsprodukte des Erythrits (d,l-Erythronsäure und d,l-Oxyerythronsäure). Biochemische Zeitschrift 24 (1910), 166–170.

222. Neuberg, Carl/Lachmann, Samuel: Zur Kenntnis der Stachyose. Biochemische Zeitschrift 24 (1910), 171–177.

223. Neuberg, Carl/Lachmann, Samuel: Über ein neues Verfahren zur Gewinnung von Glucuronsäure (und Mentholglucuronsäure). Biochemische Zeitschrift 24 (1910), 416–422.

224. Neuberg, Carl: Kleinere Mitteilungen verschiedenen Inhalts: I. Über Polarisation. II. Über Klärung und Entfärbung. III. Über einige Reaktionen vergorener Zuckerlösungen. IV. Wismutjodidjodwasserstoffsäure als Basenfällungsmittel. V. Zur Ausführung der Kjeldahlbestimmung. VI. Erfahrungen über die Naphthoresorcinreaktion. VII. Über eine Verschärfung der Tryptophanprobe. Biochemische Zeitschrift 24 (1910), 423–442.

225. Neuberg, Carl: Weitere Beiträge zur Chemie der Geschwülste. Chemisches zur Carcinomfrage VII. Biochemische Zeitschrift 26 (1910), 344–350.

226. Neuberg, Carl/Pollak, Hermann: Über Phosphorsäure- und Schwefelsäureester von Kohlenhydraten. Biochemische Zeitschrift 26 (1910), 514–528.

227. Neuberg, Carl/Pollak, Hermann: Über die Phosphorylierung von Eiweiß und die Bindung des Phosphors in den natürlichen Phosphorproteinen. Biochemische Zeitschrift 26 (1910) 529–534.

228. Neuberg, Carl: Beobachtungen an Jodproteinen. Biochemische Zeitschrift 27 (1910), 261–270.
229. Neuberg, Carl: Chemische Umwandlungen durch Strahlenarten III. Übergang von Benzoesäure in Salicylsäure im Sonnenlicht. Biochemische Zeitschrift 27 (1910), 271–272.
230. Neuberg, Carl/Hirschberg, Else: Abbauversuche in der Kohlenhydratreihe. Biochemische Zeitschrift 27 (1910), 327–338.
231. Neuberg, Carl/Hirschberg, Else: Über die γ-Naphthylisocyanatverbindungen einiger physiologisch wichtiger Substanzen. Biochemische Zeitschrift 27 (1910), 339–347.
232. Neuberg, Carl: Über eine neue Bildungsweise von Carbonylsäuren in der Kohlenhydratreihe. Biochemische Zeitschrift 28 (1910), 355–358.
233. Neuberg, Carl/Hildesheimer, A[rnold]: Die Bestimmung der Phenole im Rinderharn. Biochemische Zeitschrift 28 (1910), 525–528.
234. Neuberg, Carl: Chemische Umwandlungen durch Strahlenarten IV. Katalytische Wirkungen des Sonnenlichtes in Gegenwart anorganischer Substanzen. Biochemische Zeitschrift 29 (1910), 279–293.
235. Neuberg, Carl: Bemerkungen zur voranstehenden Mitteilung von H. L. Wheeler und L. B. Mendel. Biochemische Zeitschrift 29 (1910), 420–421.
236. Neuberg, Carl: Zur Frage der Pigmentbildung. Anhang: Beobachtungen über das Sepiamelanin. Zeitschrift für Krebsforschung 8 (1910), 195–205.

1911

237. Neuberg, Carl/Rewald, Bruno: Glykogen. In: Biochemisches Handlexikon, hrsg. von Emil Abderhalden. Bd. 2. Berlin: Springer 1911, 255–264.
238. Neuberg, Carl/Rewald, Bruno: Die einfachen Zuckerarten; in: Biochemisches Handlexikon, hrsg. von Emil Abderhalden. Bd. 2. Berlin: Springer 1911, 265–526.
239. Neuberg, Carl: Vorwort. In: Der Harn sowie die übrigen Ausscheidungen und Körperflüssigkeiten von Mensch und Tier. Ihre Untersuchung und Zusammensetzung in normalem und pathologischem Zustande. Ein Handbuch für Ärzte, Chemiker und Pharmazeuten sowie zum Gebrauche an Landwirtschaftlichen Versuchsstationen, hrsg. von Carl Neuberg. Teil 1,. Berlin: Springer 1911, V–VIII.
240. Neuberg, Carl: Die Untersuchung der organischen stickstofffreien Substanzen des Harns; in: Der Harn sowie die übrigen Ausscheidungen und Körperflüssigkeiten von Mensch und Tier. Ihre Untersuchung und Zusammensetzung in normalem und pathologischem Zustande. Ein Handbuch für Ärzte, Chemiker und Pharmazeuten sowie zum Gebrauche an Landwirtschaftlichen Versuchsstationen, hrsg. von Carl Neuberg. Teil 1. Berlin: Springer 1911, 197–527.

241. Neuberg, Carl/Wastenson, H.: Oxydative Umwandlung von Azeton. Archiv für Anatomie und Pathologie. Verhandlungen der Physiologischen Gesellschaft zu Berlin 36 (1911), 6.
242. Neuberg, Carl/Karczag, Lázló: Die Gärung der Brenztraubensäure und der Oxalessigsäure als Vorlesungsversuch. Berichte der Deutschen Chemischen Gesellschaft 44 (1911), 2477–2479.
243. Neuberg, Carl: Über den Harn nach Salicylsäuregebrauch. Berliner klinische Wochenschrift 48 (1911), 798–799.
244. Neuberg, Carl: Über die Erkennung von enzymatischer Nucleinsäurespaltung durch Polarisation. Biochemische Zeitschrift 30 (1911), 505–506.
245. Neuberg, Carl/Hildesheimer, A[rnold]: Über zuckerfreie Hefegärungen. I. Biochemische Zeitschrift 31 (1911), 170–176.
246. Neuberg, Carl/Tir, L[asar]: Über zuckerfreie Hefegärungen II. Biochemische Zeitschrift 32 (1911), 323–331.
247. Neuberg, Carl/Kretschmer, Erich: Weiteres über künstliche Kohlenhydratphosphorsäureestern und Glycerinphosphorsäure. Biochemische Zeitschrift 36 (1911), 5–14.
248. Neuberg, Carl/Kretschmer, Erich: Über p-Kresolglucuronsäure. Biochemische Zeitschrift 36 (1911), 15–21.
249. Neuberg, Carl/Saneyoshi, Sumio: Über das Verhalten der stereoisomeren Weinsäuren im Organismus des Hundes. Biochemische Zeitschrift 36 (1911), 32–36.
250. Neuberg, Carl/Miura, Soichiro: Über die hydrolysierende Wirkung des Wasserstoffsuperoxyds. Biochemische Zeitschrift 36 (1911), 37–43.
251. Neuberg, Carl/Saneyoshi, Sumio: Über den Nachweis kleiner Mengen von Disacchariden. Biochemische Zeitschrift 36 (1911), 44–55.
252. Neuberg, Carl/Saneyoshi, Sumio: Über den Nachweis kleiner Mengen Glucuronsäure als Osazon. Biochemische Zeitschrift 36 (1911), 56–59.
253. Neuberg, Carl/Karczag, Lázló: Über zuckerfreie Hefegärungen III. Biochemische Zeitschrift 36 (1911), 60–67.
254. Neuberg, Carl/Karczag, Lázló: Über zuckerfreie Hefegärungen IV. Carboxylase, ein neues Enzym der Hefe. Biochemische Zeitschrift 36 (1911), 68–75.
255. Neuberg, Carl/Karczag, Lázló: Über zuckerfreie Hefegärungen V. Zur Kenntnis der Carboxylase. Biochemische Zeitschrift 36 (1911), 76–81.
256. Neuberg, Carl/Ishida, Migaku: Die Bestimmung der Zuckerarten in Naturstoffen. Biochemische Zeitschrift 37 (1911), 142–169.
257. Neuberg, Carl/Karczag, Lázló: Über zuckerfreie Hefegärungen VI. Biochemische Zeitschrift 37 (1911), 170–176.
258. Neuberg, Carl: Biochemische Umwandlung von γ-Pyrrolidincarbonsäure in n-Valeriansäure und γ-Aminovaleriansäure. Biochemische Zeitschrift 37 (1911), 490–500.
259. Neuberg, Carl: Über die Herkunft der optisch-aktiven Valeriansäure bei der Eiweißfäulnis. Biochemische Zeitschrift 37 (1911), 501–506.

260. Neuberg, Carl: Wird d-Ornithin bei der Fäulnis racemisiert? Biochemische Zeitschrift 37 (1911), 507–509.
261. Neuberg, Carl: Versuche mit Jodocitin. Therapie der Gegenwart 52 (1911), 359–360.
262. Neuberg, Carl: Wirkungen von Lichtstrahlen auf den Abbau des Zuckers. Zeitschrift für Balneologie, Klimatotherapie und Kurort-Hygiene 3 (1911), 525–528.

1912

263. Neuberg, Carl: Zur Frage der Reaktion zwischen Traubenzucker und Methyl-phenyl-hydrazin. Berichte der Deutschen Chemischen Gesellschaft 45 (1912), 1853.
264. Neuberg, Carl/Caspary, W[ilhelm]/Löhe, H.: Weiteres über Heilversuche an geschwulstkranken Tieren mittels tumoraffiner Substanzen, Teil 1: Grundlagen der Versuche. Berliner Klinische Wochenschrift 49 (1912), 1405–1408.
265. Neuberg, Carl: Chemische Umwandlungen durch Strahlenarten. Biochemische Zeitschrift 39 (1912), 158–166.
266. Neuberg, Carl/Kerb, Johannes: Über ein Fällungsmittel für Aminosäuren. Biochemische Zeitschrift 40 (1912), 498–512.
267. Neuberg, Carl: Über zuckerfreie Hefegärungen VII. Bildung von ß-Oxybuttersäurealdehyd (Aldol) bei der Vergärung von Brenztraubensäure. Biochemische Zeitschrift 43 (1912), 491–493.
268. Neuberg, Carl/Kerb, Johannes: Über zuckerfreie Hefegärungen VIII. Entstehung von Acetaldehyd bei der sog. Selbstgärung. Biochemische Zeitschrift 43 (1912), 494–499.
269. Neuberg, Carl: Kleinere Mitteilungen verschiedenen Inhalts: 1. Über die Jodoformreaktion der Milchsäure. 2. Darstellung von d-Glucosamin. 3. Polarimetrie oder reduktionsanalytische Bestimmung von Zucker. 4. Über die Substanzen, die am Zustandekommen der so genannten Cammidgeschen Reaktion beteiligt sind. Biochemische Zeitschrift 43 (1912), 500–507.
270. Neuberg, Carl/Schewket, Omar: Polarimetrische Bestimmung des Glucosamingehaltes von Ovomucoid und Pseudomucin. Biochemische Zeitschrift 44 (1912), 491–494.
271. Neuberg, Carl/Schewket, Omar: Veränderungen einiger Arzneimittel im Licht. Biochemische Zeitschrift 44 (1912), 495–501.
272. Neuberg, Carl/Schewket, Omar: Über einen einfachen Nachweis des Vorkommens von gepaarter Glucoronsäure im normalen Harn. Biochemische Zeitschrift 44 (1912), 502–504.
273. Neuberg, Carl/Kerb, Johannes: Über zuckerfreie Hefegärungen IX. Vergärung von Ketosäuren durch Weinhefen. Biochemische Zeitschrift 47 (1912), 405–412.

274. Neuberg, Carl/Kerb, Johannes: Über zuckerfreie Hefegärungen X. Die Gärung der γ-Ketobuttersäure. Biochemische Zeitschrift 47 (1912), 413–420.

275. Neuberg, Carl/Caspary, W[ilhelm]: Tumoraffine Substanzen. Deutsche Medizinische Wochenschrift 38 (1912), 375–376.

276. Neuberg, Carl/Kerb, Johannes: Entsteht bei zuckerfreien Hefegärungen Äthylalkohol? Zeitschrift für Gärungsphysiologie 1 (1912), 114–120.

277. Neuberg, Carl: Chemische Pathologie der Krebse und Dyskrasie. Zeitschrift für Krebsforschung 10 (1910), 55–74.

1913

278. Neuberg, Carl: Chemische sowie physikalisch-chemische Wirkungen radioaktiver Substanzen und deren Beziehungen zu biologischen Vorgängen. Wiesbaden: Bergmann 1913.

279. Neuberg, Carl: Der Zuckerumsatz der Zelle. In: Handbuch der Biochemie des Menschen und der Tiere, hrsg. von Carl Oppenheimer. Ergänzungsband. Jena: Fischer 1913, 569–609.

280. Neuberg, Carl: Beziehungen des Lebens zum Licht. Berlin: Allgemeine Medizinische Verlagsanstalt 1913.

281. Neuberg, Carl/Kerb, Johannes: Über die Vorgänge bei der Hefegärung. Berichte der Deutschen Chemischen Gesellschaft 46 (1913), 2225–2228.

282. Neuberg, Carl: Über die Zerstörung von Milchsäurealdehyd und Methylglyoxal durch tierische Organe. Biochemische Zeitschrift 49 (1913), 502–506.

283. Neuberg, Carl/Rosenthal, Paul: Über zuckerfreie Hefegärungen XI. Weiteres zur Kenntnis der Carboxylase. Biochemische Zeitschrift 51 (1913), 128–142.

284. Neuberg, Carl: Weitere Untersuchungen über die biochemische Umwandlung von Methylglyoxal in Milchsäure nebst Bemerkungen über die Entstehung der verschiedenen Milchsäuren in der Natur. Biochemische Zeitschrift 51 (1913), 484–508.

285. Neuberg, Carl/Steenbock, Harry: Über die Bildung höherer Alkohole aus Aldehyden durch Hefe I. Übergang von Valerylaldehyd in Amylalkohol. Biochemische Zeitschrift 52 (1913), 494–503.

286. Neuberg, Carl/Kerb, Johannes: Über zuckerfreie Hefegärungen XII. Über die Vorgänge bei der Hefegärung. Biochemische Zeitschrift 53 (1913), 406–419.

287. Neuberg, Carl/Oertel, Walter: Studien über Methylglyoxalbildung. Biochemische Zeitschrift 55 (1913), 495–503.

288. Neuberg, Carl: Kleinere Mitteilungen verschiedenen Inhalts: I. Über die Beständigkeit der Invertase. II. Über die Beständigkeit von Carboxylase. III. Über das optische Verhalten von Hefemacerationssaft (mit Paul Rosenthal). IV. Beobachtungen über die Triketohydrindenreaktion. V. Bemerkungen über den Zucker in Pentosurieharn. Biochemische Zeitschrift 56 (1913), 495–507.

289. Neuberg, Carl/Karczag, Lázló: Über das Verhalten von Lezithin gegenüber Radiumemanation und Thorium X. Radium in Biologie und Heilkunde 2 (1913), 116–122.
290. Neuberg, Carl: Über photokatalytische Wirkungen der Pyrmonter Eisenwässer. Zeitschrift für Balneologie, Klimatotherapie und Kurort-Hygiene 5 (1913), 561–566.
291. Neuberg, Carl: Einiges über die Bedeutung des Lichtes für die Organismen. Zeitschrift für Balneologie, Klimatotherapie und Kurort-Hygiene 5 (1913), 549–653; 689–694; 716–719.

1914

292. Neuberg, Carl/Kerb, Johannes: Über Gärungen in der 3-Kohlenstoff-Reihe. Berichte der Deutschen Chemischen Gesellschaft 47 (1914), 1308–1315.
293. Neuberg, Carl/Nord, F[riedrich] F[ranz]: Phytochemische Bildung von Äthylmercaptan. Berichte der Deutschen Chemischen Gesellschaft 47 (1914), 2264–2271.
294. Neuberg, Carl/Kerb, Johannes: Zur Frage der Bildung von Acetaldehyd bei Hefegärungen. Berichte der Deutschen Chemischen Gesellschaft 47 (1914), 2730–2732.
295. Neuberg, Carl/Kerb, Johannes: Über zuckerfreie Hefegärungen XIII. Zur Frage der Aldehydbildung bei der Gärung von Hexosen sowie bei der sog. Selbstgärung. Biochemische Zeitschrift 58 (1914), 158–170.
296. Neuberg, Carl/Steenbock, Harry: Über die Bildung höherer Alkohol aus Aldehyden durch Hefe II. Weiteres über die Entstehung von Amylalkohol aus Valeraldehyd, insbesondere über die enzymatische Natur dieser Reaktion. Biochemische Zeitschrift 59 (1914), 188–192.
297. Neuberg, Carl: Bemerkung zu den vorausgehenden Auslassungen von Dakin und Dudley. Biochemische Zeitschrift 59 (1914), 194.
298. Neuberg, Carl/Welde, Ernst: Phytochemische Reduktionen I. Umwandlung der Nitrogruppe in die Aminogruppe. Biochemische Zeitschrift 60 (1914), 472–479.
299. Neuberg, Carl/Oertel, Walter: Über die Einführung von Phosphorsäure in Aminosäuren, Peptone, Albumosen und Proteine. Biochemische Zeitschrift 60 (1914), 491–508.
300. Neuberg, Carl/Rosenthal, Paul: Über zuckerfreie Hefegärungen XIV. Fortgesetzte Untersuchungen über die Carboxylase. Biochemische Zeitschrift 61 (1914), 171–183.
301. Neuberg, Carl/Kerb, Johannes: Über zuckerfreie Hefegärungen XV. Über die Bildung von n-Propylalkohol bei der Vergärung von γ-Ketobuttersäure. Biochemische Zeitschrift 61 (1914), 184–186.
302. Neuberg, Carl: Bemerkung über das Phytin. Biochemische Zeitschrift 61 (1914), 187–188.
303. Neuberg, Carl/Galambos, Antonio: Zur Biochemie der Strahlenwirkungen I. Über die Art der Beziehungen zwischen der Wirkung minerali-

scher Katalysatoren und fluorescierender Farbstoffe. Biochemische Zeitschrift 61 (1914), 315–331.
304. Neuberg, Carl/Welde, Ernst: Phytochemische Reduktionen II. Umwandlung aliphatischer Nitrokörper in Aminoverbindungen. Biochemische Zeitschrift 62 (1914), 470–476.
305. Neuberg, Carl/Welde, Ernst: Phytochemische Reduktionen III. Umwandlungen aromatischer und fettaromatischer Aldehyde in Alkohole. Biochemische Zeitschrift 62 (1914), 477–481.
306. Neuberg, Carl/Nord, F[riedrich] F[ranz]: Phytochemische Reduktionen IV. a) Über die Bildung von n-Amylalkohol durch Hefe. b) Beobachtung über natürliches Vorkommen von n-Amylalkohol. Biochemische Zeitschrift 62 (1914), 482–488.
307. Neuberg, Carl/Kerb, Johannes: Über zuckerfreie Hefegärungen XVI. Zur Frage der Bildung von Milchsäure bei der Vergärung von Brenztraubensäure durch lebende Hefen nebst Bemerkungen über die Gärungsvorgänge. Biochemische Zeitschrift 62 (1914), 489–497.
308. Neuberg, Carl/Kerb, Johannes: Über die Rolle des Acetaldehyds bei der Alkoholgärung. Bemerkung zur vorstehenden Mitteilung von Kostytschew. Biochemische Zeitschrift 64 (1914), 251–256.
309. Neuberg, Carl/Iwanoff, Nicolas [Nicolai Nikolajewitsch]: Über das ungleiche Verhalten von Carboxylase und „Zymase“ zu antiseptischen Mitteln. Biochemische Zeitschrift 67 (1914), 1–8.
310. Neuberg, Carl/Czapski, Ludwig: Über Carboxylase im Saft aus obergäriger Hefe. Biochemische Zeitschrift 67 (1914), 9–11.
311. Neuberg, Carl/Nord, F[riedrich] F[ranz]: Über die Gärwirkung frischer Hefen bei Gegenwart von Antisepticis. Biochemische Zeitschrift 67 (1914), 12–17.
312. Neuberg, Carl/Welde, Ernst: Phytochemische Reduktionen V. Zwischenstufen bei der Umwandlung der Nitrogruppe in die Aminogruppe. Biochemische Zeitschrift 67 (1914), 18–23.
313. Neuberg, Carl/Nord, F[riedrich] F[ranz]: Phytochemische Reduktionen VI. Bildung von n-Hexylalkohol durch Hefe. Biochemische Zeitschrift 67 (1914), 24–27.
314. Neuberg, Carl/Czapski, Ludwig: Kommen im Harn den Kresolen isomere Substanzen, „Urogol“ und „Urogon“ vor? Biochemische Zeitschrift 67 (1914), 28–31.
315. Neuberg, Carl/Peterson, W. H.: Die Valeraldehyd- und Amylalkoholgärung der Methyläthylbrenztraubensäure. Biochemische Zeitschrift 67 (1914), 32–45.
316. Neuberg, Carl/Nord, F[riedrich] F[ranz]: Phytochemische Reduktionen VII. Die enzymatische Umwandlung des Thioacetaldehyds in Äthylmercaptan. Biochemische Zeitschrift 67 (1914), 46–50.
317. Neuberg, Carl/Czapski, Ludwig: Über den Einfluss einiger biologisch wichtiger Säuren (Brenztraubensäure, Milchsäure, Äpfelsäure, Weinsäu-

re) auf die Vergärung des Traubenzuckers. Biochemische Zeitschrift 67 (1914), 51–55.

318. Neuberg, Carl: Über die Triketohydrinreaktion. Biochemische Zeitschrift 67 (1914), 56–58.

319. Neuberg, Carl: Zur Biochemie der Strahlenwirkungen II. Über eigenartige Bildung von Acetaldehyd aus verschiedenen Säuren der aliphatischen Reihe durch photokatalytische Vorgänge. Biochemische Zeitschrift 67 (1914), 59–62.

320. Neuberg, Carl/Peterson, W. H.: Zur Biochemie der Strahlenwirkungen III. Über die Bildung von Alkalicarbonat aus neutralen Salzen im Licht. Biochemische Zeitschrift 67 (1914), 63–70.

321. Neuberg, Carl: Über den Abbau der Bernsteinsäure und der γ-Crotonsäure zu Acetaldehyd. Biochemische Zeitschrift 67 (1914), 71–76.

322. Neuberg, Carl/Rubin, Olga: Umwandlung von Fumarsäure, Glycerinsäure und Weinsäure in Acetaldehyd. Biochemische Zeitschrift 67 (1914), 77–81.

323. Neuberg, Carl/Rubin, Olga: Über die Bildung von Thioschwefelsäure und Schwefelsäure aus Ätherschwefelsäuren und Sulfonsäuren. Biochemische Zeitschrift 67 (1914), 82–89.

324. Neuberg, Carl: Das Verhalten der γ-Ketosäuren zu Mikroorganismen I. Die Fäulnis von Brenztraubensäure und Oxalessigsäure. Biochemische Zeitschrift 67 (1914), 90–101.

325. Neuberg, Carl: Ein einfacher Polarisationsapparat für Mikro- und Makrobestimmungen bei weißem Licht. Biochemische Zeitschrift 67 (1914), 102–103.

326. Neuberg, Carl/Welde, Ernst: Phytochemische Reduktionen VIII. Die Überführung des Formaldehyds in Methylalkohol. Biochemische Zeitschrift 67 (1914), 105–110.

327. Neuberg, Carl/Welde, Ernst: Phytochemische Reduktionen IX. Die Umwandlung von Thiosulfat in Schwefelwasserstoff und Sulfit durch Hefen. Biochemische Zeitschrift 67 (1914), 111–118.

328. Neuberg, Carl/Kerb, Johannes: Weiteres über die Fällung von Aminosäuren mit Mercuriacetat und Soda. Biochemische Zeitschrift 67 (1914), 119–121.

329. Neuberg, Carl: Das Verhalten der γ-Ketosäuren zu Mikroorganismen II. Die Fäulnis von γ-Ketobuttersäure. Biochemische Zeitschrift 67 (1914), 22–126.

330. Neuberg, Carl/Rewald, Bruno: Einfache Umlagerungen in der Reihe der Glykole und ihrer stickstoffhaltigen Abkömmlinge I. Die Bildung von Acetaldehyd aus Äthylenglykol, Äthylendiamin, Colamin, Serin und Isoserin. Biochemische Zeitschrift 67 (1914), 127–136.

331. Neuberg, Carl: Ernst Salkowski zum 70. Geburtstag. Deutsche Medizinische Wochenschrift 40 (1914), 1870–1871.

332. Neuberg, Carl: Über photokatalytische Wirkungen des Mondorfer Mineralwassers. Zeitschrift für Balneologie, Klimatotherapie und Kurort-Hygiene 7 (1914), 285–289.
333. Mandel, J[ohn] A./Neuberg, Carl: Über die Wirkungen des Eisens im Lobensteiner Mineralwasser. Zeitschrift für Balneologie, Klimatotherapie und Kurort-Hygiene 7 (1914), 293–296.
334. Neuberg, Carl: <review> C. Doelter: Handbuch der Mineralchemie, Band 2, Dresden, Leipzig 1914; in: Zeitschrift für Balneologie, Klimatotherapie und Kurort-Hygiene 7 (1914), 173–174.

1915

335. Neuberg, Carl: Fortgesetzte Untersuchungen über Carboxylase und andere Hefefermente. Biochemische Zeitschrift 71 (1915), 1–103.
336. Neuberg, Carl/Schwenk, Erwin: Die Gärung der Dioxymaleinsäure. Biochemische Zeitschrift 71 (1915), 104–113.
337. Neuberg, Carl/Schwenk, Erwin: Phytochemische Reduktionen X. Reduktion von Glykolaldehyd zu Äthylenglykol. Biochemische Zeitschrift 71 (1915), 114–117.
338. Neuberg, Carl/Schwenk, Erwin: Phytochemische Reduktionen XI. Die Umwandlung von Äthylendisulfid in Äthylenmercaptan. Biochemische Zeitschrift 71 (1915), 118–121.
339. Neuberg, Carl/Rewald, Bruno: Das Verhalten der γ-Ketosäuren zu Mikroorganismen III. Die Fäulnis der d,l-Methyläthylbrenztraubensäure. Biochemische Zeitschrift 71 (1915), 122–125.
340. Neuberg, Carl/Schwenk, Erwin: Veränderungen im Alkohol- und Aldehydgehalt von Hefen bei der Aufbewahrung und bei der Autolyse. Biochemische Zeitschrift 71 (1915), 126–132.
341. Neuberg, Carl: Zur Frage der Beziehung von Carboxylase zu Zymase. Biochemische Zeitschrift 71 (1915), 133–132.
342. Neuberg, Carl/Schwenk, Erwin: Kofermentartige Wirkung von Salzen der γ-Ketosäuren. Biochemische Zeitschrift 71 (1915), 135–143.
343. Neuberg, Carl/Rewald, Bruno: Studien über Methylglyoxalbildung II. Biochemische Zeitschrift 71 (1915), 144–149.
344. Neuberg, Carl: Über Farbenreaktionen der Triosen und des Methylglyoxals. Biochemische Zeitschrift 71 (1915), 150–157.
345. Neuberg, Carl/Rewald, Bruno: Einfache Umlagerungen in der Reihe der Glykole und ihrer stickstoffhaltigen Abkömmlinge II. Die Bildung von Propionaldehyd und Aceton aus Propylenglykol, Propylendiamin, Trimethylenglykol und Trimethylendiamin. Biochemische Zeitschrift 71 (1915), 158–166.
346. Mayer, Paul/Neuberg, Carl: Phytochemische Reduktionen XII. Die Umwandlung von Citronellal in Citronellol. Biochemische Zeitschrift 71 (1915), 174–179.

347. Mandel, Joh[n] A./Neuberg, Carl: Die Umwandlung aliphatischer und aromatischer Sulfosäuren in Aldehyde bzw. Phenole. Biochemische Zeitschrift 71 (1915), 180–185.
348. Mandel, Joh[n] A./Neuberg, Carl: Darstellung einer scymnolschwefelsäureartigen Substanz. Cholesterinschwefelsäure. Biochemische Zeitschrift 71 (1915), 186–195.
349. Mandel, Joh[n] A./Neuberg, Carl: Über ein einfaches Verfahren zur Erkennung und Bestimmung von Metalloiden in organischen Verbindungen. Biochemische Zeitschrift 71 (1915), 196–213.
350. Mandel, Joh[n] A./Neuberg, Carl: Über einen einfachen Nachweis von kleinen Mengen Glycerin sowie von Alkoholen und Säuren der Kohlenhydratreihe. Biochemische Zeitschrift 71 (1915), 214–218.
351. Neuberg, Carl/Schwenk, Erwin: Zur Biochemie der Strahlenwirkungen IV. Photochemische Bildung von Indigo aus Indican. Biochemische Zeitschrift 71 (1915), 219–225.
352. Neuberg, Carl/Ringer, M.: Über das Wesen der natürlichen Bernsteinsäurebildung I. Die Bernsteinsäuregärung der γ-Ketoglutarsäure. Biochemische Zeitschrift 71 (1915), 226–236.
353. Neuberg, Carl/Ringer, M.: Über das Wesen der natürlichen Bernsteinsäurebildung II. Die Entstehung von Bernsteinsäure bei der Fäulnis von γ-Ketoglutarsäure. Biochemische Zeitschrift 71 (1915), 237–244.
354. Neuberg, Carl/Kerb, Johannes: Über die Vorgänge der natürlichen Milchsäurebildung. Biochemische Zeitschrift 71 (1915), 245–254.

1916

355. Durig, A[rnold]/Neuberg, Carl/Zuntz, Nathan: Ergebnisse der unter Führung von Prof. Pannwitz ausgeführten Teneriffaexpedition 1910 IV. Die Hautausscheidung in dem trockenen Höhenklima. Biochemische Zeitschrift 72 (1916), 253–284.
356. Neuberg, Carl: Über Hydrotropie. Sitzungsberichte der Königlich Preußischen Akademie der Wissenschaften (1916), 1034–1041.

1917

357. Neuberg, Carl: Hydrotropische Erscheinungen I. Biochemische Zeitschrift 76 (1917), 107–176.
358. Neuberg, Carl: Hornalbumose als teilweiser Ersatz von Fleischeiweiß beim Hunde. Biochemische Zeitschrift 78 (1917), 233–237.
359. Neuberg, Carl/Färber, Eduard: Über den Verlauf der alkoholischen Gärung bei alkalischer Reaktion I. Zellfreie Gärung in alkalischen Lösungen. Biochemische Zeitschrift 78 (1917), 238–263.
360. Neuberg, Carl/Färber, Eduard: Über das Vorkommen emulsinartiger von den Hefezellen abtrennbarer Fermente in den untergärigen Hefen sowie das Fehlen von Myrosin in Berliner Ober- und Unterhefen. Biochemische Zeitschrift 78 (1917), 264–272.

361. Neuberg, Carl/Färber, Eduard: Über die Wirkungsweise der Carboxylase. Biochemische Zeitschrift 79 (1917), 376–382.

362. Neuberg, Carl/Schwenk, Erwin: Über Indoxylglucoronsäure. Biochemische Zeitschrift 79 (1917), 383–388.

363. Djenab, Kemal/Neuberg, Carl: Über die Saccharophosphatase der Hefen und die Vergärung der Rohrzuckerphosphorsäure. Biochemische Zeitschrift 82 (1917), 391–411.

364. Neuberg, Carl/Färber, Eduard/Levite, Adam/Schwenk, Erwin: Über die Hexosediphosphorsäure, ihre Zusammensetzung und die Frage ihrer Rolle bei der alkoholischen Gärung sowie über das Verhalten der Dreikohlenstoffzucker zu Hefen. Biochemische Zeitschrift 83 (1917), 244–268.

1918

365. Neuberg, Carl: Über eine allgemeine Beziehung der Aldehyde zur alkoholischen Gärung nebst Bemerkung über das Koferment der Hefe. Biochemische Zeitschrift 88 (1918), 145–204.

366. Neuberg, Carl: Überführung der Fructose-diphosphorsäure in Fructose-monophosphorsäure. Biochemische Zeitschrift 88 (1918), 432–436.

367. Neuberg, Carl/Reinfurth, Elsa: Die Festlegung der Aldehydstufe bei der alkoholischen Gärung. Ein experimenteller Beweis der Acetaldehyd-Brenztraubensäuretheorie. Biochemische Zeitschrift 89 (1918), 365–414.

368. Neuberg, Carl/Ringer, M.: Phytochemische Reduktionen XIII. Der asymmetrische Verlauf der phytochemischen Hydrierung. Die Überführung von racemischem Valeraldehyd (d,l-Methyläthylacetaldehyd) in linksdrehenden Amylalkohol. Biochemische Zeitschrift 90 (1918), 388–394.

369. Neuberg, Carl/Ringer, M.: Über das Wesen der natürlichen Bernsteinsäurebildung III. Die Überführung von Aldehydopropionsäure in Bernsteinsäure mittels Hefe. Biochemische Zeitschrift 91 (1918), 131–136.

370. Neuberg, Carl/Kerb, Elisabeth: Phytochemische Reduktionen XV. Die Überführung von Acetaldol in optisch aktives ß-Butylenglykol durch Hefe nebst Anhang: Darstellung, Reaktionen und krystallisierte Abkömmlinge des Aldols. Biochemische Zeitschrift 92 (1918), 96–110.

371. Neuberg, Carl/Elisabeth Kerb: Phytochemische Reduktionen XVI. Die Umwandlung von Citral in Geraniol durch Hefe nebst Anhang: Darstellung und krystallisierte Abkömmlinge des Zyklocitrals. Biochemische Zeitschrift 92 (1918), 111–123.

372. Neuberg, Carl/Reinfurth, Elsa: Natürliche und erzwungene Glycerinbildung bei der alkoholischen Gärung. Biochemische Zeitschrift 92 (1918), 235–266.

1919

373. Neuberg, Carl/Reinfurth, Elsa: Weitere Untersuchungen über die korrelative Bildung von Acetaldehyd und Glycerin bei der Zuckerspaltung und

neue Beiträge zur Theorie der alkoholischen Gärung. Berichte der Deutschen Chemischen Gesellschaft 52 (1919), 1677–1703.

374. Neuberg, Carl/Nord, F[riedrich] F[ranz]: Die phytochemische Reduktion der Ketone. Biochemische Darstellung optisch-aktiver sekundärer Alkohole. Berichte der Deutschen Chemischen Gesellschaft 52 (1919), 2237–2248.

375. Neuberg, Carl/Nord, F[riedrich] F[ranz]: Phytochemische Reduktionen von Diketonen. Berichte der Deutschen Chemischen Gesellschaft 52 (1919), 2248–2254.

376. Neuberg, Carl: Rudolf Kobert †. Biochemische Zeitschrift 93 (1919), 127.

377. Neuberg, Carl: Nachruf auf Ivar Bang. Biochemische Zeitschrift 93 (1919), 255–256.

378. Neuberg, Carl/Nord, F[riedrich] F[ranz]: Anwendungen der Abfangmethode auf die Bakteriengärungen I. Acetaldehyd als Zwischenstufe bei der Vergärung von Zucker, Mannit und Glycerin durch Bacterium coli, durch Erreger der Ruhr und des Gasbrandes. Biochemische Zeitschrift 96 (1919), 133–157.

379. Neuberg, Carl/Nord, F[riedrich] F[ranz]: Anwendungen der Abfangmethode auf die Bakteriengärungen II. Festlegung der Aldehydstufe bei der Essiggärung. Biochemische Zeitschrift 96 (1919), 158–174.

380. Neuberg, Carl/Hirsch, Julius: Über den Verlauf der alkoholischen Gärung bei alkalischer Reaktion II. Gärung mit lebender Hefe in alkalischen Lösungen. Biochemische Zeitschrift 96 (1919), 175–202.

381. Neuberg, Carl: Franz Röhmann †. Biochemische Zeitschrift 97 (1919), 1a.

382. Neuberg, Carl/Hirsch, Julius: Wirkungsweise der Abfangmethode bei der Acetaldehyd-Glycerin-Spaltung des Zuckers. Die Korrelation von Acetaldehyd und Glycerin innerhalb der gesamten Gärführung, der zeitliche Verlauf dieser Vergärungsform und ihre Beziehung zur gewöhnlichen alkoholischen Gärung. Biochemische Zeitschrift 98 (1919), 141–158.

383. Neuberg, Carl: 100 Bände ‚Biochemische Zeitschrift'. Biochemische Zeitschrift 100 (1919), 1–2.

384. Neuberg, Carl: Die physikalisch-chemische Betrachtung der Gärungsvorgänge. Zugleich Äußerungen zu W. Ostwalds Bemerkungen über die Neubergsche Gärungstheorie. Biochemische Zeitschrift 100 (1919), 289–303.

385. Neuberg, Carl/Hirsch, Julius: Die dritte Vergärungsform des Zuckers. Biochemische Zeitschrift 100 (1919), 304–322.

1920

386. Neuberg, Carl: Max Siegfried †. Berichte der Deutschen Chemischen Gesellschaft 53 A (1920), 77–78.

387. Neuberg, Carl: Zuntz, Nathan. Berichte der Deutschen Chemischen Gesellschaft 53 A (1920), 83–88.

388. Neuberg, Carl/Reinfurth, Elsa: Über den Chemismus der alkoholischen Gärung, zugleich Bemerkungen zur gleichnamigen Mitteilung von E. Zerner. Berichte der Deutschen Chemischen Gesellschaft 53 (1920), 462–469.
389. Neuberg, Carl/Reinfurth, Elsa: Über die Vergärbarkeit der Brenztraubensäure unter den Bedingungen des Abfangverfahrens. Vergärung der Pyruvinat-Sulfite durch Hefe. Berichte der Deutschen Chemischen Gesellschaft 53 (1920), 1039–1052.
390. Neuberg, Carl/Reinfurth, Elsa: Die milchsauren Alkalien als Glycerin-Ersatz. Per- und Perka-Glycerin. Berichte der Deutschen Chemischen Gesellschaft 53 (1920), 1783–1791.
391. Neuberg, Carl/Ehrlich, Marta: Weiteres über die Beziehung der Aldehyde zur alkoholischen Gärung. Biochemische Zeitschrift 101 (1920), 239–275.
392. Neuberg, Carl/Ehrlich, Marta: Über die Beziehungen der phytochemisch reduzierbaren Substanzen zum Vorgange der alkoholischen Gärung und über die Natur der Aktivatorwirkung. Biochemische Zeitschrift 101 (1920), 278–318.
393. Neuberg, Carl: Weitere Erfahrungen über die Bildung und Bedeutung der Fructosediphosphorsäure im Stoffwechsel der Hefe. Biochemische Zeitschrift 103 (1920), 320–335.
394. Neuberg, Carl: Max Siegfried †. Biochemische Zeitschrift 105 (1920) [without page].
395. Neuberg, Carl: Nathan Zuntz †. Biochemische Zeitschrift 105 (1920), 168a.
396. Neuberg, Carl: Die physikalisch-chemische Betrachtung der Gärungsvorgänge. Biochemische Zeitschrift 105 (1920), 306.
397. Neuberg, Carl/Hirsch, Julius/Reinfurth, Elsa: Die drei Vergärungsformen des Zuckers, ihre Zusammenhänge und Bilanz. Biochemische Zeitschrift 105 (1920), 307–336.
398. Neuberg, Carl/Reinfurth, Elsa: Ein neues Abfangverfahren und seine Anwendung auf die alkoholische Gärung. Biochemische Zeitschrift 106 (1920), 276–291.
399. Neuberg, Carl/Sandberg, Marta: Weitere Mitteilungen über chemisch definierte Katalysatoren der alkoholischen Gärung. (Gärunfähige Zucker, Carbonylsäuren der Kohlenhydratreihe, aldehydische und ketonische Pflanzenbasen, Chinone und natürliche Farbstoffe, Nitro- und Nitrosokörper, Hydroxylaminderivate, organische und mineralische Disulfide, Polysulfide, Thio- und Selenosäuren, reduzierte Metallsalze sowie Elemente). Biochemische Zeitschrift 109 (1920), 290–329.
400. Neuberg, Carl/Ursum, Werner: Die dritte Vergärungsform des Zuckers als allgemeine Folge der Dismutationswirkung anorganischer und organischer Alkalisatoren. Biochemische Zeitschrift 110 (1920), 193–215.
401. Neuberg, Carl: Muneo Kumagawa †. Biochemische Zeitschrift 110 (1920) [without page].

402. Neuberg, Carl/Nord, F[riedrich] F[ranz]/Wolff, E.: Acetaldehyd als Zwischenstufe bei der Vergärung von Zucker durch B. lactis aerogenes. Biochemische Zeitschrift 112 (1920), 144–150.

1921

403. Neuberg, Carl: Über den Zusammenhang der Gärungserscheinungen in der Natur. In: Festschrift der Kaiser Wilhelm Gesellschaft zur Förderung der Wissenschaften zu ihrem zehnjährigen Jubiläum. Berlin 1921, 162–170.
404. Neuberg, Carl/Hirsch, Julius: Über ein Kohlenstoffketten knüpfendes Ferment (Carboligase). Biochemische Zeitschrift 115 (1921), 282–310.
405. Neuberg, Carl: Gino Galeotti †. Biochemische Zeitschrift 117 (1921), 117–118.
406. Neuberg, Carl/Arinstein, Bernhard: Vom Wesen der Buttersäure- und Butylalkoholgärung. Abfangung von Acetaldehyd als Umsetzungsprodukt. Übergang von Brenztraubensäure-aldol in Buttersäure. Entstehung höherer Fettsäuren aus Zucker. Biochemische Zeitschrift 117 (1921), 269–314.
407. Neuberg, Carl/Reinfurth, Elsa/Sandberg, Marta: Neue Klassen von Stimulatoren der alkoholischen Zuckerspaltung VII. Mitteilung über chemisch definierte Katalysatoren der Gärung. Biochemische Zeitschrift 121 (1921), 215–234.
408. Neuberg, Carl/Liebermann, Ludwig: Zur Kenntnis der Carboligase II. Biochemische Zeitschrift 121 (1921), 311–325.
409. Neuberg, Carl/Liebermann, Ludwig: Über Glucose- und Rohrzuckermonoschwefelsäure III. Biochemische Zeitschrift 121 (1921), 326–332.
410. Neuberg, Carl/Cohen, Clara: Über die Bildung von Acetaldehyd und die Verwirklichung der zweiten Vergärungsform bei verschiedenen Pilzen. Biochemische Zeitschrift 122 (1921), 204–224.
411. Neuberg, Carl/Sandberg, Marta: Von den Stimulatoren der alkoholischen Zuckerspaltung VIII. Mitteilung über chemisch definierte Katalysatoren der Gärung. Biochemische Zeitschrift 125 (1921), 202–219.
412. Neuberg, Carl: F. Czapek †. Biochemische Zeitschrift 125 (1921), 221.
413. Neuberg, Carl/Ohle, Heinz: Über einen Schwefelgehalt des Agars. Biochemische Zeitschrift 125 (1921), 311–313.
414. Neuberg, Carl/Sandberg, Marta: Über Stimulatoren der alkoholischen Zuckerspaltung IX. Mitteilung über chemisch definierte Katalysatoren der Gärung. Biochemische Zeitschrift 126 (1921/22), 153–178.
415. Neuberg, Carl: Gärung und Synthese. Die Naturwissenschaften 9 (1921), 334–337.

1922

416. Neuberg, Carl: Von der Chemie der Gärungserscheinungen. Berichte der Deutschen Chemischen Gesellschaft 55 (1922), 3624–3638.

417. Neuberg, Carl/Ohle, Heinz: Zur Kenntnis der Carboligase III. Der Bau der biosynthetisch verknüpften mehrgliedrigen Kohlenstoffketten. Biochemische Zeitschrift 127 (1922), 327–339.
418. Neuberg, Carl: Johannes Feigl †. Biochemische Zeitschrift 128 (1922) [without page].
419. Neuberg, Carl/Hirsch, Julius: Zur Klassifizierung der Carboligase. Biochemische Zeitschrift 128 (1922), 608–609.
420. Neuberg, Carl/Ohle, Heinz: Zur Kenntnis der Carboligase IV. Weitere Feststellungen über die biosynthetische Kohlenstoffkettenverknüpfung beim Gärungsvorgang. Biochemische Zeitschrift 128 (1922), 610–618.
421. Neuberg, Carl: W. Palladin †. Biochemische Zeitschrift 130 (1922) [without page].
422. Neuberg, Carl/Dalmer, O.: Krystallisierte Salze einiger physiologisch wichtiger Zuckerphosphorsäure-Verbindungen. Biochemische Zeitschrift 131 (1922), 188–192.
423. Neuberg, Carl: A. Koch †. Biochemische Zeitschrift 132 (1922), 1.
424. Neuberg, Carl/Hirsch, Julius/Reinfurth, Elsa: Weitere Mitteilungen über die äquivalente Bildung von Acetaldehyd und Glycerin bei der zweiten Vergärungsform. Zugleich über die Fehlerquellen in den Angaben H. Gehles. Biochemische Zeitschrift 132 (1922), 589–596.
425. Neuberg, Carl: Ergebnisse der neueren Gärungsforschung. Chemiker-Zeitung 46 (1922), 862.

1923

426. Neuberg, Carl: Ernst Salkowski †. Berichte der Deutschen Chemischen Gesellschaft 56 A (1923), 58–60.
427. Neuberg, Carl: Franz Hofmeister †. Biochemische Zeitschrift 134 (1923), 1–2.
428. Neuberg, Carl: Ernst Salkowski †. Biochemische Zeitschrift 138 (1923), 1–4.
429. Neuberg, Carl: Hartwig Franzen †. Biochemische Zeitschrift 138 (1923), 5.
430. Neuberg, Carl/Reinfurth, Elsa: Phytochemische Reduktionen XVIII. Biochemische Zeitschrift 138 (1923), 561–568.
431. Neuberg, Carl/Cohen, Reinhold: Über Zwischenprodukte des bakteriellen Abbaues von Cellulose. Biochemische Zeitschrift 139 (1923), 527–549.
432. Neuberg, Carl/Kurono, K.: Über Sulfatase. Über die enzymatische Spaltung der Phenolätherschwefelsäure. Biochemische Zeitschrift 140 (1923), 295–298.
433. Neuberg, Carl/Albert, V.: Die Bilanz der Brenztraubensäuregärung. Biochemische Zeitschrift 140 (1932), 299–314.
434. Neuberg, Carl/Linhardt, K.: Über Sulfatase II. Biochemische Zeitschrift 142 (1923), 191–194.

435. Neuberg, Carl/Rosenthal, O.: Über die Cellase der Takadiastase VI. Mitteilung über Carboligase. Biochemische Zeitschrift 143 (1923), 553–558.
436. Neuberg, Carl/Reinfurth, Elsa: Eine neue Form der Umwandlung des Acetaldehyds durch gärende Hefe VI. Mitteilung über Carboligase. Biochemische Zeitschrift 143 (1923), 558–565.
437. Neuberg, Carl/Gottschalk, A[lfred]/Strauß, H.: Das Eingreifen von Insulin in Abbauvorgänge der tierischen Zelle. Deutsche Medizinische Wochenschrift 49 (1923), 1407–1408.
438. Neuberg, Carl/Hirsch, Julius: Über Fragen der Gärungs-Chemie. Ergebnisse der Physiologie 21, 1. Abt. (1923), 400–437.
439. Neuberg, Carl/Gottschalk, A[lfred]: Vorführung der Acetaldehydbildung durch Organe des Warmblüters im Vorlesungsversuche. Klinische Wochenschrift 2 (1923), 1458.
440. Neuberg, Carl: Vom Abbau der Cellulose in der Natur. Die Naturwissenschaften 11 (1923), 657–659.

1924

441. Neuberg, Carl: Kohlenhydrate. I. Zucker; in: Handbuch der Biochemie des Menschen und der Tiere, hrsg. von Carl Oppenheimer. 2. Aufl., Bd. 1. Jena: G. Fischer 1924, 477–541.
442. Neuberg, Carl/Pincussen, L[udwig]: Von der Physiologie der Sonnenstrahlung; in: Handbuch der Balneologie, medizinischen Klimatologie und Balneographie, hrsg. von [Eduard] Dietrich und [Siegfried] Kaminer. Band 3: Klimatophysiologie und Strahlenphysiologie. Leipzig: Thieme 1924, 92–118.
443. Neuberg, Carl/Rosenthal, O.: Zusammenhang von carboligatischer Synthese mit carboxylatischem Abbau. Verwendung von Acetaldehyd als Abfangmittel. Berichte der Deutschen Chemischen Gesellschaft 57 (1924), 1436–1441.
444. Neuberg, Carl: Zur Frage der Konstitution des biochemisch gebildeten Oxy-oxo-propyl-benzols. Biochemische Zeitschrift 144 (1924), 44–46.
445. Neuberg, Carl/Rosenthal, O.: Über Taka-Lactase. Biochemische Zeitschrift 145 (1924), 186–188.
446. Neuberg, Carl: Hartog Jakob Hamburger †. Biochemische Zeitschrift 145 (1924), 379–380.
447. Neuberg, Carl/Gottschalk, Alfred: Quantitative Untersuchungen über die Bildung und Herkunft des Acetaldehyds im intermediären Zellstoffwechsel der Warmblüter. Biochemische Zeitschrift 146 (1924), 164–184.
448. Neuberg, Carl/Gottschalk, Alfred: Abtrennung und Identifizierung des im intermediären Stoffwechsel überlebender Warmblüterzellen gebildeten Acetaldehyds. Biochemische Zeitschrift 146 (1924), 185–187.
449. Neuberg, Carl: Jacques Loeb †. Biochemische Zeitschrift 146 (1924), 387–388.

450. Neuberg, Carl/Reinfurth, Elsa: Über die Beziehungen der Hexose-monophosphorsäure zur Hexose-di-phosphorsäure. Biochemische Zeitschrift 146 (1924), 589–593.
451. Neuberg, Carl/Noguchi, J.: Über die enzymatische Spaltung der Phenacetursäure. Biochemische Zeitschrift 147 (1924), 370–371.
452. Neuberg, Carl/Linhardt, K.: Die enzymatische Spaltung benzoylierter Aminosäuren und ihr asymmetrischer Verlauf. Biochemische Zeitschrift 147 (1924), 372–376.
453. Neuberg, Carl/Gottschalk, A[lfred]: Beobachtungen über den Verlauf der anaeroben Pflanzenatmung. Biochemische Zeitschrift 151 (1924), 167–168.
454. Neuberg, Carl/Gottschalk, A[lfred]: Weitere Untersuchungen über die Entstehung von Acetaldehyd in tierischen Organen (Aceton-Leber). Biochemische Zeitschrift 151 (1924), 169–176.
455. Neuberg, Carl: Einige Beobachtungen über Hefenfermente. Biochemische Zeitschrift 152 (1924), 203–206.
456. Neuberg, Carl: Egmont Münzer †. Biochemische Zeitschrift 153 (1924), 1.
457. Neuberg, Carl/Gottschalk, A[lfred]: Erfahrungen über die Vergärung des Dioxyacetons. Biochemische Zeitschrift 154 (1924), 487–491.
458. Gottschalk, A[lfred]/Neuberg, Carl: Untersuchungen über die Phosphorylierung. Biochemische Zeitschrift 154 (1924), 492–494.
459. Neuberg, Carl/Gorr, G[ünther]: Neue Untersuchungen über die Vergärung der Oxalessigsäure. Biochemische Zeitschrift 154 (1924), 495–502.
460. Neuberg, Carl: Über das neue Ferment Sulfatase. Die Naturwissenschaften 12 (1924), 797–800.
461. Neuberg, Carl: Die Verwendung des Zuckers zur biologischen Synthese. Die Deutsche Zuckerindustrie 49 (1924) [without page].

1925

462. Neuberg, Carl: Vom Zuckerumsatz der pflanzlichen Zelle; in: Handbuch der Biochemie des Menschen und der Tiere, hrsg. von Carl Oppenheimer. 2. Aufl. Bd. 2. Jena: G. Fischer 1925, 442–484.
463. Neuberg, Carl/Gottschalk, Alfred: Die Chemie der Neubildungen; in: Handbuch der Biochemie des Menschen und der Tiere, hrsg. von Carl Oppenheimer. 2. Aufl. Bd. 4. Jena: G. Fischer 1925, 441–455.
464. Neuberg, Carl: Über abnorme Löslichkeiten; in: Handbuch der praktischen und wissenschaftlichen Pharmazie, hrsg. von Hermann Thoms. Bd. 2. Berlin, Wien: Urban & Schwarzenberg 1925, 259–262.
465. Neuberg, Carl: Julius Morgenroth †. Biochemische Zeitschrift 155 (1925), 369–370.
466. Neuberg, Carl/Kobel, M[aria]: Zur Frage der künstlichen und natürlichen Phosphorylierung des Zuckers. Biochemische Zeitschrift 155 (1925), 499–506.

467. Neuberg, Carl/Simon, E[rnst]: Über Sulfatase V. Mitteilung: Tierische Sulfatase. Biochemische Zeitschrift 156 (1925), 365–373.

468. Neuberg, Carl/Simon, E[rnst]: Zur Kenntnis der biochemischen Acyloinsynthese VIII. Mitteilung über Carboligase. Biochemische Zeitschrift 156 (1925), 374–378.

469. Neuberg, Carl: Arthur Heffter †. Biochemische Zeitschrift 158 (1925), 1–2.

470. Neuberg, Carl/Gottschalk, A[lfred]: Über die Bedeutung der Acetaldehydbildung im Froschmuskel und ihre zahlenmäßige Beziehung zur Atmung. Biochemische Zeitschrift 158 (1925), 253–256.

471. Neuberg, Carl: August von Wassermann †. Biochemische Zeitschrift 158 (1925), 257–258.

472. Neuberg, Carl/Kobel, Maria: Über das physiologische Verhalten des Acetoins I. Mitteilung: Über das Verhalten des Acetoins zu Hefe. Biochemische Zeitschrift 160 (1925), 250–255.

473. Neuberg, Carl/Gottschalk, A[lfred]: Über den Nachweis von Acetaldehyd als Zwischenstufe bei der anaeroben Atmung höherer Pflanzen. Biochemische Zeitschrift 160 (1925), 256–260.

474. Neuberg, Carl/Kobel, Maria: Weiteres zur Phosphorylierung des Zuckers. Biochemische Zeitschrift 160 (1925), 464–466.

475. Neuberg, Carl/Sabetay, Sebastian: Über lösliche und unlösliche Salze der Hexose-di-phosphorsäure. Biochemische Zeitschrift 161 (1925), 240–243.

476. Neuberg, Carl/Gottschalk, A[lfred]: Über Apozymase und Cozymase. Zur Lehre von der Phosphorylierung. Biochemische Zeitschrift 161 (1925), 244–256.

477. Neuberg, Carl/Wagner, Joachim: Über die Spaltbarkeit von Ätherschwefelsäuren der verschiedenen Reihen durch Sulfatase und über den Mechanismus der Sulfatasewirkung VII. Mitteilung über Sulfatase. Biochemische Zeitschrift 161 (1925), 492–505.

478. Neuberg, Carl/Sabetay, Sebastian: Die enzymatische Spaltung der Saccharose-phosphorsäure in Fruchtzucker und Glucose-phosphorsäure. Biochemische Zeitschrift 162 (1925), 479–483.

479. Neuberg, Carl/Gottschalk, A[lfred]: Über das physiologische Verhalten des Acetoins II. Mitteilung: Über das Verhalten des Acetoins im Tierkörper. Biochemische Zeitschrift 162 (1925), 484–487.

480. Neuberg, Carl/Dalmer, O.: Notiz über das Aldehydacetal des Methylglyoxals. Biochemische Zeitschrift 162 (1925), 488–489.

481. Neuberg, Carl/Gorr, G[ünther]: Über den Mechanismus der Milchsäurebildung bei Bakterien. Biochemische Zeitschrift 162 (1925), 490–495.

482. Neuberg, Carl/Kobel, M[aria]: Über eine Reaktion der Fructose mit Alanin. Biochemische Zeitschrift 162 (1925), 496–501.

483. Neuberg, Carl/Perelman,[1090] Genia: Verlauf der alkoholischen Zuckerspaltung in Gegenwart von Schwefelwasserstoff und Cyanwasserstoff. Biochemische Zeitschrift 165 (1925), 238–244.
484. Neuberg, Carl/Gorr, Günther: Überführung des Methylglyoxals in Brenztraubensäure. Biochemische Zeitschrift 166 (1925), 442–443.
485. Neuberg, Carl/Gorr, Günther: Über die gekreuzte Dismutation zwischen Aldehyd und Keton. Biochemische Zeitschrift 166 (1925), 444–449.
486. Neuberg, Carl/Carl Oppenheimer: Zur Nomenklatur der Gärungsfermente und Oxydasen. Biochemische Zeitschrift 166 (1925), 450–453.
487. Neuberg, Carl/Windisch, Fritz: Über die Essiggärung und die chemischen Leistungen der Essigbakterien. Biochemische Zeitschrift 166 (1925), 454–481.
488. Neuberg, Carl/Gorr, Günther: Über den Mechanismus der Milchsäurebildung bei Bakterien II. Biochemische Zeitschrift 166 (1925), 482–487.
489. Neuberg, Carl/Kobel, M[aria]: Vergleichende Versuche über die zellfreie Vergärung von Hexose-di-phosphorsäure, Glucose, Fructose, Saccharose, sowie Invertzucker. Biochemische Zeitschrift 166 (1925), 488–496.
490. Neuberg, Carl/Gorr, G[ünther]: Die Bilanz der Oxalessigsäuregärung. Ergebnisse der Physiologie 24 (1925), 191–195.
491. Neuberg, Carl: Über das Wesen einfacher Oxydationsvorgänge (Essigärung). Klinische Wochenschrift 4 (1925), 598.
492. Neuberg, Carl/Windisch, Fritz: Vom Wesen der Essiggärung und von verwandten Erscheinungen. Die Naturwissenschaften 13 (1925), 993–996.
493. Neuberg, Carl/Kobel, Maria: Über die Milchsäure in ihrer Bedeutung für die Chemie und Physiologie. Zeitschrift für angewandte Chemie 38 (1925), 761–766.

1926

494. Neuberg, Carl: Gärung; in: Handbuch der praktischen und wissen.schaftlichen Pharmazie, hrsg. von Hermann Thoms. Bd. 4. Berlin, Wien: Urban & Schwarzenberg 1926, 73–80.
495. Neuberg, Carl/Behrens, Martin: Über die enzymatische Abspaltung von Rohrzucker aus Salzen der Saccharose-phosphorsäure. Biochemische Zeitschrift 170 (1926), 254–264.
496. Neuberg, Carl/Simon, Ernst: Über das Verhalten des p-Xylochinons zu Hefe. Biochemische Zeitschrift 171 (1926), 256–260.
497. Neuberg, Carl/Gorr, Günther: Über den Mechanismus der Milchsäurebildung bei Phanerogamen. Biochemische Zeitschrift 171 (1926), 475–484.
498. Neuberg, Carl/Wagner, Joachim: Zur Kenntnis der Phosphatase und über die Darstellung von sauren Estern der Pyrophosphorsäure. Biochemische Zeitschrift 171 (1926), 485–500.

1090 See also Lusman-Per(e)lman, Genia.

499. Neuberg, Carl/Gorr, Günther: Untersuchungen über den Mechanismus der Milchsäurebildung bei Phanerogamen II. Biochemische Zeitschrift 173 (1926), 358–362.

500. Neuberg, Carl/Gorr, Günther: Über die saccharogene Bildung von Milchsäure durch verschiedene Bakterien, die Methylglyoxal dismutieren, und über eine einfache Art der Isolierung von Lactat. Biochemische Zeitschrift 173 (1926), 476–481.

501. Neuberg, Carl/Simon, Ernst: Über die Dismutation des Methyl-äthyl-acetaldehyds. Biochemische Zeitschrift 174 (1926), 452–456.

502. Neuberg, Carl/Wagner, Joachim: Über die Verschiedenheit der Sulfatase und Myrosinase VIII. Mitteilung über Sulfatase. Biochemische Zeitschrift 174 (1926), 457–463.

503. Neuberg, Carl/Kobel, Maria: Weiteres über die Reaktion einzelner Zuckerarten mit verschiedenen Aminosäuren nebst Bemerkungen über die Vergärung von Aminosäuren-Zucker-Gemischen. Biochemische Zeitschrift 174 (1926), 464–479.

504. Neuberg, Carl/Kobel, Maria: Neue vergleichende Versuche über die Vergärung von Hexose-di-phosphorsäure, Glucose, Fructose, Saccharose und Invertzucker durch Hefensäfte sowie frische Hefen unter verschiedenen Bedingungen. Biochemische Zeitschrift 174 (1926), 480–492.

505. Neuberg, Carl/Kobel, Maria: Über die Wirkung wechselnder Mengen von Arsenat auf die Phosphorylierung. Biochemische Zeitschrift 174 (1926), 493–496.

506. Neuberg, Carl: L. v. Liebermann †. Biochemische Zeitschrift 179 (1926), 1–2.

507. Neuberg, Carl/Simon, Ernst: Beiträge zur Biochemie des Asymmetrieproblems. Biochemische Zeitschrift 179 (1926), 443–450.

508. Neuberg, Carl/Kobel, Maria: Fortgesetzte vergleichende Versuche über die Vergärbarkeit freier und phosphorylierter Hexosen und über eine polarimetrisch feststellbare Bindung dieser Substanzen an Inhaltsstoffen der Hefenzelle. Biochemische Zeitschrift 179 (1926), 451–458.

509. Neuberg, Carl/Kobel, Maria: Über die Vorgänge im frischen und getrockneten Tabakblatt vor und während der Fermentation. Biochemische Zeitschrift 179 (1926), 459–490.

510. Neuberg, Carl: Betrachtungen zum Fermentproblem. Hoppe-Seylers Zeitschrift für physiologische Chemie 157 (1926), 299–315.

511. Neuberg, Carl/Klopstock, Felix: Über ein neues Antigen für die Serodiagnostik der Tuberkulose. Klinische Wochenschrift 5 (1926), 1078–1079.

512. Neuberg, Carl/Gorr, Günther: Über die Bildung von Milchsäure durch die Zellen grüner Pflanzen. Die Naturwissenschaften 14 (1926), 437–439.

513. Neuberg, Carl/Molinari, Emil: Über den Verlauf der Essiggärung. Die Naturwissenschaften 14 (1926), 758–759.

514. Neuberg, Carl/Kobel, Maria: Zur Biochemie des Tabaks. Die Naturwissenschaften 14 (1926), 1182–1184.

515. Neuberg, Carl: Über die Rohrzuckerphosphorsäure und ihren chemischen sowie biochemischen Abbau. Zeitschrift des Vereins der deutschen Zuckerindustrie 76 (1926), 463–465.

516. Neuberg, Carl: Gärung. Zentralblatt für Bakteriologie, Parasitenkunde und Infektionskrankheiten 97 (1926), I. Abt. (Originale), Beiheft, 2–14.

1927

517. Neuberg, Carl: Albrecht Kossel †. Berichte der Deutschen Chemischen Gesellschaft 60 (1927), 159–160.

518. Neuberg, Carl/Simon, Ernst: Nachweis von Zucker neben Proteinen und vermeintlichen Kondensationsprodikten von Kohlenhydraten mit Eiweiß. Berichte der Deutschen Chemischen Gesellschaft 60 (1927), 817–824.

519. Neuberg, Carl: Über die Vergärungsgeschwindigkeit der Brenztraubensäure und damit zusammenhängende Fragen. Biochemische Zeitschrift 180 (1927), 471–490.

520. Neuberg, Carl/Kobel, Maria: Verhalten der Hexosen sowie Hexose-Mono-Phosphorsäureester zu Serum und Aminosäuren. Biochemische Zeitschrift 182 (1927), 273–284.

521. Neuberg, Carl/Komarewsky, Wassily: Über die Umwandlung des Methyl-Benzoylcarbinols durch gärende Hefe. Biochemische Zeitschrift 182 (1927), 285–290.

522. Neuberg, Carl/Kobel, Maria: Über die Bildung von reiner d-Milchsäure durch frische Hefen und Trockenhefe sowie von d, l-Milchsäure durch Hefenmacerationssaft. Biochemische Zeitschrift 182 (1927), 470–487.

523. Neuberg, Carl/Kitasato, Torao: Synthese der γ-Keto-d-gluconsäure. Biochemische Zeitschrift 183 (1927), 485–488.

524. Neuberg, Carl/Leibowitz, J[esaia]: Untersuchungen über das durch Gärung gewonnene Hexose-Monophosphat. Biochemische Zeitschrift 184 (1927), 489–507.

525. Neuberg, Carl/Kobel, Maria: Über den Abbau von Aminosäuren und Aminopurinen durch Methylglyoxal und verwandte Substanzen. Biochemische Zeitschrift 185 (1927), 477–479.

526. Neuberg, Carl/Simon, Ernst: Über die quantitative Dismutation des Methylglyoxals zu Milchsäure durch den Bacillus Delbrücki sowie durch das Bacterium aerogenes nebst Versuchen über das abweichende stereochemische Verhalten des Phenylglyoxals bei dieser Reaktion. Biochemische Zeitschrift 186 (1927), 331–336.

527. Neuberg, Carl/Simon, Ernst: Vom Wesen der Brenztraubensäurevergärung. Biochemische Zeitschrift 187 (1927), 220–253.

528. Neuberg, Carl/Leibowitz, J[esaia]: Über die enzymatische Umwandlung von Hexose-di-phosphat in Hexose-Monophosphorsäure und die enzymatische Synthese von Hexose-di-Phosphat aus Hexose-Monophosphat. Biochemische Zeitschrift 187 (1927), 481–490.

529. Neuberg, Carl/Kobel, Maria: Quantitative Untersuchungen über den Abbau von Aminosäuren und Aminopurinen durch Methylglyoxal und verwandte Substanzen. Biochemische Zeitschrift 188 (1927), 197–210.
530. Neuberg, Carl/Kobel, Maria: Weiteres über Abfangverfahren. Biochemische Zeitschrift 188 (1927), 211–216.
531. Neuberg, Carl/Ottenstein, Berta: Übertritt von Methylalkohol in den Tabakrauch. Biochemische Zeitschrift 188 (1927), 217–226.
532. Neuberg, Carl/Wagner, Joachim/Jacobsohn, Kurt P.: Über die asymmetrische Wirkungsweise der Phosphatase und eine Methode zur biochemischen Darstellung der beiden entgegengesetzt drehenden Alkohole aus ihren Racematen. Biochemische Zeitschrift 188 (1927), 227–240.
533. Neuberg, Carl/Simon, Ernst: Die Verwirklichung der phytochemischen Reduktionen durch Bakterien. Biochemische Zeitschrift 190 (1927), 226–231.
534. Neuberg, Carl/Kobel, Maria: Über die enzymatische Abspaltung von Methylalkohol aus Pektin durch ein Ferment des Tabaks. Biochemische Zeitschrift 190 (1927), 232–240.
535. Neuberg, Carl/Leibowitz, J[esaia]: Über die partielle De-Phosphorylierung der Hexose-di-Phosphorsäure durch Hefe. Biochemische Zeitschrift 191 (1927), 450–455.
536. Neuberg, Carl/Leibowitz, J[esaia]: Abbau von Zymo-di-phosphat mittels tierischer Phosphatase zu Hexose-Mono-phosphorsäure-ester. Biochemische Zeitschrift 191 (1927), 456–459.
537. Neuberg, Carl/Leibowitz, J[esaia]: Arsenat-Aktivierung und die Spezifität von Phosphatase. Biochemische Zeitschrift 191 (1927), 460–465.
538. Neuberg, Carl/Schou, Svend Aage: Die Struktur der Hexose-Phosphorsäureester und des Methylglyoxyls nach ihrem spektrographischen Verhalten. Biochemische Zeitschrift 191 (1927), 466–471.
539. Neuberg, Carl/Kobel, Maria: Vermeintliche Abfangung von Methylglyoxal bei der alkoholischen Gärung. Biochemische Zeitschrift 191 (1927), 472–477.
540. Neuberg, Carl/Kobel, Maria: Über die Bildung von optisch aktiver Milchsäure auf biochemischen Wege und über eine einfache Methode zur Bereitung des Zinksalzes von d-Milchsäure. [Komppa-Festschrift: Suomalaisen Tiedeakatemian Toimituksia, Helsinki], Annales Academiae Scientiarum Fennicae Ser. A 29 (1927), 9.
541. Neuberg, Carl/Simon, E[rnst]: Zur Brenztraubensäurefrage. Hoppe-Seylers Zeitschrift für physiologische Chemie 171 (1927), 1–3.
542. Neuberg, Carl/Wagner, Joachim: Über die Zerlegung des myronsauren Kaliums durch animalische Sulfatase IX. Mitteilung über Sulfatase. Zeitschrift für die gesamte experimentelle Medizin einschließlich experimentelle Chirurgie 56 (1927), 334–343.

1928

543. Neuberg, Carl/Kobel, Maria: Milchsäurebildung in Blättern grüner Pflanzen; in: Festschrift anlässlich des siebzigsten Geburtstages von Julius Stoklasa, hrsg. von Ernst Gustav Doerell et al. Berlin: Parey 1928, 269–278.

544. Neuberg, Carl/Leibowitz, J[esaia]: Chemische Charakteristik der einfachen Kohlehydrate (Zuckerarten); in: Grafes Handbuch der organischen Warenkunde mit Einschluss der mechanischen Technologie und technischen Warenprüfung, hrsg. von Victor Grafe. Bd. 2,1: Kohlehydrate und die Waren der Kohlehydratgruppe. Stuttgart: Poeschel 1928, 1–60.

545. Neuberg, Carl/E[rnst] Simon: Von den chemischen Wirkungen des Lichtes; in: Handbuch der gesamtem Strahlenheilkunde, Biologie, Pathologie und Therapie. Bd. 1. München: Bergmann 1928, 393–403.

546. Neuberg, Carl: Carboxylase; in: Die Fermente und ihre Wirkungen, hrsg. von Carl Oppenheimer und Ludwig Pincussen. Bd. 3: Die Methodik der Fermente. Leipzig: Thieme 1928, 1317–1221.

547. Neuberg, Carl/Kobel, Maria: Zuckerphosphate und Phytin; in: Die Fermente und ihre Wirkungen, hrsg. von Carl Oppenheimer und Ludwig Pincussen. Bd. 3: Die Methodik der Fermente. Leipzig: Thieme 1928, 407–422.

548. Neuberg, Carl/Kobel, Maria: Abbau der Hexosen; in: Die Fermente und ihre Wirkungen, hrsg. von Carl Oppenheimer und Ludwig Pincussen. Bd. 3: Die Methodik der Fermente. Leipzig: Thieme 1928, 1160–1182.

549. Neuberg, Carl/Kobel, Maria: Pflanzliche Phosphatbindung; in: Die Fermente und ihre Wirkungen, hrsg. von Carl Oppenheimer und Ludwig Pincussen. Bd. 3: Die Methodik der Fermente. Leipzig: Thieme 1928, 1183–1184.

550. Neuberg, Carl/Kobel, Maria: Carboligase; in: Die Fermente und ihre Wirkungen, hrsg. von Carl Oppenheimer und Ludwig Pincussen. Bd. 3: Die Methodik der Fermente. Leipzig: Thieme 1928, 1329–1337.

551. Neuberg, Carl/Simon, Ernst: Einfache Phosphorsäure-ester der aliphatischen, aromatischen und hydroaromatischen Reihe; in: Die Fermente und ihre Wirkungen, hrsg. von Carl Oppenheimer und Ludwig Pincussen. Bd. 3: Die Methodik der Fermente. Leipzig: Thieme 1928, 423–427.

552. Neuberg, Carl/Simon, Ernst: Phosphatasen; in: Die Fermente und ihre Wirkungen, hrsg. von Carl Oppenheimer und Ludwig Pincussen. Bd. 3: Die Methodik der Fermente. Leipzig: Thieme 1928, 743–755.

553. Neuberg, Carl/Simon, Ernst: Ketonaldehydmutase; in: Die Fermente und ihre Wirkungen, hrsg. von Carl Oppenheimer und Ludwig Pincussen. Bd. 3: Die Methodik der Fermente. Leipzig: Thieme 1928, 1311–1316.

554. Neuberg, Carl/Wagner, Joachim: Sulfatase; in: Die Fermente und ihre Wirkungen, hrsg. von Carl Oppenheimer und Ludwig Pincussen. Bd. 3: Die Methodik der Fermente. Leipzig: Thieme 1928, 760–764.

555. Neuberg, Carl/Wagner, Joachim: Alkoholdehydrase; in: Die Fermente und ihre Wirkungen, hrsg. von Carl Oppenheimer und Ludwig Pincussen. Bd. 3: Die Methodik der Fermente. Leipzig: Thieme 1928, 1338–1344.
556. Neuberg, Carl/Kobel, Maria: Phosphorylierung und alkoholische Zuckerspaltung. Justus Liebigs Annalen der Chemie 465 (1928), 272–282.
557. Neuberg, Carl: Sur les éthers hexose-phosphoriques, leur formation, leurs transformations, leur signification. Conférence faite au Congrès International de la Vigne et du Pin Maritime, Bordeaux, 1928, Annales de la Brasserie et de la Distillerie 27 (1928), 1–9.
558. Neuberg, Carl/Leibowitz, J[esaia]: Biochemische Darstellung eines Disaccharid-mono-phosphorsäure-esters. Biochemische Zeitschrift 193 (1928), 237–244.
559. Neuberg, Carl/Kobel, Maria: Zur Frage des Nachweises von Methylglyoxal als Zwischenprodukt der Glykolyse. Biochemische Zeitschrift 193 (1928), 464–467.
560. Neuberg, Carl/Simon, Ernst: Alkoholische Zuckerspaltung durch Essigbakterien. Biochemische Zeitschrift 197 (1928), 259–260.
561. Neuberg, Carl/Ottenstein, Berta: Bildung von Methylalkohol bei der Autolyse frischer Tabakblätter. Biochemische Zeitschrift 197 (1928), 491–501.
562. Neuberg, Carl/Kobel, Maria: Ergänzung zur vorstehenden Veröffentlichung und zu unserer früheren Mitteilung über vermeintliche Abfangung von Methylglyoxal bei der alkoholischen Gärung. Biochemische Zeitschrift 199 (1928), 230–231.
563. Neuberg, Carl/Simon, Ernst: Über Verschiedenheit der Vorgänge bei der alkoholischen Zuckerspaltung und der Acetaldehyd-dismutation. Biochemische Zeitschrift 199 (1928), 232–247.
564. Neuberg, Carl/Weinmann, Fritz/Vogt, Marthe: Synthese von Glycerinsäure-mono-phosphorsäure. Biochemische Zeitschrift 199 (1928), 248–252.
565. Neuberg, Carl/Jacobsohn, Kurt P.: Fortgesetzte Untersuchungen über den Wirkungsbereich der Phosphatase. Biochemische Zeitschrift 199 (1928), 498–517.
566. Neuberg, Carl/Simon, Ernst: Weitere Untersuchungen über die stereochemische Spezifität der Ketonaldehydmutase. Biochemische Zeitschrift 200 (1928), 468–472.
567. Neuberg, Carl/Kobel, Maria: Über die Erscheinungen beim Zusammentreffen von Dipeptid bzw. Pepton mit Zucker, Hexose-di-phosphat sowie Methylglyoxal. Biochemische Zeitschrift 200 (1928), 459–467.
568. Neuberg, Carl/Fritz Weinmann: Strukturchemische Spezifität der Carboxylase. Biochemische Zeitschrift 200 (1928), 473–476.
569. Neuberg, Carl/Kobel, Maria: Über die biochemische Umwandlung von Di-oxyaceton in Hexosen auf dem Gärungswege und über die Vergärungsgeschwindigkeit des Di-oxyacetons im Zusammenhang mit der

Verbrennungswärme dieser Triose I. Biochemische Zeitschrift 203 (1928), 452–462.

570. Neuberg, Carl/Kobel, Maria: Die desmolytische Bildung von Methylglyoxal durch Hefeenzym. Biochemische Zeitschrift 203 (1928), 463–468.

571. Klopstock, Felix/Neuberg, Carl: Weitere Untersuchungen zur Serodiagnostik der Tuberkulose. Klinische Wochenschrift 7 (1928), 537–539.

572. Neuberg, Carl/Klopstock, Felix: Serodiagnostik der Tuberkulose. Die Medizinische Welt 1928, Nr. 15, 1–5.

573. Neuberg, Carl/Kobel, Maria: Chemische und enzymatische Prozesse im Lichte der stereochemischen Forschung. Die Naturwissenschaften 16 (1928), 392–396.

574. Neuberg, Carl/Kobel, Maria: Über die Frage nach der Identität von Mutase und Ketonaldehydmutase. Kinetische Untersuchungen. Zeitschrift für physikalische Chemie. Abt. A 139 (1928), 631–646.

1929

575. Neuberg, Carl: Hefe in Milchsäuregärung. Biochemische Zeitschrift 264 (1933), 452–455.

576. Neuberg, Carl/Lüdtke, Max: Die Umwandlungen der Kohlenhydrate durch Gärungsvorgänge; in: Handbuch der Ernährung und des Stoffwechsels der landwirtschaftlichen Nutztiere als Grundlage der Fütterungslehre, hrsg. von Ernst Mangold. Berlin: Springer 1929, 73–82.

577. Neuberg, Carl: John A. Mandel †. Berichte der Deutschen Chemischen Gesellschaft 62 A (1929), 80–81.

578. Neuberg, Carl/Kobel, Maria: Isolierung von Methylalkohol aus Tabakrauch. Biochemische Zeitschrift 206 (1929), 240–244.

579. Kitasato, T[orao]/Neuberg, Carl: Gewinnung von Osonlösungen zwecks Synthese von Keturonsäuren. Biochemische Zeitschrift 207 (1929), 230–231.

580. Neuberg, Carl/Kobel, Maria: Die Isolierung von Methylglyoxal bei der Milchsäuregärung. Biochemische Zeitschrift 207 (1929), 232–262.

581. Neuberg, Carl: W. L. Omeliansky †. Biochemische Zeitschrift 209 (1929), 247.

582. Neuberg, Carl: Emil Fromm †. Biochemische Zeitschrift 209 (1929), 248.

583. Neuberg, Carl: Theodor Paul †. Biochemische Zeitschrift 209 (1929), 249–250.

584. Neuberg, Carl/Kobel, Maria: Weiteres über die Vorgänge bei der desmolytischen Bildung von Methylglyoxal durch Hefe. Biochemische Zeitschrift 210 (1929), 466–488.

585. Neuberg, Carl: Über das Verhalten des glucose-schwefligsauren Natriums zu Hefe und damit zusammenhängende Fragen. Biochemische Zeitschrift 212 (1929), 477–489.

586. Neuberg, Carl/Collatz, Herbert: Wirkung des Sulfits bei der zweiten Vergärungsform. Biochemische Zeitschrift 216 (1929), 233–237.
587. Neuberg, Carl: Wilhelm Wiechowski †. Biochemische Zeitschrift 216 (1929), 241.
588. Neuberg, Carl: John A. Mandel †. Biochemische Zeitschrift 216 (1929), 242.
589. Neuberg, Carl/Kobel, Maria: Die Bildung von Brenztraubensäure als Durchgangsglied bei der alkoholischen Zuckerspaltung. Ihre Isolierung als Hauptprodukt der Gärung. Biochemische Zeitschrift 216 (1929), 493–496.
590. Neuberg, Carl/Jacobsohn, K[urt] P./Wagner, J[oachim]: Bildung und Spaltung von Glukosiden als Methode zur chemischen und biochemischen Trennung razemischer Alkohole in ihre optisch aktiven Formen, Fermentforschung (Berlin/Wien), (1929), 491–531.
591. Neuberg, Carl/Scheuer, Max: Nachweis und Isolierung des biochemisch gebildeten Methylglyoxals als Dioxim. Sitzungsberichte der Akademie der Wissenschaften in Wien. Mathem.-naturwiss. Klasse, Abt. II B Chemie 138 (1929), 1031–1035.
592. Neuberg, Carl/Scheuer, Max: Nachweis und Isolierung des biochemisch gebildeten Methylglyoxals als Dioxim. Monatshefte für Chemie 53/54 (1929), 1031–1035. (= Festschrift Rudolf Wegscheider zum siebzigsten Geburtstag dargebracht).[1091]

1930

593. Max Lüdtke/Neuberg, Carl: Alloiomorphe Zucker (am-Zucker); in: Handbuch der Biochemie des Menschen und der Tiere, hrsg. von Carl Oppenheimer. 2. Aufl. Ergänzungsband. Jena: G. Fischer 1930, 117–132.
594. Neuberg, Carl: Adolf Harnack. Berichte der Deutschen Chemischen Gesellschaft 63 A (1930), 1431.
595. Neuberg, Carl/Kobel, Maria: Vorführung der Methylglyoxal- und Brenztraubensäure-Stufe bei der alkoholischen Zuckerspaltung als Demonstrationsversuch. Berichte der Deutschen Chemischen Gesellschaft 63 (1930), 1986–1989.
596. Neuberg, Carl/Scheuer, Max: Neue Art des Nachweises von Methylglyoxal bei der biochemischen Spaltung des Zuckers. Berichte der Deutschen Chemischen Gesellschaft 63 (1930), 3068–3072.
597. Neuberg, Carl/Kobel, Maria: Über den Verlauf der Brenztraubensäurebildung bei der Hefengärung. Biochemische Zeitschrift 219 (1930), 490–494.
598. Neuberg, Carl/Collatz, Herbert: Über die individuelle Existenz und Beständigkeit von Dioxyaceton und Glycerinaldehyd in wässeriger Lösung. Biochemische Zeitschrift 223 (1930), 494–501.

1091 Identical to “Supplement der Sitzungsberichte“, see No. 591.

599. Neuberg, Carl/Hofmann, Eduard: Über einfache Darstellung von Methylglyoxal-lösungen. Biochemische Zeitschrift 224 (1930), 491–497.
600. Neuberg, Carl: Photochemische und phytochemische Acyloinbildung (carboligatische Wirkung). Biochemische Zeitschrift 225 (1930), 238–241.
601. Neuberg, Carl/Collatz, Herbert: Über die Dismutation der Methylglyoxalyl-essigsäure zu d-γ-Oxyglutarsäure. Biochemische Zeitschrift 225 (1930), 242–246.
602. Hofmann, Eduard/Neuberg, Carl: Weiteres über einfache Gewinnung von Methylglyoxallösungen. Biochemische Zeitschrift 226 (1930), 489–491.
603. Neuberg, Carl: Versuche zur biochemischen Darstellung von optisch aktivem Glycerinaldehyd. Biochemische Zeitschrift 228 (1930), 259–261.
604. Neuberg, Carl/Kobel, Maria: Weiteres über die Bildung von Methylglyoxal sowie Brenztraubensäure durch Hefen unter der Wirkung verschiedener plasmolytischer Stoffe. Biochemische Zeitschrift 229 (1930), 255–262.
605. Neuberg, Carl/Kobel, Maria: Über den Kohlenhydratstoffwechsel der höheren Pflanzen. Die Bildung von Methylglyoxal durch Enzyme der gekeimten Samen. Biochemische Zeitschrift 229 (1930), 433–442.
606. Neuberg, Carl/Hofmann, Eduard: Über eine einfache Darstellung von Phenylglyoxal. Biochemische Zeitschrift 229 (1930), 443–445.
607. Neuberg, Carl/Kobel, Maria: Die Zerlegung von nicht phosphoryliertem Zucker durch Hefe unter Bildung von Glycerin und Brenztraubensäure. Biochemische Zeitschrift 229 (1930), 446–454.
608. Neuberg, Carl/Kobel, Maria: Studien an Tabakenzymen. Biochemische Zeitschrift 229 (1930), 455–463.
609. Neuberg, Carl/Clara Ostendorf: Über ein Modell für die Pectase. Vorläufiger Bericht. Biochemische Zeitschrift 229 (1930), 464–466.
610. Neuberg, Carl/Fritz Weinmann: Zur Theorie der Hydrotropie. Biochemische Zeitschrift 229 (1930), 467–479.
611. Neuberg, Carl/Kobel, Maria: Die vierte und fünfte Vergärungsform des Zuckers. Die Naturwissenschaften 18 (1930), 427–429.
612. Neuberg, Carl: Die Wege des Zuckerzerfalls bei der alkoholischen Gärung. Jahrbuch der Versuchs- und Lehranstalt für Brauerei 21 (1930), 208–211.
613. Neuberg, Carl/Kobel, Maria/Hans Laser: Über den Mechanismus der Zuckerspaltung im Tumor und embryonalen Gewebe. Zeitschrift für Krebsforschung 32 (1930), 92–98.

1931

614. Neuberg, Carl: Biochemie. In: Wissenschaft und Landwirtschaft. Festschrift zum 50jährigen Bestehen der Landwirtschaftlichen Hochschule Berlin. Berlin 1931, 155–160.
615. Neuberg, Carl: Friedrich Wilhelm Semmler †. Berichte der Deutschen Chemischen Gesellschaft 64 (1931), 108–109.
616. Simon, Ernst/Neuberg, Carl: Quantitative Bestimmung von Methylglyoxal, Brenztraubensäure nebst Acetaldehyd für sich und im Gemisch sowie gleichzeitige Bestimmung von Milchsäure in Gegenwart der genannten Stoffe und in einer Materialprobe. Biochemische Zeitschrift 232 (1931), 479–484.
617. Neuberg, Carl/Hofmann, Eduard/Kobel, Maria: Über die Verbrennungswärme des d,l-Glycerinaldehyds. Biochemische Zeitschrift 234 (1931), 341–344.
618. Neuberg, Carl/Hofmann, E[duard]: Enzymatischer Abbau von Chondroitinschwefelsäure. Biochemische Zeitschrift 234 (1931), 345–346.
619. Neuberg, Carl/Kobel, Maria: Kleinere Mitteilungen verschiedenen Inhalts: I. Fehlerquellen bei Zuckerbestimmungen, die auf der Reduktion von Cuprisalzlösungen beruhen. II. Über fehlerhafte Anwendung der Hagedorn-Jensen-Methode zur Zuckerbestimmung. III. Fehlerquellen bei Verwendung von Jodessigsäure als gärungs hemmende Agens. IV. Über die Beständigkeit der Glykolase. V. Über die Wirksamkeit einer 19 Jahre lang aufbewahrten Invertaselösung. Biochemische Zeitschrift 238 (1931), 226–252.
620. Neuberg, Carl/Hofmann, Eduard: Schnellverfahren zur billigen Darstellung von Phenylglyoxal. Biochemische Zeitschrift 239 (1931), 495–497.
621. Neuberg, Carl/Hans von Euler: Zur Nomenklatur der an der alkoholischen Gärung beteiligten Katalysatoren. Biochemische Zeitschrift 240 (1931), 245.
622. Falkenhausen, F[riedrich] Freiherr von/Neuberg, Carl: Synthetische Herstellung von reinem optisch aktiven Amylalkohol (Methyl-äthyl-carbincarbinol). Biochemische Zeitschrift 242 (1931), 482–486.
623. Neuberg, Carl/Kobel, Maria: Über die Ausführung der Reaktion von Alduronsäuren mit Naphthoresorcin. Biochemische Zeitschrift 243 (1931), 435–450.
624. Neuberg, Carl/Kobel, Maria: Über den Mechanismus des Abbaus von Zucker durch das Thermobakterium mobile Lindneri. Biochemische Zeitschrift 243 (1931), 451–460.
625. Neuberg, Carl/Scheuer, Max: Pektin des Tabaks. Biochemische Zeitschrift 243 (1931), 461–471.
626. Neuberg, Carl/Joseph Burkard: Über neue stickstofffreie Bestandteile des Tabakrauches. Biochemische Zeitschrift 243 (1931), 472–484.
627. Neuberg, Carl/Hofmann, Eduard: Über Chondro-sulfatase, ein neues Enzym. Die Naturwissenschaften 19 (1931), 484–485.

628. Neuberg, Carl: Les processus de dégradation des sucres par fermentation. (Conférence faite devant la Société de Chimie Biologique et la Société Chimique de France, Paris, Juin 1931). Bulletin de la Société de Chimie Biologique 13 (1931), 1294–1309.

1932

629. Falkenhausen, Friedrich Freiherr von/Neuberg, Carl: Alkohole; in: Handbuch der Pflanzenanalyse, hrsg. von Gustav Klein. Bd. 2: Spezielle Analyse. Teil 1: Anorganische Stoffe: Organische Stoffe 1 . Wien: Springer 1932, 205–245.

630. Simon, Ernst/Neuberg, Carl: Aldehyde und Ketone; in: Handbuch der Pflanzenanalyse, hrsg. von Gustav Klein. Bd. 2: Spezielle Analyse, Teil 1: Anorganische Stoffe; Organische Stoffe 1. Wien: Springer 1932, 255–286.

631. Kobel, Maria/Neuberg, Carl: Phosphorsäure-ester des Kohlehydratstoffwechsels; in: Handbuch der Pflanzenanalyse, hrsg. von Gustav Klein. Bd. 2: Spezielle Analyse, Teil 1: Anorganische Stoffe; Organische Stoffe 1. Wien: Springer 1932, 546–577.

632. Neuberg, Carl: La dégradation et la transformation des hydrates de carbone par des cellules différentes et leurs ferments. Leur chimie et leur énergie calorique. (Conférence faite devant le IVme Congrès de la Section Italienne de la Société Internationale de Microbiologie, Milan, Octobre 1932.) 1932, 3–15.

633. Neuberg, Carl: Über die biochemischen Zusammenhänge der wichtigsten Leistungen von tierischen und pflanzlichen Zellen. Annalen der Tomarkin-Foundation. Organ für experimentell-therapeutische Forschung und angewandte Medizin 2 (1932), 130–133.

634. Neuberg, Carl/Kobel, Maria: Untersuchungen über die Glycolase und ihre Haltbarkeit. Australian Journal of Experimental Biology and Medical Science 9 (1932), 127–133.

635. Neuberg, Carl: Hermann Thoms †. Biochemische Zeitschrift 245 (1932), 1.

636. Neuberg, Carl/Kobel, Maria: Weiteres über biochemische Leistungen des Thermobakterium mobile Lindneri. Biochemische Zeitschrift 247 (1932), 246–248.

637. Neuberg, Carl/Scheuer, Max: Verbindungen der Fructose-di-phosphorsäure. Biochemische Zeitschrift 249 (1932), 478–482.

638. Neuberg, Carl: Hans Murschhauser †. Biochemische Zeitschrift 251 (1932), 1.

639. Neuberg, Carl/Kobel, Maria: Über die Oxydation des Methylglyoxals zu Brenztraubensäure mit molekularem Sauerstoff. Biochemische Zeitschrift 252 (1932), 215–230.

640. Neuberg, Carl/Hofmann, Eduard: Über enzymatische Spaltung der Malto- und Lacto-bionsäure. Biochemische Zeitschrift 252 (1932), 434–439.

641. Neuberg, Carl/Hofmann, Eduard: Über die Verbrennungswärme des Methylglyoxals und damit zusammenhängende Fragen. Biochemische Zeitschrift 252 (1932), 440–450.
642. Neuberg, Carl/Burkard, Josef: Beobachtungen über die momentane Reaktion des Methylglyoxals mit stickstofffreien Verbindungen. Biochemische Zeitschrift 253 (1932), 222–224.
643. Neuberg, Carl/Simon, Ernst: Über isoliertes Vorkommen von Carboxylase und über enzymatische Wirkungen des Essigbakterium Bordeaux. Biochemische Zeitschrift 253 (1932), 225–230.
644. Neuberg, Carl: [S. N.] Salaskin †. Biochemische Zeitschrift 256 (1932), 1.
645. Neuberg, Carl/Hofmann, Eduard: Neue Beobachtungen über ß-Glucosidase. Biochemische Zeitschrift 256 (1932), 450–461.
646. Neuberg, Carl/Kobel, Maria: Oxydation des Glyoxalbisulfits durch molekularen Sauerstoff zu Glyoxylsäure. Anhang: Oxydation des Glyoxals durch Brom zu Glyoxylsäure und Oxalsäure. Biochemische Zeitschrift 256 (1932), 475–484.
647. Neuberg, Carl/Simon, Ernst: Dismutation des Glyoxals. Biochemische Zeitschrift 256 (1932), 485–491.
648. Neuberg, Carl/Simon, Ernst: Sulfatasen und ihre Substrate. Ergebnisse der Physiologie 34 (1932), 896–906.
649. Neuberg, Carl/Hofmann, Eduard: Die Wärmequelle der anaerob vollzogenen Energie liefernden Reaktionen. Die Naturwissenschaften 20 (1932), 379–381.
650. Neuberg, Carl/Kobel, Maria: Über Autooxydation. Die Naturwissenschaften 20 (1932), 953–954.
651. Neuberg, Carl/Collatz, Herbert: Über die Einwirkung von Glukuronsäure und Galakturonsäure auf Alanin. Schweizerische Medizinische Wochenschrift 62 (1932), 1–4.

1933

652. Neuberg, Carl/Simon, Ernst: Holozymase und Teilfermente, Aldehydrasen, Alkoholdehydrasen; in: Handbuch der Biochemie des Menschen und der Tiere, hrsg. von Carl Oppenheimer. 2. Aufl. Ergänzungswerk Bd. 1, 1. Jena: G. Fischer 1933, 508–517.
653. Neuberg, Carl/Kobel, Maria: Abbau der Zucker. A) Allgemeines. B) Die Gärungen; in: Handbuch der Biochemie des Menschen und der Tiere, hrsg. von Carl Oppenheimer. 2. Aufl. Ergänzungswerk Bd. 1, 2. Jena: G. Fischer 1933, 905–913.
654. Neuberg, Carl/Simon, Ernst: Die Endoxydation; in: Handbuch der Biochemie des Menschen und der Tiere, hrsg. von Carl Oppenheimer. 2. Aufl. Ergänzungswerk Bd. 1, 2. Jena: G. Fischer 1933, 929–931.
655. Kobel, Maria/Neuberg, Carl: Untersuchung von Gärflüssigkeiten; in: Handbuch der Pflanzenanalyse, hrsg. von G. Klein. Bd. 4: Spezielle Analyse. Teil 3: Organische Stoffe 3. Wien: Springer 1932, 1253–1344.

656. Neuberg, Carl/Kobel, Maria: Über Darstellung und Bedeutung der Glycerin-mono-phosphorsäure. Angewandte Chemie 46 (1933), 711–713.
657. Neuberg, Carl/Kobel, Maria: Über Vergärung von Methylglyoxal. Biochemische Zeitschrift 258 (1933), 365–370.
658. Neuberg, Carl/Kobel, Maria: Überführung der synthetischen Glycerin-mono-phosphorsäure in Brenztraubensäure mittels Hefe und Milchsäurebakterien. Biochemische Zeitschrift 260 (1933), 241–246.
659. Neuberg, Carl/Kobel, Maria: Über das Verhalten der Glycerin-mono-phosphorsäure gegen Hefe II. Biochemische Zeitschrift 263 (1933), 219–227.
660. Auhagen, Ernst/Neuberg, Carl: Über Umschaltung der alkoholischen Zuckerspaltung durch Herstellung von Phosphoglycerinsäure. Biochemische Zeitschrift 264 (1933), 456–460.
661. Neuberg, Carl/Simon, Ernst: Chemische Vorgänge und energetische Verhältnisse bei physiologischen Ab- und Umbau der Kohlenhydrate und ihrer Spaltungsprodukte. Ergebnisse der Enzymforschung 2 (1933), 118–138.
662. Neuberg, Carl/Schoenebeck, Otto von: Über die Teilfermente der Myrosinase. Biochemische Zeitschrift 265 (1933), 223–236.
663. Neuberg, Carl/Schoenebeck, Otto von: Über Aufteilung der Myrosinase. Die Naturwissenschaften 21 (1933), 404–405.

1934

664. Neuberg, Carl/Kobel, Maria: Gärung. Alkohol; in: Handwörterbuch der Naturwissenschaften, hrsg von Rudolf Dittler et al. 2. Aufl., Bd. 4. Jena: Fischer 1934, 603–614.
665. Kobel, Maria/Neuberg, Carl: Zur Frage des Auftretens von Triose beim desmolytischen Hexosenabbau. Biochemische Zeitschrift 269 (1934), 441–446.
666. Burkard, J[osef]/Neuberg, Carl: Zur Frage nach der Entstehung des Rohrzuckers. Biochemische Zeitschrift 270 (1934), 229–234.
667. Neuberg, Carl/Schuchardt, W[erner]/Vercellone, A[lberto]: Umwandlung von Glycerinsäure-di-phosphorsäure in Brenztraubensäure. Biochemische Zeitschrift 271 (1934), 229–232.
668. Cattaneo, Carlo/Neuberg, Carl: Umstellung der Coli-Gärung auf reine Milchsäure-Gärung. Biochemische Zeitschrift 272 (1934), 441–444.
669. Neuberg, Carl/Kobel, Maria: Weiteres über Auftreten von Triose beim desmolytischen Hexosenabbau durch Mikroben und höhere Pflanzen. Biochemische Zeitschrift 272 (1934), 445–456.
670. Neuberg, Carl/Kobel, Maria: Umwandlung von Phosphoglycerinsäure durch die Fermente gekeimter Erbsen und Bohnen. Biochemische Zeitschrift 272 (1934), 457–458.
671. Neuberg, Carl/Kobel, Maria: Rein chemische Überführung der Glycerinsäure und Glycerinsäure-mono-phosphorsäure in Brenztraubensäure. Biochemische Zeitschrift 272 (1934), 459–460.

672. Neuberg, Carl/Kobel, M[aria]: Kristallisierte und gelatinöse Salze der Glycerinsäure-mono-phosphorsäure. Biochemische Zeitschrift 272 (1934), 461–464.
673. Neuberg, Carl/Kobel, Maria: Present status of the problem of sugar fermentation. Journal of Bacteriology 28 (1934), 461–471.

1935

674. Neuberg, Carl: Sur la fermentation des hydrates de carbone. Academia das Cînecias de Lisboa. Biblioteca de Altos Estudos 40 (1935) [without page].
675. Neuberg, Carl/Cahill, W[illiam M.]: L'idrolisi enzimatica totale degli acidi condroitin-solforico e mucoitinsolforico nei loro componenti. (Atti Reale Accademia nazionale dei Lincei, Rendiconti) Classe di Scienze fisiche, matematiche e naturali (1935), 149–155.
676. Neuberg, Carl: Über die Anwendung der Doppelbindungsregel auf Fragen der Zucker-Chemie. Berichte der Deutschen Chemischen Gesellschaft 68 (1935), 505–506.
677. Neuberg, Carl/Cahill, W[illiam M.]: Über den enzymatischen Abbau von Mucoitinschwefelsäure. Biochemische Zeitschrift 275 (1935), 328–329.
678. Schoenebeck, Otto von: Stabilisierung von Carboxylaselösungen. Neuberg, Carl: Fortgesetzte Untersuchungen über die Carboxylase. Biochemische Zeitschrift 275 (1935), 330–338.
679. Neuberg, Carl/Kobel, Maria: Über Bildung von Diketonen aus Bestandteilen des Tabaks. Biochemische Zeitschrift 275 (1935), 339–343.
680. Neuberg, Carl/Vercellone, A[lberto]: Über die phytochemische Reduktion des d,l-Milchsäurealdehyds. Biochemische Zeitschrift 279 (1935), 140–142.
681. Neuberg, Carl/Hofmann, E[duard]: Notiz über einfache Gewinnung von kristallisiertem Dioxyaceton auf biochemischem Wege. Biochemische Zeitschrift 279 (1935), 318–320.
682. Neuberg, Carl/Ostendorf, Cl[ara]: Dismutation des p-Tolylglyoxals. Biochemische Zeitschrift 279 (1935), 459–462.
683. Vercellone, A[lberto]/Neuberg, Carl: Weitere Vereinfachung der Darstellung von d-3-Phosphoglycerinsäure. Biochemische Zeitschrift 280 (1935), 161–162.
684. Neuberg, Carl: Phosphorylierung mit lebender Hefe. Biochemische Zeitschrift 280 (1935), 163–166.
685. Neuberg, Carl/Hofmann, Eduard: Über Vergärung von Triosen durch Hefen. Biochemische Zeitschrift 280 (1935), 293–296.
686. Neuberg, Carl/Kobel, Maria: Über den spezifischen Einfluss des Zinks auf die Dismutation des Methylglyoxals. Biochemische Zeitschrift 280 (1935), 297–306.
687. Neuberg, Carl/Hofmann, Eduard: Über die Hydrolyse des Milchzuckers durch Fermente aus Leber und Niere. Biochemische Zeitschrift 281 (1935), 431–437.

688. Neuberg, Carl/Hofmann, Eduard: Über Triosen vergärende Hefen. Die Naturwissenschaften 23 (1935), 484–486.
689. Neuberg, Carl/Kobel, Maria: Über die Umwandlung des Rutins durch Tabakenzyme in braune Farbstoffe. Die Naturwissenschaften 23 (1935), 800–801.
690. Kobel, Maria/Neuberg, Carl: Über Vorkommen von Saponinen im Tabak und über einen Begleitstoff. Journal für praktische Chemie N.F. 143 (1935), 29–41.

1936

691. Neuberg, Carl: Kaiser Wilhelm-Institut für Biochemie in Berlin-Dahlem; in: 25 Jahre Kaiser-Wilhelm-Gesellschaft zur Förderung der Wissenschaften, hrsg. von Max Planck. Bd. 2: Naturwissenschaften. Berlin: Springer 1936, 336–342.
692. Neuberg, Carl: Carboxylase; in: Handbuch der biologischen Arbeitsmethoden, hrsg. von Emil Abderhalden. Abt. IV, Teil 1: Methoden der Fermentforschung. Berlin, Wien: Urban & Schwarzenberg 1936, 565–582.
693. Neuberg, Carl/Wagner, Joachim: Sulfatase; in: Handbuch der biologischen Arbeitsmethoden, hrsg. von Emil Abderhalden. Abt. IV, Teil 1: Methoden der Fermentforschung. Berlin, Wien: Urban & Schwarzenberg 1936, 583–592.
694. Neuberg, Carl/Kobel, Maria: Abfangverfahren; in: Handbuch der biologischen Arbeitsmethoden, hrsg. von Emil Abderhalden. Abt. IV, Teil 1: Methoden der Fermentforschung. Berlin, Wien: Urban & Schwarzenberg 1936, 593–614.
695. Neuberg, Carl/Simon, Ernst: Phosphatasen; in: Handbuch der biologischen Arbeits-methoden, hrsg. von Emil Abderhalden. Abt. IV, Teil 1: Methoden der Fermentforschung. Berlin, Wien: Urban & Schwarzenberg 1936, 615–624.
696. Neuberg, Carl/Kobel, Maria: Carboligase; in: Handbuch der biologischen Arbeits-methoden, hrsg. von Emil Abderhalden. Abt. IV, Teil 1: Methoden der Fermentforschung. Berlin, Wien: Urban & Schwarzenberg 1936, 625–644.
697. Neuberg, Carl/Gorr, Günther: Phytochemische Reduktionen; in: Handbuch der biologischen Arbeitsmethoden, hrsg. von Emil Abderhalden. Abt. IV, Teil 1: Methoden der Fermentforschung. Berlin, Wien: Urban & Schwarzenberg 1936, 645–670.
698. Neuberg, Carl/Simon, Ernst: Ketonaldehydmutase; in: Handbuch der biologischen Arbeitsmethoden, hrsg. von Emil Abderhalden. Abt. IV, Teil 2. Berlin, Wien: Urban & Schwarzenberg 1936, 2225–2248.
699. Neuberg, Carl: Abbau der d-Fructose-6-phosphorsäure zu d-Arabonsäure-5-phos-phorsäure und deren enzymatische Spaltung. Cellulosechemie 17 (1936), 125–128.
700. Neuberg, Carl: Darstellung von d-Arabinose. Cellulosechemie 17 (1936), 128.

701. Neuberg, Carl/Cahill, W[illiam] M.: Über die enzymatische Aufspaltung der Chon-droitin-Schwefelsäure und Mucoitin-Schwefelsäure in ihre Bausteine. Enzymologia 1 (1936/37), 22–38.
702. Neuberg, Carl/Cahill, W[illiam] M.: Verwirklichung der phytochemischen Reduktion von Triketopentan. Enzymologia 1 (1936/37), 142–144.
703. Neuberg, Carl/Minard, Gisèle: Verhalten der γ-Chinolin-brenztraubensäure zu Hefe; ihr Einfluss auf die Vergärung von Brenztraubensäure und Glucose. Enzymologia 1 (1936/37), 161–167.
704. Neuberg, Carl/Kobel, Maria: Umwandlung des Rutins durch Tabakenzyme in braune Farbstoffe und Einwirkung von Tabakenzymen auf andere Phenolkörper. Enzymo-logia 1 (1936/37), 177–182.
705. Neuberg, Carl/Schuchardt, W[erner]: Über die Synthese der Phospho-d-weinsäure und ihre phosphatatische Spaltung. Enzymologia 1 (1936/37), 39–47.
706. Neuberg, Carl/Kobel, Maria: Isolierung von Rutin aus Tabak. Zeitschrift für Untersuchung der Lebensmittel 72 (1936), 113–116.
707. Neuberg, Carl/Kobel, Maria: Vorkommen von Sorbit im Tabak. Zeitschrift für Unter-suchung der Lebensmittel 72 (1936), 116–121.

1937

708. Neuberg, Carl: Ein Fall von freiwilliger Trennung isomerer Verbindungen. Biochimija (russ.) 2 (1937), 383.
709. Neuberg, Carl/Fischer, H[ans] A[rmin]: Enzymatische Spaltung anorganischer Tri-phos-phorsäure. Enzymologia 2 (1937/38), 191.
710. Neuberg, Carl/Fischer, H[ans] A[rmin]: Über die enzymatische Spaltung von Tri-phos-phorsäure II. Enzymologia 2 (1937/38), 241–257.

1938

711. Neuberg, Carl/Fischer, Hans Armin: Enzymatische Spaltung von Triphosphorsäure. I. Comptes Rendus des Travaux du Laboratoire Carlsberg, Ser. chim. 22 (1938), 366–374.
712. Neuberg, Carl/Fischer, H[ans] A[rmin]: Über die enzymatische Spaltung von Tri-phosphorsäure III. Enzymologia 2 (1937/38), 360–366.
713. Neuberg, Carl/Ziffer, Rudolf: Enzymatische Spaltung von Acetalen. Enzymologia 5 (1938/39), 389.
714. Neuberg, Carl/Schwietzer, C. H.: Untersuchungen über Agar. Monatshefte für Chemie 71 (1938), 46–66.

1939

715. Neuberg, Carl/Ziffer, Rudolf: Neues über Abfangung von Acetaldehyd bei der alko-holischen Gärung. Enzymologia 7 (1939), 225–227.
716. Neuberg, Carl/Peiser, Elisabeth: Über die vermeintliche optische Inaktivierung der d-Weinsäure durch Pankreas. Enzymologia 7 (1939), 228–230.

1940

717. Neuberg, Carl/Fischer, Hans Armin: Vergleichende Untersuchungen über die hydro-tropischen Wirkungen correspondierender und isomerer Verbindungen der C7- und C8-Reihe. Recueil des Travaux chimiques des Pays-Bas 59 (1940), 77–86.

1941

718. Feigenbaum, J./Neuberg, Carl: Simplified method for the preparation of aromatic sulfuric acid esters. Journal of the American Chemical Society 63 (1941), 3529–3530.

1942

719. Neuberg, Carl: Development of chemistry of fermentation. American Brewery 75 (1942), 22–30.
720. Neuberg, Carl/Lustig, Hilda: Preparation of d-Fructose-1, 6-diphosphate by Means of Baker's Yeasts. Journal of the American Chemical Society 64 (1942), 2722–2723.

1943

721. Neuberg, Carl/Lustig, Hilda: Preparation of active zymase extracts from top yeast. Archives of Biochemistry 1 (1943), 191–196.
722. Neuberg, Carl: d-3-Phosphoglyceric acid. Archives of Biochemistry 1 (1943), 311–318.
723. Neuberg, Carl/Lustig, Hilda/R. N. Cagan: Enzymatic reduction of furil and furoin. Archives of Biochemistry 1 (1943), 391–395.
724. Neuberg, Carl/Lustig, Hilda: Question of the significance of phosphorglyceric acids under pathological conditions. Experimental Medicine and Surgery 1 (1943), 14–21.
725. Neuberg, Carl/Lustig, H[ilda]: Pancreas and the problem of asymmetry. The non-existence of transformation of d-tartaric to meso-tartaric acid by pancreas. Experimental Medicine and Surgery 1 (1943), 314–335.
726. Frankenthal, Lea/Neuberg, Carl: Triphosphate and triphosphatase I. The enzymatic hydro-lysis of triphosphoric acid by triphosphatase of tumors and vegetable objects. II. Roberts, Irene Stephanie/Neuberg, Carl: Specific chemical action of triphosphates. Experimental Medicine and Surgery 1 (1943), 386–401.
727. Neuberg, Carl: <review> Dictionary of biochemistry and related subjects, ed. by William M. Malisoff. New York 1943. Experimental Medicine and Surgery 1 (1943), 413.

1944

728. Neuberg, Carl/Lustig, H[ilda]/Rothenberg, M[ortimer] A.: Fructose-1, 6-Diphosphoric acid and Fructose-6-Monophosphoric Acid. Archives of Biochemistry 3 (1944), 33–44.
729. Neuberg, Carl: Synthesis of d(-)3-Phosphoglyceric Acid and d(+)2-Phosphoglyceric Acid. Archives of Biochemistry 3 (1944), 105–112.
730. Neuberg, Carl/Strauss, E[duard]/Lipkin, L. E.: Convenient Method for Deproteinization. Archives of Biochemistry 4 (1944), 101–104.
731. Neuberg, Carl: <review> Fritz Feigl, Laboratory manual of spot tests, New York 1943. Experimental Medicine and Surgery 2 (1944), 192.
732. Neuberg, Carl: <review> Edwin J.Cohn/John T. Edsall, Proteins, Amino Acids and Peptides as Ions and Dipolar Ions, New York 1943. Experimental Medicine and Surgery 2 (1944), 271.

1945

733. Neuberg, Carl/Strauss, Eduard: Microdetermination of Dicarbonyl Compounds. On the Coloring of Nitro Compounds in Alkaline Solutions. Archives of Biochemistry 7 (1945), 211–230.
734. Neuberg, Carl/Strauss, E[duard]: The deproteinization by perchlorid acid. Experimental Medicine and Surgery 3 (1945), 39–43.
735. Neuberg, Carl: <review> Richard J. Block/Diana Bolling, The Amino Acid Composition of Proteins and Foods, Springfield (Ill.) 1945. Archives of Biochemistry 7 (1945), 273–274.
736. Neuberg, Carl: <review> George F. Gause, Optical Acitivity and Living Matter, Normandy (Mo) 1941. Archives of Biochemistry 7 (1945), 491–493.

1946

737. Neuberg, Carl/Roberts, Irene Stephanie: Invertase. New York: New York Academic Press 1946.
738. Neuberg, Carl/Mandl, Ines: The Identification of Fructose. Archives of Biochemistry 11 (1946), 451–456.
739. Neuberg, Carl/Strauss, Eduard: Quantitative Formation of Osazones. Archives of Biochemistry 11 (1946), 457–465.
740. Neuberg, Carl: The Biochemistry of Yeast. Annual Review of Biochemistry 15 (1946), 435–474.
741. Neuberg, Carl: Emil Starkenstein 1884–1942. Experimental Medicine and Surgery 4 (1946), 184–185.
742. Neuberg, Carl: <review> Die Methoden der Fermentforschung, hrsg. von Eugen Bamann und Karl Myrbäck, New York 1945. Archives of Biochemistry 9 (1946), 302–304.
743. Neuberg, Carl: <review> Israel Kleiner, Human Biochemistry, St. Louis 1945. Experimental Medicine and Surgery 4 (1946), 285–286.
744. Neuberg, Carl: <review> Georgia Leffingwell/Milton Lesser, Glycerin, Brooklyn 1945. Experimental Medicine and Surgery 4 (1946), 286.

1947

745. Neuberg, Carl: <review> C. S. Hudson, The Collected Papers, Vol. 1, New York 1946. Archives of Biochemistry 12 (1947), 326.

746. Neuberg, Carl: <review> Pierre Thomas, Manuel de Biochimie, Paris 1946. Archives of Biochemistry 13 (1947), 303–304.

747. Neuberg, Carl: <review> Methods of Vitamin Assay, New York and London 1947. Experimental Medicine and Surgery 5 (1947), 446.

748. Neuberg, Carl: <review> Samuel Soskin/Rachmiel Levine, Carbohydrate metabolism, Chicago. Experimental Medicine and Surgery 5 (1947), 448–449.

1948

749. Neuberg, Carl/Mandl, Ines: An unknown effect of amino acids. Archives of Biochemistry 19 (1948), 149–161.

750. Neuberg, Carl/Lustig, Hilda/Dresel, Rita: Dismutation in the heterocyclic series. Dismutation of furfural by yeast and related problems. Archives of Biochemistry 19 (1948), 163–169.

751. Neuberg, Carl/Mandl, Ines: Beachtliche Wirkung von Salzen organischer Säuren auf unlösliche anorganische Verbindungen (Mineralysis). Zeitschrift für Vitamin-, Hormon- und Fermentforschung 2 (1948/49), 479–492.

752. Neuberg, Carl: <review> Emil Abderhalden, Lehrbuch der physiologischen Chemie, 26. Aufl., Basel. Experimental Medicine and Surgery 6 (1948), 119.

753. Neuberg, Carl: <review> Encyclopedia of chemical technology ed. by Raymond E. Kirk/Donald F. Othmer. Vol. 1, New York 1947. Experimental Medicine and Surgery 6 (1948), 324–325.

754. Neuberg, Carl: <review> Jesse P. Greenstein, Biochemistry of Cancer, Bethesda 1947. Archives of Biochemistry 17 (1948), 366.

1949

755. Neuberg, Carl: Biochemical reductions at the expense of sugars. Advances in Carbohydrate Chemistry 74 (1949), 75–117.

756. Neuberg, Carl/Roberts, Irene S[tephanie]: Remarkable Properties of Nucleic Acids and Nucleotides. Archives of Biochemistry 20 (1949), 185–210.

757. Neuberg, Carl/Mandl, Ines: An Unknown Effect of Amino Acids II. Interaction of Nitrogenous Polycarboxylic Acids (N-Substituted Amino Acids) and Insoluble Metal Sulfides and Mercaptides. Archives of Biochemistry 23 (1949), 271–274.

758. Neuberg, Carl/Mandl, Ines: Functions of ATP and other Phosphoric Acid Derivatives. Archives of Biochemistry 23 (1949), 499–501.

759. Neuberg, Carl: <review> C. S. Hudson, The Collected Papers. Vol. 2. Bethesda 1948. Archives of Biochemistry 20 (1949), 181–182.

760. Neuberg, Carl: <review> Bioquimica Analitica Cuantitativa ed. Agustin D. Marenzi et al.. Buenos Aires 1949. Archives of Biochemistry 20 (1949), 184.

761. Neuberg, Carl: <review> Elvin A. Kabat/Manfred M. Mayer, Experimental immunochemistry, Springfield (Ill.) 1949. Experimental Medicine and Surgery 7 (1949), 70–71.

1950

762. Neuberg, Carl/Mandl, Ines: Invertase; in: The Enzymes ed. by James B. Sumner/Karl Myrbaeck. Bd. 1.1. New York: Academic Press 1950, 527–550.

763. Neuberg, Carl: Carbohydrates in Yeast. American Brewery 83 (1950) [without page].

764. Neuberg, Carl/Lustig, Hilda/Mandl, Ines: Heavy-Metal Hydroxides in Statu Nascendi as Reagents for the Purification of Amino Acid Mistures and the Preparation of Pure Heavy-Metal Salts of Individual Amino Acids. Archives of Biochemistry 26 (1950), 77–84.

765. Neuberg, Carl: Gärung und phytochemische Reduktion. Biochimica et Biophysica Acta (Amsterdam) 4 (1950), 170–178.

766. Neuberg, Carl/Mandl, Ines: Preparation of d- and l-amino acids by enzymatic resolution. Enzymologia 14 (1950/51), 128–133.

767. Neuberg, Carl/Grauer, A[melie]/Mandl, I[nes]: The formation of pyrophosphate by enzymatic breakdown of inorganic triphosphate. V. Communication on enzymatic cleavage of triphosphoric acid. Enzymologia 14 (1950/51), 157–163.

768. Kirsch, Charlotte/Neuberg, Carl: Differentiation of free and bound tryptophane. Note on tryptamine. Experimental Medicine and Surgery 8 (1950), 295–300.

769. Grauer, A[melie]/Mandl, I[nes]/Strauss, E[duard]/Neuberg, Carl: Deproteinization with perchloric acid. (Institute of Polymer Research, Polytechnic Institute of Brooklyn, Brooklyn, N.Y.). Experimental Medicine and Surgery 8 (1950), 301–307.

770. Neuberg, Carl: <review> Outlines of biochemistry, 3rd. ed. by Ross Aiken Gortner, jr./Willis Alway Gortner, New York, London 1949. Archives of Biochemistry 25 (1950), 471–474.

1951

771. Mandl, Ines/Neuberg, Carl: 5-Phospho-D-arabonic Acid. Archives of Biochemistry and Biophysics 33 (1951), 191–194.

772. Neuberg, Carl: <review> Biochemistry of Glucuronic Acid ed. by Neal E. Artz/Elizabeth N. Osman New York 1950. Archives of Biochemistry and Biophysics 31 (1951), 157.

773. Neuberg, Carl: <review> Amino acids, Cold Spring Harbor Symposia on Quantitative Biology, vol. XIV. Experimental Medicine and Surgery 9 (1951), 157.

1952

774. Mandl, Ines: Conversion of a phosphorylated ketohexose (D-fructose-6-phosphate) to a phosphorylated aldopentose derivative (D-arabonic acid-6-phosphate). Acta Biochemica et Biophysica <Shanghai> B 37 (1952), 83–84.

775. Neuberg, Carl/Grauer, A[melie]/Pisha, B. V.: The precipitation of carbonyl compounds with 2, 4-dinitrophenylhydrazine. Formation of isomeric dinitrophenylhydrazones. Analytica chimica Acta 7 (1952), 238–242.

776. Mandl, Ines/Neuberg, Carl: Characterization of Some Sugars of Interest to Biochemistry. Archives of Biochemistry and Biophysics 35 (1952), 326–334.

777. Mandl, Ines/Grauer, Amelie/Neuberg, Carl: Solubilization of Insoluble Matter in Nature I. The Part played by Salts of Adenosinetriphosphate. Biochimica et Biophysica Acta <Amsterdam> 8 (1952), 654–663.

778. Neuberg, Carl/Grauer, A[melie]/Mandl, I[nes]: Enzymatic cleavage of conjugated glucuronic acids in relation to the problem of mineralysis. Enzymologia 15 (1952/53), 115–121.

779. Mandl, I[nes]/Grauer, A[melie]/Neuberg, Carl: Transport of insoluble matter in nature: dependence upon complex formation. Federation of American Societies for Experimental Biology. Federation Proceedings 11 (1952), 253.

780. Neuberg, Carl/Grauer, Amelie: Biosynthese von Mercaptanen. Hoppe-Seylers Zeitschrift für Physiologische Chemie 289 (1952), 253–256.

781. Neuberg, Carl: Aufnahme und Wanderung unlöslicher Stoffe in der Natur. Separata da Revista da Faculdada al Cîencias de Lisboa Serie 2B (1952), 165–176.

782. Neuberg, Carl: <review> Israel Kleiner, Human biochemistry, St. Louis 1951. Experimental Medicine and Surgery 10 (1952), 172.

783. Neuberg, Carl: <review> M. G. Sevag, Immuno-catalysis and related fields of bacteriology and biochemistry, 2nd ed., Springfield Illnois. 1951. Experimental Medicine and Surgery 10 (1952), 245–246.

1953

784. Grauer, A[melie]/Neuberg, Carl: The naphthoresorcinol reaction of uronic acids. Analytica Chimica Acta 8 (1953), 422–425.

785. Neuberg, Carl/Forrest, Irene S[tephanie]: Action of Isonicotinic Acid in Microbial Metabolism. Archives of Biochemistry and Biophysics 45 (1953), 237–238.

786. Mandl, Ines/Grauer, Amelie/Neuberg, Carl: Solubilization of insoluble matter in nature II. The part played by salts of organic and inorganic acids occurring in nature. Biochimica et Biophysica Acta <Amsterdam> 10 (1953), 540–569.

787. Forrest, Irene S[tephanie]/Neuberg, Carl: Preparation and analytical data of pure D(-)-3-phosphoglyceric acid. Biochimica et Biophysica Acta <Amsterdam> 11 (1953), 588–589.

788. Neuberg, Carl/Grauer, A[melie]: Precipitability of small quantities of cell constituents and metabolites as insoluble radioactive compounds. Biochimica et Biophysica Acta <Amsterdam> 12 (1953), 265–272.

789. Neuberg, Carl/Forrest, Irene S[tephanie]: Einfache Darstellung aktiver Zymaselösungen aus Bäckerhefe. Hoppe-Seylers Zeitschrift für Physiologische Chemie 295 (1953), 110–113.

790. Neuberg, Carl: <review> Michel Polonovski, Biochimie Medicale, 5. ed., Paris 1952. Archives of Biochemistry and Biophysics 44 (1953), 256–257.

791. Neuberg, Carl: <review> Joseph Fruton/Sofia Simmonds, General Biochemistry, New York 1953. Archives of Biochemistry and Biophysics 45 (1953), 472–473.

792. Neuberg, Carl: <review> Virgilio Bolcato; La Chimica delle Fermentazioni, Bologna 1952. Archives of Biochemistry and Biophysics 46 (1953), 259–260.

793. Neuberg, Carl: <review> Fritz Lieben, Vorstellungen vom Aufbau der Materie im Wandel der Zeiten, Wien 1953. Archives of Biochemistry and Biophysics 46 (1953), 491–492.

1954

794. Neuberg, Carl: Bioreduction of trimethylamine oxide. Bulletin of the Research Council of Israel 4 (1954), 12–14.

795. Forrest, I[rene] S[tephanie]/Grauer, A[melie]/Neuberg, Carl: Thionophosphatase. Enzymologia 16 (1953/54), 305–310.

796. Forrest, I[rene] S[tephanie]/Grauer, A[melie]/Kreidl, M./Neuberg, Carl: Enzymatic formation of insoluble phosphates. Suggested Mechanism for Redissolution of Phosphates and Carbonates in Nature. Enzymologia 17 (1954/56), 97–109.

797. Forrest, I[rene] S[tephanie]/Grauer, A[melie]/Neuberg, Carl: Thionophosphatase. Federation of American Societies for Experimental Biology. Federation Proceedings 13 (1954) Nr. 1, 210.

798. Neuberg, Carl: <review> Leland A. Underkofler, Industrial Fermentation, New York 1954; Archives of Biochemistry and Biophysics 53 (1954), 523–524.

1955

799. Mandl, Ines/Neuberg, Carl: Metaphosphatase. Methods in Enzymology Bd. 2 (1955), 577–580.

800. Mandl, Ines/Neuberg, Carl: Triphosphatase. Methods in Enzymology Bd. 2 (1955), 580–582.

801. Forrest, I[rene] S[tephanie]/Grauer, A[melie]/Kreidl, M./Neuberg, Carl: Reaction between Neutral Compounds Involving Major Changes in Hydrogen-Ion Concentration. Formation of Free Mineral Acids. Archives of Biochemistry and Biophysics 55 (1955), 555–565.

802. Neuberg, Carl/Grauer, A[melie]/Kreidl, M.: The Basis of the Carbamate Reaction. Archives of Biochemistry and Biophysics 58 (1955), 169–177.

803. Forrest, I[rene] S[tephanie]/Grauer, A[melie]/Kreidl, M./Neuberg, Carl: Modification and regulation of H-ion concentration by simple reactions. Formation of free mineral acid from neutral substances. Experimental Medicine and Surgery 13 (1955), 22–32.

804. Neuberg, Carl: <review> Fritz Feigl, Spot tests, Amsterdam 1955, Archives of Biochemistry and Biophysics 54 (1955), 256–257.

805. Neuberg, Carl: <review> O. Hoffmann-Ostenhof, Enzymologie, Wien 1954, Archives of Biochemistry and Biophysics 56 (1955), 279–280.

1956

806. Mandl, Ines/Neuberg, Carl: Solubilization, Migration, and Utilization of Insoluble Matter in Nature. Advances in Enzymology 17 (1956), 135–158.

1957

[published post hum]

807. Mandl, Ines/Neuberg, Carl: Ketohexose phosphates. Methods in Enzymology Vol. 3 (1957), 162–172

808. Mandl, Ines/Neuberg, Carl: Preparation of D(-)3-phosphoglyceric acid in Enzymology. Methods in Enzymology <New York> 3 (1957), 208–214.

809. Mandl, Ines/Neuberg, Carl: Preparation of D(+)2-phosphoglyceric acid in Enzymology. Methods in Enzymology <New York> 3 (1957), 214–216.

810. Neuberg, Carl: Pro Memoria Carl Neuberg. Lecture-Demonstration. Experientia 13 (1957), 451–453.

11.2. DOCTORAL THESES SUPERVISED BY CARL NEUBERG

1. Wolff, Hans Georg: Zur Kenntnis des Glucosamin. Diss. phil. Berlin 1903.

2. Neimann, Wilhelm: 1. Zur Kenntnis der Glucuronsäure und über die Synthese gepaarter Glucuronsäuren. 2. Über eine Methode zur Isolierung von Aldehyden und Ketonen. Diss. phil. Berlin 1904.

3. Neimann, Ernst: I. Synthese von Diaminokorksäure und Diaminosebacinsäure. II. Neue Synthese von Diaminen. III. Über gelatinöse anorganische Baryumsalze. Diss. phil. Berlin 1905.

4. Silbermann, Martin: Untersuchungen in der 3-Kohlenstoffreihe. Über ein neues Verfahren zur Darstellung von Oxyaminosäuren. Diss. phil. Berlin 1905.

5. Federer, Max: Über γ-γ1-Diaminoazelainsäure, γ-ß-Diaminobuttersäure und γ-Oxy-ß-aminobuttersäure. Über eine neue Methode der Spaltung von Racemkörpern. Diss. phil. Berlin 1906.

6. Ascher, Erich: I. Untersuchungen in der 3-Kohlenstoffreihe. II. Chemische Untersuchungen über den Krebs. Diss. phil. Berlin 1907.
7. Marx, Fritz: Untersuchungen in der Zuckerreihe. Diss. phil. Berlin 1907.
8. Rosenberg, Egon: Über Eiweißfäulnis. Diss. phil. Berlin 1907.
9. Rewald, Bruno: Über l- und d-Capronsäure. Über kolloidale und gelatinöse Erdalkaliverbindungen. Diss. phil. Berlin 1908.
10. Scott, Leonard: Abbau in der Kohlehydratreihe durch Elektrolyse. II. Über Jodospongin. Diss. phil. Berlin 1908.
11. Kansky, Eugen: Reduktion von Aminosäuren zu Aminoaldehyden. Über die Isolierung aliphatischer Alkohole. Diss. phil. Berlin 1909.
12. Lachmann, Samuel: Untersuchungen aus der Kohlenhydratreihe. Diss. phil. Berlin 1909.
13. Hirschberg, Else: Abbauversuche in der Kohlenhydratreihe. Über die γ-Naphthylisocyanatverbindungen einiger physiologisch wichtiger Substanzen. Diss. phil. Berlin 1910.
14. Pollak, Hermann: Über Phosphorsäure- und Schwefelsäureester von Kohlenhydraten und über die Phosphorylierung von Eiweiß. Diss. phil. Berlin 1910.
15. Kretschmer, Erich: Über die Einführung der Phosphorsäure in Kohlenhydrate und Glycerin. Über p-Kresolätherschwefelsäure und p-Kresolglucuronsäure. Diss. phil. Berlin 1911.
16. Oertel, Walter: I. Über die Einführung der Phosphorsäure in Aminosäuren, Peptone, Albumosen und Proteine. II. Studien über die Methylglyoxalbildung. III. Über den Einfluss des ultravioletten Lichtes auf halogensauerstoffsaure Alkalien. Diss. phil. Berlin 1914.
17. Rosenthal, Paul: Zur Kenntnis der Karboxylase und der Gärungsvorgänge. Diss. phil. Berlin 1914.
18. Kerb, Johannes: Physiologisch-chemische Untersuchungen in der Kohlenhydratreihe. I. Über den Verlauf der alkoholischen Gärung bei Gegenwart von kohlensaurem Kalk. II. Über eine Verbindung der Stärke mit Phosphorsäure. Diss. med. Berlin 1919.
19. Kerb, Elisabeth: Zur Kenntnis der phytochemischen Reduktionsprozesse. A. Die Überführung von Acetaldol in optisch-aktives ß-Butylenglykol durch Hefe. Nebst Anhang: Darstellung, Reaktionen und krystallisierte Abkömmlinge des Aldols. B. Die Umwandlung von Citral in Geraniol durch Hefe. Nebst Anhang: Darstellung und krystallisierte Abkömmlinge des Zyklocitrals. Diss. phil. Berlin 1919.
20. Reinfurth, Elsa: Die Festlegung der Aldehydstufe bei der alkoholischen Gärung. Ein experimenteller Beweis der Acetaldehyd-Brenztraubensäuretheorie. Diss. ing. Karlsruhe 1919.
21. Ursum, Werner: Die dritte Vergärungsform des Zuckers als allgemeine Folge der Wirkung anorganischer und organischer Alkalisatoren. Diss. phil. Berlin 1920.
22. Arinstein, Bernhard: Vom Wesen der Buttersäure- und Butylalkoholgärung. Diss. phil. Berlin 1921.

23. Grab, Max von: Brenztraubensäure als Zwischenprodukt der alkoholischen Zuckerspaltung. Diss. phil. Berlin 1921.
24. Liebermann, Ludwig: I. Zur Kenntnis der Carboligase. II. Über Glucose- und Rohrzuckermonoschwefelsäure. Diss. phil. Berlin 1921
25. Cohen, Clara: Über die Bildung von Acetaldehyd und die Verwirklichung der zweiten Vergärungsform bci verschiedenen Pilzen. Diss. phil. Berlin 1922.
26. May, Albert von: Die Bilanz der Brenztraubensäuregärung. Diss. phil. Berlin 1923.
27. Simon, Ernst: I. Untersuchungen über die Sulfatase. II. Zur Kenntnis der biochemischen Acyloinkondensation. Diss. phil. Berlin 1925.
28. Wagner, Joachim: I. Über die Spaltbarkeit von Ätherschwefelsäuren der verschiedenen Reihen durch Sulfatase und über den Mechanismus der Sulfatasewirkung. II. Über eine neue Methode zur Trennung des Äthylalkohols von Acetaldehyd und Aceton. Diss. phil. Berlin 1925.
29. Lusmann-Perlmann, Genia: Untersuchungen über die alkoholische Gärung und die Milchsäurebildung. Diss. phil. Berlin 1926.
30. Windisch, Friedrich [Fritz]: Über das Wesen der Essiggärung und die chemischen Leistungen der Essigbakterien. Diss. phil. Berlin 1926.
31. Jacobsohn, Kurt: Bildung und Spaltung von Glukosiden als Methode zur chemischen und biochemischen Trennung razemischer Alkohole in ihre optisch aktiven Komponenten. Diss. phil. Berlin 1929.
32. Vogt, Marthe: Untersuchungen über Bildung und Verhalten einiger biologisch wichtiger Substanzen aus der Dreikohlenstoffreihe. Diss. phil. Berlin 1929.
33. Molinari, Emil: Über die biochemische Dismutation nebst Untersuchungen über die Essiggärung. Diss. phil. Berlin 1930.
34. Neuberg, Irene Stephanie: Untersuchungen in der 3-Kohlenstoffreihe und Gewinnung von Zuckern aus ihren Hydrazonen. Diss. phil. Berlin 1932.
35. Bergmann, Felix: Untersuchungen über das chemische und enzymatische Verhalten einiger natürlich vorkommender Glukuronsäureverbindungen. Mit Anhang: Über die Zucker der Frauenmilch. Diss. phil. Berlin 1933.
36. Cahill, William M.: Über die enzymatische Aufspaltung von Chondroitin-schwefelsäure und Mucoitin-schwefelsäure in ihre Bausteine. Diss. phil. Berlin 1936.

11.3. PUBLICATIONS BY OTHER STAFF OF CARL NEUBERG'S RESEARCH DEPARTMENTS, INSTITUTES AND LABORATORIES, 1903–1955

1903–1906

1. Euler, Hans von: Chemische Dynamik der zellfreien Gärung. Hoppe–Seylers Zeitschrift für physiologische Chemie 44 (1905), 53–73.
2. Manasse, A[rmand]: Über die quantitative Bestimmung des Zuckers im Harn. In: Arbeiten aus dem Pathologischen Institut zu Berlin. Zur Feier der Vollendung der Instituts-Neubauten, hrsg. von Johannes Orth. Berlin: Hirschwald 1906, 608–617.

1907–1914

3. Rewald, Bruno: Über die Pentose aus Gesamtpankreas. Erwiderung auf die Mitteilung von Levene und Jacobs. Berichte der Deutschen Chemischen Gesellschaft 43 (1910), 3502–3503.
4. Karczag, Lázló: In welcher Weise wird die Weinsäure durch Hefe angegriffen. Biochemische Zeitschrift 43 (1912), 44–46.
5. Róna, Elisabeth: I. Über die Reduktion des Zimtaldehyds durch Hefe. II. Vergärung von Benzylbenztraubensäure. Biochemische Zeitschrift 67 (1914), 137–142.

1915–1919

6. Czapski Ludwig: Zur Methodik der Bestimmung von Milchsäure neben Brenztraubensäure. Biochemische Zeitschrift 71 (1915), 167–168.
7. Karczag, Lázló: Notizen über die Brenztraubensäure. Biochemische Zeitschrift 84 (1917), 225–230.
8. Nord, F[riedrich] F[ranz]: Biochemische Bildung von Aminoäthylalkohol aus Serin. Biochemische Zeitschrift 95 (1919), 281–285.
9. Kerb, Johannes: Über eine Verbindung der Stärke mit Phosphorsäure. Biochemische Zeitschrift 100 (1919), 3–14.
10. Nord, F[riedrich] F[ranz]: Der Acetaldehyd in der Natur. Ergebnisse des Abfangverfahrens. Die Naturwissenschaften 7 (1919), 685–690.

1920

11. Ehrlich, Marta: Über die angebliche Inaktivierung des d-Mannits. Biochemische Zeitschrift 103 (1920), 312–314.
12. Nord, F[riedrich] F[ranz]: Phytochemische Reduktion von o-Nitrobenzaldehyd. Biochemische Zeitschrift 103 (1920), 315–319.
13. Nord, F[riedrich] F[ranz]: Gemischte Dismutation der Aldehyde. Biochemische Zeitschrift 106 (1920), 275–280.
14. Róna, Elisabeth: Über die Wirksamkeit der Fermente unter abnormen Bedingungen und über die angebliche Aldehydnatur der Enzyme. Biochemische Zeitschrift 109 (1920), 279–289.

15. Cohen, Clara: Über die Bildung von Acetaldehyd bei den Umsetzungen von Zucker durch Pilze. Biochemische Zeitschrift 112 (1920), 139–143.

16. Färber, Eduard/Nord, F[riedrich] F[ranz]: Die phytochemische Reduktion des Acetols zu optisch-aktivem Propylenglykol. Nebst Bemerkungen und Versuchen zur Frage des Auftretens sowie der Verarbeitung von Racemkörpern bei Tier und Pflanze von Carl Neuberg. Biochemische Zeitschrift 112 (1920), 313–323.

17. Färber, Eduard: Die Zwischenstufen bei chemischen Umwandlungen. Die Naturwissenschaften 8 (1920), 321–327.

1921

18. Hirsch, Julius: Zur Kenntnis des oxydativen Zuckerabbaus im Tierkörper. Biochemische Zeitschrift 117 (1921), 113–116.

19. Grab, Max von: Brenztraubensäure als Zwischenprodukt der alkoholischen Zuckerspaltung. Biochemische Zeitschrift 123 (1921), 69–89.

1922

20. Sandberg, Marta: Über den Verlauf der alkoholischen Gärung in Gegenwart von Harnstoff. Biochemische Zeitschrift 128 (1922), 76–79.

21. Kumagawa, H.: Erzielung der zweiten und dritten Vergärungsform mit Saccharomycetes, Zygosaccharomyces major und Zygosaccharomyces salsus. Biochemische Zeitschrift 131 (1922), 148–156.

22. Kumagawa, H.: Über die Zerlegung des meso-Inosits und Glycerins nach Art der wahren Zucker durch den Bacillus lactis aerogenes. Biochemische Zeitschrift 131 (1922), 157–160.

23. Tomita, M.: Zur Kenntnis der Phosphatasen I: Saccharo-phosphatase. Biochemische Zeitschrift 131 (1922), 161–169.

24. Tomita, M.: Zur Kenntnis der Phosphatasen II: Hexose-mono-phosphatase. Biochemische Zeitschrift 131 (1922), 170–174.

25. Tomita, M.: Über den Einfluss des Thyroxins auf die alkoholische Gärung. Biochemische Zeitschrift 131 (1922), 175–177.

26. Hirsch, Julius: Über eine biosynthetische Kohlenstoffkettenverknüpfung in der aliphatischen Reihe. Zur Kenntnis der Carboligase. Biochemische Zeitschrift 131 (1922), 178–187.

27. Ohle, Heinz: Über Zuckerschwefelsäuren IV. Biochemische Zeitschrift 131 (1922), 601–610.

28. Ohle, Heinz: Zur Konstitution des Vacciniins. Biochemische Zeitschrift 131 (1922), 611–613.

1923

29. Hirsch, Julius: Acetaldehyd im intermediären Stoffwechsel überlebender Muskulatur. Biochemische Zeitschrift 134 (1923), 415–423.

30. Kurono, K.: Über die Umwandlung tertiärer γ-Aminosäuren durch Hefe. Ein Beitrag zur Lehre vom biochemischen Abbau der Aminosäuren. Biochemische Zeitschrift 134 (1923), 424–433.

31. Kurono, K.: Synthese zweier neuer Leucine. Biochemische Zeitschrift 134 (1923), 434–436.
32. Soda, T.: Zur Kenntnis der Glucose-mono-schwefelsüre und Rohrzucker-mono-schwefelsäure. V. Mitteilung. Biochemische Zeitschrift 135 (1923), 621–628.
33. Ohle, Heinz: Über Zuckerschwefelsäuren VI. Biochemische Zeitschrift 136 (1923), 428–448.
34. Sen, K. H.: Über die Gärung der γ-Keto-n-capronsäure. Biochemische Zeitschrift 140 (1923), 447–452.
35. Nagai, K.: Über die Bildung von Acetaldehyd bei der Vergärung von Fructose, Galaktose, Saccharose, Maltose und Lactose durch Bacterium coli und Bacillus lactis aerogenes. Biochemische Zeitschrift 141 (1923), 261–265.
36. Nagai, K.: Über die Bildung von Acetaldehyd beim bakteriellen Abbau von Säuren der Kohlenhydratreihe und verwandten Säuren. Biochemische Zeitschrift 141 (1923), 266–273.
37. Tamba, R.: Über einen Schwefelsäure-ester der Stärke. Biochemische Zeitschrift 141 (1923), 274–277.
38. May, Albert von: Über die Einwirkung von Metallsalzen auf den Verlauf der alkoholischen Gärung. Biochemische Zeitschrift 141 (1923), 447–457.
39. Akamatsu, S.: Über Galaktose-schwefelsäure. Biochemische Zeitschrift 142 (1923), 181–183.
40. Akamatsu, S.: Über das Vorkommen von Glycero-phosphatase in der "Takadiastase". Biochemische Zeitschrift 142 (1923), 184–185.
41. Akamatsu, S.: Über Lecithinspaltung durch "Takadiastase". Biochemische Zeitschrift 142 (1923), 186–187.
42. Akamatsu, S.: Phytochemische Reduktion in der Cyclohexanreihe. Biochemische Zeitschrift 142 (1923), 188–190.
43. Noguchi, J.: Über Sulfatase III. Über die enzymatische Spaltung von im Pferde-, Hammel- und Kaninchenharn enthaltenen aromatischen Ätherschwefelsäure-Verbindungen. Biochemische Zeitschrift 143 (1923), 186–189.
44. Noguchi, J.: Über die Hexose-mono-Phosphatase der Takadiastase. Biochemische Zeitschrift 143 (1923), 190–194.
45. Sen, H. K.: Über die carboxylatische Spaltung der Di-methyl-brenztraubensäure und die Herstellung der γ-Keto-iso-valeriansäure. Biochemische Zeitschrift 143 (1923), 195–200.

1924

46. Linhardt, K.: Stickstoffhaltige Kohlenhydrate; in: Handbuch der Biochemie des Menschen und der Tiere, hrsg. von Carl Oppenheimer. 2. Aufl. Bd. 1, Jena: G. Fischer 1924, 144–188.

47. Noguchi, J.: Über Sulfatase IV. Mitteilung. Über die enzymatische Spaltung der ß-Naphthol-schwefelsäure. Biochemische Zeitschrift 144 (1924), 138–140.
48. Takahashi, Y.: Über das Vorkommen von Inulase in der Takadiastase. Biochemische Zeitschrift 144 (1924), 199–202.
49. Takahashi, Y.: Über die Spaltung von Hexose-di-phosphorsäure durch verschiedene Organe. Biochemische Zeitschrift 145 (1924), 178–181.
50. Hatano, J.: Über Gelatine-schwefelsäure und Casein-schwefelsäure. Biochemische Zeitschrift 145 (1924), 182–185.
51. Tamba, R.: Hydrotropische Erscheinungen. Biochemische Zeitschrift 145 (1924), 415–421.
52. Takahashi, Y.: Über die enzymatische Zerlegung von Hexose-monophosphorsäure durch Extrakte des Femur. Biochemische Zeitschrift 146 (1924), 161–163.
53. Gottschalk, Alfred: Über tierische Carboxylase. Biochemische Zeitschrift 146 (1924), 582–588.
54. Gottschalk, Alfred: Der Acetaldehyd im intermediären Zellstoffwechsel. Klinische Wochenschrift 3 (1924), 713–715.
55. Gottschalk, Alfred: Die Beeinflussung des oxydativen Kohlenhydratabbaues durch Inkrete. Klinische Wochenschrift 3 (1924), 1356–1357.
56. Noguchi, J.: Über den Abbau von Nucleinsäuren durch Takadiastase. Biochemische Zeitschrift 147 (1924), 255–257.
57. Hatano, J.: Über „Taka-lab“. Biochemische Zeitschrift 149 (1924), 228–231.
58. Santomauro, Pietro: Über die phytochemische Reduktion des Methyl-γ-chlor-äthyl-ketons. Biochemische Zeitschrift 151 (1924), 48–50.
59. Sen, H. K.: Die biochemische Umwandlung von unsymmetrischem Dichlor-aceton in optisch-aktiven γ, γ-Di-chlor-iso-propyl-alkohol. Biochemische Zeitschrift 151 (1924), 51–53.
60. Hatano, J.: Über die Spaltung von Seidenfibroinpepton durch Takadiastase. Biochemische Zeitschrift 151 (1924), 335–336.
61. Kerb, Johannes/Kerb-Etzdorf, Elisabeth: Zur Kenntnis der Glucosane. Nachtrag zu der früheren Mitteilung. Biochemische Zeitschrift 151(1924), 435–437.
62. Hatano, J.: Amygdalinspaltung durch Takadiastase. Biochemische Zeitschrift 151 (1924), 498–500.
63. Hatano, J.: Über die Spaltung verschiedener ß-Glucoside durch Takadiastase. Biochemische Zeitschrift 151 (1924), 501–503.
64. Gottschalk, Alfred: Über den Wirkungsmechanismus von Glykokinin aus Hefe auf den Stoffumsatz von Leberzellen. Deutsche Medizinische Wochenschrift 50 (1924), 538.
65. Herzog, R. O./Kobel, Maria. Proteinstudien. II. Versuche von Molekulargewichtsbestimmungen an Seidenfibroin. Hoppe-Seyler 134 (1924), 296–299.

1925

66. Gottschalk, Alfred: Der Kohlenhydratumsatz in tierischen Zellen; in: Handbuch der Biochemie des Menschen und der Tiere, hrsg. von Carl Oppenheimer. 2. Aufl. Bd. 2, Jena: G. Fischer 1925, 485-521.
67. Gottschalk, Alfred: Neuere Ergebnisse über die ersten Stufen des biochemischen Kohlenhydratabbaues. Klinische Wochenschrift 4 (1925), 2454–2457.
68. Gottschalk, Alfred: Untersuchungen über die hormonale Regulation des intermediären Kohlenhydratstoffwechsels I. Mitteilung: Die Bedeutung von Adrenalin und Insulin für die Verwertung von Traubenzucker durch die Warmblüterzelle. Biochemische Zeitschrift 155 (1925), 348–355.
69. Rosenfeld, L.: Über die phytochemische Reduktion des γ, γ, β-Trichlorbutylaldehyds zum 2,2,3-Trichlorbutanol. Biochemische Zeitschrift 156 (1925), 54–57.
70. Yabusoe, Muneo: Über Eisen- und Blutfarbstoffbestimmungen in normalen Geweben und in Tumorgewebe. Biochemische Zeitschrift 157 (1925), 388–395.
71. Tanakai, Kanichi: Versuche zur Prüfung der Wielandschen Atmungstheorie. Biochemische Zeitschrift 157 (1925), 425–433.
72. Rosenfeld, L.: Über das Vorkommen und Verhalten der Sulfatase in menschlichen Organen VI. Mitteilung über Sulfatase. Biochemische Zeitschrift 157 (1925), 434–437.
73. Blaschko, Hermann: Über die Verbrennungswärme der Brenztraubensäure und ihre physiologische Bedeutung. Biochemische Zeitschrift 158 (1925), 428–434.
74. Hatano, J.: Über die partielle Hydrolyse der Rohrzucker-phosphorsäure zu d-Fructose und d-Glucose-phosphorsäure. Biochemische Zeitschrift 159 (1925), 175–178.
75. Gottschalk, Alfred: Untersuchungen über die hormonale Regulation des intermediären Kohlenhydratstoffwechsels II. Mitteilung: Demonstration des Insulin-Adrenalin-Antagonismus am lebenden Frosch. Biochemische Zeitschrift 159 (1925), 502–506.
76. Gorr, Günther/Wagner, Joachim: Über eine neue Methode zur Trennung des Äthylalkohols von Acetaldehyd und Aceton. Biochemische Zeitschrift 161 (1925), 488–491.
77. Sabetay, S[ebastian]/Rosenfeld, L.: Über Glucose-phosphorsäure. Biochemische Zeitschrift 162 (1925), 469–478.
78. Iwatsuru, Ryuzo: Über Polysaccharid-spaltung durch Hefe. Biochemische Zeitschrift 166 (1925), 409–413.

1926

79. Iwatsuru, Ryuzo: Über die Vergärung der γ-Ketoglutarsäure durch das Bacterium Xylinum. Biochemische Zeitschrift 168 (1926), 34–35.
80. Elion, L.: Zur Kenntnis der Acetoinbildung bei der Vergärung von Zucker durch Hefe. Biochemische Zeitschrift 169 (1926), 471–477.

81. Behrens, Martin/Nikolai, Nikolajewitsch Iwanoff: Neue Versuche über die Carboligase. Biochemische Zeitschrift 169 (1926), 478–481.

82. Iwatsuru, R[yuzo]: Über die Spaltung der Mono-phenyl-phosphorsäuren und Mono-Äthyl-phosphorsäuren Salze durch pflanzliche und tierische Phosphatase. Biochemische Zeitschrift 173 (1926), 348–357.

83. Gorr, Günther/Lusmann-Perlmann, Genia: Über die Einwirkung des Sauerstoffs auf den Verlauf der alkoholischen Zuckerspaltung. Biochemische Zeitschrift 174 (1926), 425–432.

84. Gorr, Günther/Lusmann-Perlmann, Genia: Weitere Untersuchungen über die Bildung von Milchsäure aus Methylglyoxal durch Ketonaldehydmutase tierischer und pflanzlicher Herkunft. Biochemische Zeitschrift 174 (1926), 433–439.

85. Binder-Kotrba, G.: Über die Vergärung der Phenylglyoxylsäure. Biochemische Zeitschrift 174 (1926), 440–442.

86. Binder-Kotrba, G.: Die Umwandlung von Phenylglyoxal in Mandelsäure durch die Ketonaldehydmutase grüner Pflanzen. Biochemische Zeitschrift 174 (1926), 443–447.

87. Binder-Kotrba, G.: Über die Dismutation des Acetaldols durch Bacterium ascendens. Biochemische Zeitschrift 174 (1926), 448–451.

88. Nagelschmidt, Günther: Gestufte phytochemische Reduktionen. Biochemische Zeitschrift 186 (1927), 317–321.

1927

89. Nagelschmidt, Günther: Über die Fällbarkeit der Diaminosäuren durch Mercuriacetat und Soda. Biochemische Zeitschrift 186 (1927), 322–326.

90. Klar, Ludwig: Ketonaldehydmutase in Weizen- und Roggenkörnern. Biochemische Zeitschrift 186 (1927), 327–330.

91. Maurer, Kurt: Über die biochemische Überführung von Oximinobrenztraubensäure in Alanin. Biochemische Zeitschrift 189 (1927), 216–219.

92. Kitasato, Torao: Die partielle Hydrolyse des Populins zu Saligenin und Benzoylglucose durch ein Enzym der Takadiastase. Biochemische Zeitschrift 190 (1927), 109–113.

93. Kageura, Naomi: Einwirkung des Bacteriums lactis aerogenes und des Bacteriums coli auf Hexose-Monophosphorsäure. Biochemische Zeitschrift 190 (1927), 181–187.

94. Maurer, Kurt: Beobachtungen über die Zuckerspaltung durch das Bacterium propionicum. Biochemische Zeitschrift 191 (1927), 83–87.

1928

95. Leibowitz, Jesaia: Vergleichende Versuche über die Vergärbarkeit der Zymohexosen, des Glykogens und der Stärke. Hoppe-Seylers Zeitschrift für Physiologische Chemie 173 (1928), 84–88.

96. Wagner, Joachim: Die quantitative Bestimmung von Acetaldehyd nach verschiedenen Methoden. Biochemische Zeitschrift 194 (1928), 441–452.

97. Kitasato, Torao: Acyloinaufbau mittels Enzymen der Essigbakterien. Biochemische Zeitschrift 195 (1928), 118–127.
98. Kitasato, Torao: Über Meta-phosphatase. Biochemische Zeitschrift 197 (1928), 257–258.
99. Kobel, Maria/Tychowski, Alexander: Biochemische Spaltung des Zukkers nach der zweiten Vergärungsform unter dem Einfluss von Carbaminsäurehydrazid und Thio-carbaminsäurehydrazid. Isolierung von Acetaldehyd und Glycerin. Biochemische Zeitschrift 199 (1928), 218–229.
100. Kitasato, Torao: Weitere Untersuchungen über die Metaphosphatase. Biochemische Zeitschrift 201 (1928), 206–211.
101. Kobel, Maria/Roth, Walther A.: Über die Verbrennungs- und Lösungswärme des Dioxyacetons. Biochemische Zeitschrift 203 (1928), 159–163.
102. Kobel, Maria: Neuere Forschungen über die Assimilation. Die Naturwissenschaften 16 (1928), 457–460.

1929

103. Lüdtke, Max: Zucker; in: Handbuch der Ernährung und des Stoffwechsels der landwirtschaftlichen Nutztiere als Grundlage der Fütterungslehre, hrsg. von Ernst Mangold. Berlin: Springer 1929, 10–30.
104. Lüdtke, Max: Andere stickstofffreie Extraktstoffe; in: Handbuch der Ernährung und des Stoffwechsels der landwirtschaftlichen Nutztiere als Grundlage der Fütterungslehre, hrsg. von Ernst Mangold. Berlin: Springer 1929, 30–43.
105. Lüdtke, Max: Die Substanzen der pflanzlichen Zellmembran; in: Handbuch der Ernährung und des Stoffwechsels der landwirtschaftlichen Nutztiere als Grundlage der Fütterungslehre, hrsg. von Ernst Mangold. Berlin: Springer 1929, 44–72.
106. Andreadis, Thales: Über den Nitratgehalt des Tabaks und über eine Fehlerquelle der bisherigen Bestimmungsmethode. Biochemische Zeitschrift 204 (1929), 484–492.
107. Weinmann, Fritz: Darstellung von Silbersalzen der Hexosen-monophosphorsäure-ester. Biochemische Zeitschrift 204 (1929), 493–494.
108. Weinmann, Fritz: Über Sulfatase X. Biochemische Zeitschrift 205 (1929), 214–218.
109. Gualdi, Antonio: Über gerichtete enzymatische Dismutation durch die Ketonaldehyfmutase vom B. subtilis. Biochemische Zeitschrift 205 (1929), 318–319.
110. Gualdi, Antonio: Die Wirkungsweise tierischer Phosphatase in stereochemischer Beziehung. Biochemische Zeitschrift 205 (1929), 320–324.
111. Hayashi, Shigeru: Experimentelle Untersuchungen über die sterischen Verhältnisse bei der Dismutation von Phenylglyoxal-hydrat durch verschiedene Bakterien (B. proteus, B. fluorescens, B. pyocyaneum, B. prodigiosum und B. coli). Biochemische Zeitschrift 206 (1929), 223–227.
112. Kitasato, Torao: Weitere Synthesen von γ-Ketosäuren der Kohlenhydratreihe (γ-Keto-d-galaktonsäure oder d-Tagaturonsäure sowie γ-

Keto-maltobionsäure oder d-Glucosido-d-fructuronsäure). Biochemische Zeitschrift 207 (1929), 217–229.

113. Fromageot, Claude: Die stereochemische Spezifität der Sulfatase. (Über Sulfatase XI.). Biochemische Zeitschrift 208 (1929), 482–489.

114. Fromageot, Claude: Über die asymmetrische Verwertung des p-sek.-Butylphenols im Tierkörper. Biochemische Zeitschrift 208 (1929), 490–492.

115. Vogt, Marthe: Über Glycerin-mono-phosphorsäure, Biochemische Zeitschrift 211 (1929), 1–16.

116. Tychowski, Alexander/Maria Kobel: Über eine schnell verlaufende Umwandlung von Hexose-di-phosphat in Lactat unter dem Einfluss des typischen Milchsäurebildners Bacillus Delbrücki, Biochemische Zeitschrift 209 (1929), 134–141.

117. Vogt, Marthe: Bildung und Isolierung von Methylglyoxal bei der Glykolyse durch tierisches Enzym. Biochemische Zeitschrift 211 (1929), 17–39.

118. Andreadis, Thales: Untersuchungen über Vorgänge bei der Tabakfermentation. Biochemische Zeitschrift 211 (1929), 378–394.

119. Lüdtke, Max: Untersuchungen über Agar. Biochemische Zeitschrift 212 (1929), 419–429.

120. Lüdtke, Max: Über Inulin-phosphorsäure. Biochemische Zeitschrift 212 (1929), 475–476.

121. Pi-Suner Bayo, César: Gewinnung von Methylglyoxal bei der Spaltung von Zucker durch Hefenmazerationssaft. Biochemische Zeitschrift 213 (1929), 489–494.

122. Pi-Suner Bayo, César: Studien über die Dismutation von Methylglyoxal und Phenylglyoxal durch das Enzym grüner Blätter. Versuche mit Lindenblättern. Biochemische Zeitschrift 213 (1929), 495–500.

123. Molinari, Emil: Über die biochemische Dismutation nebst Untersuchungen über die Essiggärung. Biochemische Zeitschrift 216 (1929), 187–215.

124. Kobel, Maria/Scheuer, Max: Über den Kohlenhydratumsatz im Tabakblatt. Nachweis von Methylglyoxal als Zwischenprodukt im Stoffwechsel grüner Blätter. Biochemische Zeitschrift 216 (1929), 216–223.

125. Fromageot, Claude: Über den Kohlenhydratabbau durch Bacterium coli. Die Isolierung von Methylglyoxal. Biochemische Zeitschrift 216 (1929), 467–474.

126. Widmann, Ernst: Quantitative Umwandlung von Methyl-glyoxal in reine d(-) Milchsäure durch das Bacterium fluorescens. Biochemische Zeitschrift 216 (1929), 475–478.

127. Widmann, Ernst: Methylglyoxal als Produkt der Glykolyse durch Blutkörperchen. Biochemische Zeitschrift 216 (1929), 479–492.

128. Vogt, Marthe: Die Isolierung von Methylglyoxal als Zwischensubstanz bei der Glykolyse. Klinische Wochenschrift 8 (1929), 1–4.

1930

129. Schwaibold, Julius: Über den Jodgehalt frischer, dachreifer und fermentierter Tabakblätter. Biochemische Zeitschrift 218 (1930), 318–320.
130. Yamasaki, Izue: Über das desmolytische System von Torula colliculosa. 2. Mitt.: Carboxylase. Biochemische Zeitschrift 218 (1930), 252–254.
131. Yamasaki, Izue: Über das desmolytische System von Torula colliculosa. 1. Mitt.: Glycolase und Bildung von Methylglyoxal. Biochemische Zeitschrift 218 (1930), 468–471.
132. Neuberg, Irene Stephanie: Zur Frage nach der Enolstruktur der Brenztraubensäure und ihrer aliphatischen Abkömmlinge. Biochemische Zeitschrift 219 (1930), 165–170.
133. Falkenhausen, Friedrich Freiherr von: Über das Verhalten des Methylbenzoyl-carbinols zu gärender Hefe. Biochemische Zeitschrift 219 (1930), 241–247.
134. Neuberg, Irene Stephanie/Ostendorf, Clara: Über das besondere Verhalten der d-Mannose bei der biochemischen Phosphorylierung. Biochemische Zeitschrift 221 (1930), 154–165.
135. Neuberg, Irene Stephanie: Über eine neue Darstellung von Glycerinaldehyd und Glycerin. Biochemische Zeitschrift 221 (1930), 492–493.
136. Simon, Ernst: Über das zymatische System und die Wirkungen der Essigbakterien. Biochemische Zeitschrift 224 (1930), 253–291.
137. Kuroya, Masahiko: Über willkürlich beeinflusste asymmetrische Spaltung der d, l-Borneolphosphorsäure durch Hefen- sowie Takaphosphatase. Biochemische Zeitschrift 225 (1930), 452–463.
138. Yamasaki, Izue: Über die Wirkung von mono-jod-essigsaurem Natrium auf die Teilfermente der Zymase und die Vergärung von Hexose-di-phosphat. Biochemische Zeitschrift 228 (1930), 123–126.
139. Yamasaki, Izue: Das zymatische System von Saccharomyces Johannisberg. Biochemische Zeitschrift 228 (1930), 127–134.
140. Kobel, Maria/Scheuer, Max: Die Bilanz der 4. Vergärungsform bei der zellfreien Hefegärung. Biochemische Zeitschrift 229 (1930), 238–247.

1931

141. Veibel, Stig: Über die Methylglyoxalyl-essigsäure und ihre Dismutation durch Bacterium coli. Biochemische Zeitschrift 232 (1931), 435–441.
142. Sym, Ernst: Die quantitativen Verhältnisse bei der Bildung und Umwandlung von Methylglyoxal durch Kaltblütermuskulatur. Biochemische Zeitschrift 233 (1931), 251–257.
143. Kuroya, Masahiko: Bildung und Umwandlung von Methylglyoxal durch Fermente der Tuberkel- und der Timothee-Bazillen. Biochemische Zeitschrift 235 (1931), 438–443.
144. Kuroya, Masahiko: Bildung von Methylglyoxal durch das wasserlösliche Ferment des Rattensarkoms. Biochemische Zeitschrift 235 (1931), 444–446.

145. Kuroya, Masahiko: Nachweis und quantitative Bestimmung der Lactose im Harn. Biochemische Zeitschrift 235 (1931), 447–450.
146. Fujise, Shin-ichiro: Über die phytochemische Reduktion der Oxalessigsäure zu Äpfelsäure. Biochemische Zeitschrift 236 (1931), 231–236.
147. Fujise, Shin-ichiro: Dismutative Umwandlung von Methylglyoxalylessigsäure in d-γ-Oxy-glutarsäure mit dem Enzym tierischer Zellen. Biochemische Zeitschrift 236 (1931), 237–240.
148. Fujise, Shin-ichiro: Über das Thienyl-glyoxal und seine biochemische Dismutation. Biochemische Zeitschrift 236 (1931), 241–246.
149. Neuberg, Irene Stephanie: Mono-methyläther und Mono-äthyläther des Dioxyacetons. Biochemische Zeitschrift 238 (1931), 459–460.
150. Veibel, Stig: Über die Vorgänge bei der Phosphorylierung von Zucker durch Hefe, insbesondere über die Veränderungen während der Periode der totalen Phosphorylierung. Biochemische Zeitschrift 239 (1931), 350–373.
151. Veibel, Stig: Phytochemische Darstellung von 4,5-Octandiol aus Butyroin. Biochemische Zeitschrift 239 (1931), 456–460.
152. Falkenhausen, Friedrich Freiherr von: Über Versuche zur Trennung des synthetischen racemischen Amylalkohols [2-Methyl-butanol-(1)] in die optischen Antipoden. Biochemische Zeitschrift 242 (1931), 472–481.
153. Simon, Ernst: Aldehyd-dismutation und Essiggärung. Biochemische Zeitschrift 243 (1931), 401–405.
154. Kobel, Maria: Die Bildung äquimolekularer Mengen von Glycerin und Brenztraubensäure bei der zellfreien Vergärung von Glucose. Biochemische Zeitschrift 243 (1931), 406–415.
155. Hofmann, Eduard: Über den Abbau von Glucose-ureid durch Bakterien. Biochemische Zeitschrift 243 (1931), 416–422.
156. Hofmann, Eduard: Über die Bildung von Oxalsäure aus Uronsäuren durch Aspergillus niger. Biochemische Zeitschrift 243 (1931), 423–428.
157. Hofmann, Eduard: Über die Verhältnisse bei der carboxylatischen Spaltung der γ-Oxo-n-valeriansäure. Biochemische Zeitschrift 243 (1931), 429–434.

1932

158. Simon, Ernst: Überführung von Milchsäure in Brenztraubensäure mittels Bakterium Delbrücki. Biochemische Zeitschrift 245 (1932), 488–493.
159. Simon, Ernst: Bestimmung von Furfurol und Methyl-furfurol als 2,4-Dinitrophenylhydrazon. Biochemische Zeitschrift 247 (1932), 171–177.
160. Imanishi, Yoshio: Neue Versuche über die Wirkung der Phosphatase gegenüber verschiedenen Substraten. Biochemische Zeitschrift 247 (1932), 406–417.
161. Tankó, Béla: Über die Bildung von Acetoin durch Termobakterium mobile Lindneri. Biochemische Zeitschrift 247 (1932), 482–485.
162. Tankó, Béla: Spaltung der Glucose-schwefelsäure und Saccharoseschwefelsäure durch Bakteriensulfatase. Biochemische Zeitschrift 247 (1932), 486–490.

163. Falkenhausen, Friedrich Freiherr von: Darstellung und enzymatische Spaltung verschiedener Amylphosphorsäuren. Biochemische Zeitschrift 253 (1932), 152–160.
164. Ochiai, Eiji: Gewinnung und enzymatische Hydrolyse von l- und d,l-Difenchyl-pyro-phosphorsäure. Biochemische Zeitschrift 253 (1932), 185–192.
165. Simon, Ernst: Beobachtungen über Gärungshemmung durch Halogen-Verbindungen. Biochemische Zeitschrift 253 (1932), 218–221.
166. Hofmann, Eduard: Über das physiologische Verhalten der Ureido-lactose und der Ureiodo-maltose, insbesondere über die enzymatische Spaltung der beiden Disaccharid-ureide. Biochemische Zeitschrift 253 (1932), 462–469.
167. Brücke, Franz von: Über die phosphatische Spaltung stellungsisomerer Di-aryl-phosphorsäuren. Biochemische Zeitschrift 253 (1932), 470–481.
168. Neuberg, Irene Stephanie: Untersuchungen in der 3-Kohlenstoffreihe: I. Neue Synthese des Glycerinaldehyds und Glycerins. II. Chemische und biochemische Gewinnung von Dioxyaceton-mono-alkyläthern. III. Anhang: Über kristallisiertes γ-Phenyl-glycerin, über einen Nachweis von Acrolein sowie über Methylglyoxal. Biochemische Zeitschrift 255 (1932), 1–26.
169. Collatz, Herbert/Neuberg, Irene Stephanie: Über ein einfaches Verfahren der Gewinnung von Zuckern aus ihren Hydrazonen. Biochemische Zeitschrift 255 (1932), 27–37.
170. Hofmann, Eduard: Die ß-Glykosidase der Milchzuckerhefen. Biochemische Zeitschrift 256 (1932), 462–474.
171. Klopstock, Felix/Simon, Ernst: Zur Frage des Vorkommens tumorspezifi-scher Antigene in bösartigen Geschwülsten (Versuche am Rattensarkom). Zeitschrift für Krebsforschung 37 (1932), 127–136.
172. Simon, Ernst: Ketonaldehydmutase und Glykolase bei echten Milchsäurebakterien. Zentralblatt für Bakteriologie, Parasitenkunde und Infektionskrankheiten. II. Abt. 85 (1932), 269–276.

1933

173. Hagenguth, K[urt]: Summarische Analyse der unbekannten Harnsubstanzen. Biochemische Zeitschrift 257 (1933), 484–491.
174. Grzycki, Stefan: Teilfermente im Enzymsystem der Milchzuckerhefen. Biochemische Zeitschrift 265 (1933), 191–194.
175. Grzycki, Stefan: Teilfermente im Enzymsystem der Milchzuckerhefen. Biochemische Zeitschrift 265 (1933), 195–198.
176. Collatz, Herbert: Über Nachweis und Bestimmung von Tri-bromäthylalkohol. Biochemische Zeitschrift 265 (1933), 199–202.
177. Hofmann, Eduard: Über die Phosphorylierung durch Lactose-Hefe. Biochemische Zeitschrift 265 (1933), 203–208.
178. Hofmann, Eduard: Über Lactase in untergärigen Hefen. Biochemische Zeitschrift 265 (1933), 209–212.

179. Auhagen, Ernst: Die Umsetzung zugefügten Methylglyoxals durch Blut und Muskulatur. Biochemische Zeitschrift 265 (1933), 213–216.

180. Auhagen, Ernst/Grzycki, Stefan: Der Einfluss von Fluorid auf die Phosphatasewirkung. Biochemische Zeitschrift 265 (1933), 217–222.

181. Hofmann, Eduard: Über das Emulsin der Hagebutten (rosa canina). Biochemische Zeitschrift 267 (1933), 309–312.

182. Antoniani, C[laudio]: Über die Umwandlung der optisch aktiven Phosphoglycerinsäure durch Bacterium coli. Biochemische Zeitschrift 267 (1933), 376–379.

183. Antoniani, C[laudio]: Über den Abbau der d,l-Glycerinsäure durch B. coli und B. lactis aerogenes. Biochemische Zeitschrift 267 (1933), 380–382.

184. Baba, Tamezi: Vergleichende Untersuchungen über die biochemische Bildung von optisch aktiver Phosphoglycerinsäure aus verschiedenen Zuckern. Biochemische Zeitschrift 267 (1933), 452–455.

185. Cattaneo, Carlo: Gewinnung von Phosphoglycerinsäure aus Galaktose und Spaltung von Phosphoglycerinsäure durch Galaktosehefen. Biochemische Zeitschrift 267 (1933), 456–459.

1934

186. Windisch, Fritz: Gärung. Bildung von organischen Säuren; in: Handwörterbuch der Naturwissenschaften, hrsg. von Rudolf Dittler. 2. Aufl., Bd. 4, Jena: Fischer 1934, 643–653.

187. Simon, Ernst: Gärung. Eiweiß; in: Handwörterbuch der Naturwissenschaften, hrsg. von Rudolf Dittler. 2. Aufl., Bd. 4, Jena: Fischer 1934, 654–664.

188. Kobel, Maria/Collatz, Herbert: Zur Frage des Übergangs von Glycerinaldehyd-3-phosphorsäure in Methylglyoxal und Milchsäure. Biochemische Zeitschrift 268 (1934), 202–204.

189. Auhagen, E[rnst]/Auhagen, T[rude]: Über Umschaltung der alkoholischen Zuckerspaltung durch Hefe in Milchsäuregärung II. Biochemische Zeitschrift 268 (1934), 247–252.

190. Gaffron, Hans: Über die Kohlensäure-Assimilation der roten Schwefelbakterien. I. Biochemische Zeitschrift 269 (1934), 447–453.

191. Cattaneo, Carlo: Über die Einwirkung von Timotheebakterien auf die links-drehende Glycerinsäure-mono-phosphorsäure. Biochemische Zeitschrift 270 (1934), 382–385.

192. Schoenebeck, Otto von: Versuche zur Reinigung von Carboxylase. Biochemische Zeitschrift 272 (1934), 42–50.

193. Hofmann, Eduard: Über das Vorkommen von Glucosidasen bzw. Galaktosidasen und Disaccharide spaltenden Enzymen in Bakterien. Biochemische Zeitschrift 272 (1934), 133–143.

194. Hofmann, Eduard: Vorkommen von Maltase und Saccharase bei Schizo-Saccharomyces octosporus (Beijerinck) und deren Trennung. Biochemische Zeitschrift 272 (1934), 417–425.

195. Hofmann, Eduard: Untersuchungen über verschiedene Pflanzenemulsine. Biochemische Zeitschrift 272 (1934), 426–429.
196. Cattaneo, Carlo: Über die enzymatische Spaltung der Cellobiose-carbonsäure. Biochemische Zeitschrift 272 (1934), 430–433.
197. Schuchardt, W[erner]/Vercellone, A[lberto]: Abbau der linksdrehenden Mono-phosphoglycerinsäure durch frische untergärige Hefe. Biochemische Zeitschrift 272 (1934), 434.
198. Schuchardt, W[erner]/Vercellone, A[lberto]: Weiteres über die Vergärung der Glycerinsäure-di-phosphorsäure durch Hefe. Biochemische Zeitschrift 272 (1934), 435–436.
199. Schuchardt, W[erner]/Vercellone, A[lberto]: Verhalten von optisch aktiver Mono- und Diphosphoglycerinsäure gegenüber Erythrozyten verschiedener Herkunft. Biochemische Zeitschrift 272 (1934), 437–440.
200. Hofmann, Eduard: Untersuchungen über Glykoside und Disaccharide spaltende Enzyme von Schimmelpilzen. Biochemische Zeitschrift 273 (1934), 198–206.
201. Baba, Tamezi: Über die besonderen Verhältnisse bei der Phosphorylierung von Maltose durch frische untergärige Hefe, Trockenhefe und Mazerationssaft. Biochemische Zeitschrift 273 (1934), 207–218.
202. Antoniani, Claudio: Über die enzymatische Spaltung der Cellobionsäure. Biochemische Zeitschrift 273 (1934), 219–222.
203. Hofmann, Eduard: Neues zur Frage nach der Spezifität der Glykosidasen, insbesondere bei Schimmelpilzen und Bakterien. Die Naturwissenschaften 22 (1934), 406–409.

1935

204. Baba, Tamezi: Über den Abbau des Hexose-di-phosphats durch ein Enzym der Tabakblätter. Biochemische Zeitschrift 275 (1935), 248–252.
205. Baba, Tamezi: Die Fällung von Zuckern mit methylalkoholischem Bariumhydroxyd. Biochemische Zeitschrift 275 (1935), 253–260.
206. Schuchardt, W[erner]/Vercellone, A[lberto]: Weiteres über das Verhalten von Mono- und Di-phosphoglycerinsäure gegenüber Hämolysaten und Organsäften. Biochemische Zeitschrift 275 (1935), 261.
207. Gaffron, Hans: Über den Stoffwechsel der Purpurbakterien. II. Biochemische Zeitschrift 275 (1935), 301–319.
208. Hofmann, Eduard: Über Beziehungen zwischen Wasserstoffzahl und Abkunft verschiedener Carbohydrasen. Biochemische Zeitschrift 275 (1935), 320–327.
209. Schoenebeck, Otto von: Stabilisierung von Carboxylaselösungen. Neuberg, Carl: Fortgesetzte Untersuchungen über die Carboxylase. Biochemische Zeitschrift 275 (1935), 330–338.
210. Klopstock, Felix/Cattaneo, Carlo: Chemischer Aufbau und antigenes Verhalten des Tuberkelbazillus. Zeitschrift für Immunitätsforschung 84 (1934), 34–45.

211. Schoenebeck, Otto von: Zur Frage der Vergärbarkeit der Glycerinsäure. Biochemische Zeitschrift 276 (1935), 421–424.
212. Schoenebeck, Otto von: Nachträge zur Abhandlung „Stabilisierung von Carboxylaselösungen". Biochemische Zeitschrift 277 (1935), 451–452.
213. Schuchardt, Werner: Über Spaltung von γ- und ß-Glycero-phosphaten durch frische Hefe und gelöstes Hefeferment. Biochemische Zeitschrift 278 (1935), 164–172.
214. Collatz, Herbert: Die enzymatische Hydrolyse der Dioxyaceton-monophosphorsäure. Biochemische Zeitschrift 278 (1935), 364–371.
215. Vercellone, A[lberto]: Über die phytochemische Reduktion von Anthradi-chinon. Biochemische Zeitschrift 279 (1935), 137–139.
216. Hofmann, Eduard: Über die Hydrolyse der ß-Glykolyse und der gepaarten Glucuronsäuren durch Fermente der Leber und Niere. Biochemische Zeitschrift 281 (1935), 438–441.
217. Schuchardt, Werner: Beitrag zur Spezifität der Phosphatasen. Biochemische Zeitschrift 283 (1936), 433–434.
218. Hofmann, Eduard: Zur Spezifität der Glykosidasen. Biochemische Zeitschrift 285 (1936), 429–447.

1936–1955

219. Schuchardt, Werner: Über die Phosphatasen von Hefe. Biochemische Zeitschrift 285 (1936), 448–456.
220. Gaffron, Hans: Metastabiler Sauerstoff und Kohlensäure-Assimilation. Biochemische Zeitschrift 287 (1936), 130–139.
221. Hofmann, Eduard: Über die Glykosidasen der Spalthefen. Biochemische Zeitschrift 287 (1936), 271 275.
222. Neuberg-Rabinowitsch, Irene: Vergärung von Maltose-carbonsäure und Melibionsäure. Enzymologia 3 (1937), 41–43.
223. Forrest, I[rene] S[tephanie]: Enzymatic Hydrolysis of O, O-Diisopropyl Thionothiolphosphate. Archives of Biochemistry and Biophysics 58 (1955), 178–180.

12. INDEX OF SOURCES AND LITERATURE

12.1. ARCHIVE MATERIALS AND UNPUBLISHED SOURCES

Archives of the Academy of Sciences, Göttingen:
- Pers. 20, Neuberg, Carl (1052).

Archives of the Czechoslovak Academy of Agriculture, Prague:
- CAZ, inv. C. 27/N47, k. 109.

Archives of the Free University, Berlin (AFU):
- HSA FUB: Med. Fak./Dekanat, Minutes of Faculty Meetings (meeting on 16.06.1952).
- HSA FUB: Rektorat/Aussenamt, file "Aussenkommission: Gastprofessoren. Buchstaben Na-Nh", 1950s and 1960s, Neuberg correspondence.

Archives of the German Academy of Sciences Leopoldina (Deutsche Akademie der Naturforscher Leopoldina), Halle:
- M No. 3476.

Archives of Hanover Medical School
- Akte 1/Gründungsgeschichte (1995/96): Die Bauplaner der Medizinischen Hochschule Hannover (File 1/foundation (1995/96): The planners of the buildings of Hanover Medical School). Jan W. Prendel, Karl Dieckmann, Konstanty Gutschow, Godber Nissen.

Archives of the Humboldt University, Berlin (AHU):
- Charité, 933.
- Neuberg's Personnel File, UK-N 29, Vols. 1–3.
- Faculty of Philosophy, 357: Doctorate documents.
- Faculty of Philosophy, 1226: *habilitation* documents.
- Faculty of Philosophy, 1438.
- Faculty of Philosophy, 1439.
- Faculty of Philosophy, 1468.

Archives of the "Kongelige Danske Videnskabernes Selskab", Copenhagen:
- Prot. No. 213 (1929).

Archives of New York University (ANYU):
- Refugee Appointments – 1940–1941, Office of Vice President Harold O. Voorhis, Administrative Subject Files, 1925–1963.
- The NYU Album 1942.
- New York University Directory (1941–1942).
- New York University Directory (1942–1943).
- New York University Directory (1949–1950).

Archives of Springer Verlag, Heidelberg (ASV):
- B. N. 32.
- B. B. 158.
- D. 38.7, Neuberg, Carl.
- E. 72.1, Neuberg, Carl.

Archives on the History of the Max Planck Society, Berlin (MPG Archives):
- Sec. I, Rep. 1A, 1445–1449, 1455, 1459, 1463a, 2031–2032, 2035, 2036, 2040, 2041.
- Sec.. II, Rep. 1A, Neuberg's Personnel File.
- Sec.. III, Rep. 84/1.
- Collection of pictures.

Archivum Panstwowe w Ponaniu, Poznan:
- Kartoteka mieszkancow Poztnania z lat 1870–1930 [civil status register].

Carl Neuberg Papers (Recatalogued by rsc 2002) (CNP). American Philosophical Society. Philadelphia. Series I. Correspondence: *

Box 1
- Abderhalden, Emil [1945–1949]
- Adams, Roger [1946–1948]
- Addinal, C. R. [1946]*
- Adenauer, Konrad [1954]
- American Cancer Society [1949–1956]
- Ammon, Robert [1953–1956]

* Initials are given complete wherever possible, and otherwise correspond to the form in which they are found in the Carl Neuberg Papers [www.amphilsoc.org/library/mole/n/neuberg.htm, 02/21/2006]. See also Kay (1989).

Box 2
- Anschütz, Ludwig [1948–1955]
- Antoniani, Claudio [1947–1955]
- Aron, Hans C. S. [1942–1955, 2 folders]
- Ascorbic Acid [1922–1940]
- Barken, G. [1943–1944]
- Beitcher, Kurt [1955]

Box 3
- Bergmann, Ernst [1942–1955]
- Berlak, Millie [1941–1955]
- Bersin, Theodor [1952–1955]
- Bloch-Frankenthal, Leah [1945–1956]
- Böhm, Ernest [1951]
- Burk, Dean [1952–1956]
- Butenandt, Adolf [1947–1956]
- Carl Neuberg Society for International Scientific Relations [1964]

Box 4
- Collatz, Herbert [1951–1955]
- Corbiere, Henri [1953]
- Dazian Foundation for Medical Research [1943–1950, 2 folders]
- Deuticke, Hans Joachim [1951–1956]
- Deutsche Physiologisch-Chemische-Gesellschaft [1953]
- Durig, Arnold [1946–1954]
- Ella Sachs Plotz Foundation [1942–1950, 2 folders]

Box 5
- Fodor, Andor [1942–1954]
- Frank, Alfred [1946–1947]
- Fritzsching, Erwin T. [1942–1944]
- Fromageot, Henri Claude [1940–1953]
- Fürst, Walter M. [1947]
- Gaffron, Hans [1943–1956]
- Gisel, Father Eugene A. [1948]
- Glum, Friedrich [1947–1948]
- Glycerine - Misc. Correspondence [1916]

Box 6
- Goldenberg, Geromont & Co. – Chemische Fabrik [1916–1918]
- Goldschmiedt, Henry [1942–1956]
- Häusler, H. [1952–1955]
- Hahn, Otto [1947–1956]
- Hamburger, Adolf [1941–1954]
- Hebrew University of Jerusalem [1940–1956]
- Heubner, Wolfgang [1952–1954]
- Hofmann, Eduard [1948–1955]

Box 7
- Jacobsohn, Kurt [1929–1956, 5 folders]
- Jander, G. [1952]
- Java Pacific Line [1940]
- Jewish Cultural Reconstruction, Inc. [1951]
- Kaiser-Wilhelm-Gesellschaft [1925–1949, 2 folders]
- Kaiser-Wilhelm-Institut für Biochemie [1927]
- KWI für Physikalische Chemie und Elektrochemie [1916]
- Karpen, M. [1952–1953]
- Kimura, T. [1954]
- Kindler, Hilde [1942]
- Kisch, Bruno [1946–1956]
- Kobel, Maria [1948–1956]
- Krebs, Hans A. [1954–1955]
- Krekeler, Heinz L. [1951–1954]
- Kriegschemikalien Aktiengesellschaft [1916]
- Kriegsministerium-Allgemeines-Kriegs-Department [1917]
- Krolof-Becker, Mignon [1946–1948]
- Kuhn, Richard [1948]

Box 8
- Lang, K. [1947]
- Lederer, Marianne [1943–1956]
- Linke, Agnes [1947–1953]
- Loewi, Otto [1951–1956]
- Loon, Hendrik von [1943]
- Lynen, Fedor [1951–1955]
- MacTavish, W. C. [1942–1948]
- Madison Foundation for Biochemical Research [1947–1954, 2 folders]
- Mark, Hermann F. [1947–1956]

Box 9
- Max-Planck-Gesellschaft [1948–1956, 3 folders]
- Mayer, Paul [1942–1943]
- Mazia, Daniel [1955–1956]
- Micheel, Fritz [1947–1956]

Box 10
- Nathan, Blanche [1942–1947]
- National Vitamin Foundation [1951]
- Neuberg, Carl:
  - Awards and Diplomas [1926–1956, 4 folders, see also oversize and extra-oversize folders]
  - Biographical Material

Box 11
- Neuberg, Carl:
  - Birthday Congratulations [1952, 3 folders]
  - Manuscript "Jews in Biochemistry". Also includes biographical notes on Alfred Gottschalk (1p.), autographical notes on Neuberg (1p.), and 2 TLSs relating to publication of the article [1947, 12pp.]
  - Manuscripts – misc. [2 folders]

Box 12
- Neuberg, Carl:
  - Patents
  - Personal [1898–1956, 2 folders]
- Neuberg, Ernst [1947–1954]
- Neuberg, Julius [1946]
- New York Medical College
- New York University [1942–1955, 2 folders]
- Nord, (Frans) Friedrich [1942–1956]
- Overberger, Charles G. [1942]
- Ochoa, Severo [1947–1955]

Box 13
- Photographs
- Plesch, Johannes [1947]
- Polytechnic Institute of Brooklyn [1951–1952, 2 folders]
- Potter, Van R. [1943]
- Reindel, F. [1952–1955]
- Reinold, Marianne [1948–1956]
- Reiter, Melchior [1951–1955]
- Roberts, Irene Neuberg (see also Forrest, Irene S.) [1943]

Box 14
- Roth, Walter [1946–1947]
- Sabalitschka, Theodor [1947–1952]
- Sachs, Helena [1943–1956]
- Scheminzky, F. [1951–1953]
- Schenley Research Institute [1943]
- Schlitter, E. [1949]
- Schönebeck, Otto von [1947–1956]

Box 15
- Simon, Ernst [1941–1956]
- Society for the Protection of Science and Learning [1942–1947]
- Sugar Research Foundation [1944]
- Swift & Company [1944–1945]
- Szemzö, Bela [1952–1955]
- Thilo, Erich [1951–1956]
- Thomas, Karl [1949–1954]
- Tobacco-Misc. [1940s]
- Tomita, Masaji [1953–1956]

Box 16
- U. S. Atomic Energy Commission [1949–1953, 4 folders]
- U. S. Coordinator of Information [1942]
- U. S. Department of the Navy. Office of Naval Research [1950–1955, 4 folders]
- U. S. Public Health Services (folders 1–4) [1943–1955, 4 folders]

Box 17
- U. S. Public Health Services (folders 5–8) [1943–1955, 4 folders]
- Voorhis [1940]
- Wastel, H. [1946]
- Warburg, Otto [1948–1956]
- Warwick Chemical Company [1943]
- Wertheimer, E. [1943–1945]

Box 18
- Wieland, Heinrich O. [1946–1955]
- Windaus, Adolf [1947–1952]
- Windisch, Fritz [1948–1956]
- Zeitschrift für die Zuckerindustrie [1945]
- Ziffer, Rudolf [1947–1949]
- Zondek, Marie [1942–1943]

Central Archives of the Hebrew University of Jerusalem:
- Copy of the summary of Carl Neuberg's relations with the university from 29/11/1940 by D. W. Senator.

Deutsche Bank (1938). Annual Report for the year 1938: [www.bankgeschichte.de/downloads/gb/gb_1938.pdf, 03/21/2006].

Hanover City Archives:
- Mortgage files, A 24 101, Karlstrasse 3 – Fernrode.
- Verbal communication from Herr Schulze to Hinderk Conrads, July 1997.

Jewish Cemetery, Weissensee, Berlin:
- Certificate of an entry of death of February 4, 1921 [Johanna Lewinski].
- Certificate of an entry of death of January 20, 1928 [Sigismund Lewinski].
- Death certificate of March 25, 1929 [Hela Neuberg].
- Interment registration of September 30, 1936 [Hela Neuberg].

Leo Baeck Institute Library, New York (LBIL):
- Membership list of the Greater Berlin Association of the Centralverein, as of 10/24/1931, MfW 12-III.

Lower Saxony Central State Archives, Hanover (LSCSA):
- Hanover 74, Hildesheim, 1841–1871.
- Hanover 83b,
  - [A] Jewish Civil Status Registry, Dankelshausen synagogue congregation.
  - [B] Jewish Civil Status Registry, Hanover synagogue congregation.
  - [C] Jewish Civil Status Registry, Hildesheim synagogue congregation.
  - [D] Jewish Civil Status Registry, Pyrmont synagogue congregation.

City of Poznań, Civil Status Register (1870–1930) (Kartoteka mieszkancow Poztnania z lat 1870–1930), Archivum Panstwowe w Poznaniu.

Register of deaths of the Jewish Cemetry, Weissensee, Berlin.

Reports of proceedings of the Finnish Academy of Sciences (1932), no page numbers given.

Staatsbibliothek zu Berlin, Preussischer Kulturbesitz (State Library of the Prussian Cultural Heritage, Berlin) (SBPK):
- Manuscripts department: Kisch papers, K2 and K3 (a–c).

Swedish Academy of Sciences. Svenska Läkaresällskapets förhandlingar. Stockholm (1931).

University of St. Andrews. University Senatus Minutes (1933).

Zehlendorf District Registry, Zehlendorf Local Court, Berlin (ZLC):
- Land registry, Berlin-Dahlem: Vol. 20.

Zehlendorf District Registry, Zehlendorf Town Hall, Berlin (ZTH):
- Building records, Hittorfstrasse 18, Vol. 1.

12.2. LITERATURE

Adressbücher der Stadt Berlin (1896–1909).

Adressbücher der Stadt Hannover (1874–1892).

Albrecht, Helmuth/Hermann, Armin (1990): Die Kaiser-Wilhelm-Gesellschaft im Dritten Reich (1933–1945). In: Vierhaus/Brocke (1990), 356–406.

Albu, A[lbert/lbrecht] (1916): Therapeutische Erfahrungen mit einem Glycerinersatzmittel. Berliner Klinische Wochenschrift 19 (1916), 496–497.

Albu, Albert/Neuberg, Carl (1906): Physiologie und Pathologie des Mineralstoffwechsels. Nebst Tabellen über die Mineralstoffzusammensetzung der menschlichen Nahrungs- und Genussmittel, sowie der Mineralbrunnen und -bäder. Berlin: Springer.

American Men & Women of Science (1995–1996): A biographical directory of today's leaders in physical, biological and related sciences. 19. Ed. (1994). New Providence, NJ: Bowker.

American Society of European Chemists and Pharmacists (ed.) (1947): The ASEC Bulletin (1947). New York.

Ammon, Robert/Dirscherl, Wilhelm (1956): In memoriam Carl Neuberg. Arzneimittelforschung 6 (1956), 411–412.

Artelt, Walter/Heischkel, Edith/Steudel, Johannes (ed.) (1953): Festschrift zum fünfundsiebzigsten Geburtstag von Paul Diepgen. Wiesbaden: Franz Steiner Verlag.

Barkai, Avraham (2002): "Wehr dich!" Der Centralverein deutscher Staatsbürger jüdischen Glaubens (C.V.) 1893–1938. München: Beck-Verlag.

Bayer-AG (ed.) (1967): Unser Werk [Zeitschrift], Nr. 6.

Becke-Goering, Margot (1990): Freunde in der Zeit des Aufbruchs der Chemie. Der Briefwechsel zwischen Theodor Curtius und Carl Duisberg. Sitzungsberichte der Heidelberger Akademie der Wissenschaften Mathematisch-Naturwissenschaftliche Klasse. Jahrgang 1990. 1. Abhandlung. Berlin: Springer.

Beneke, Klaus (2005): Hermann Franz Mark (3. Mai 1895 Wien Österreich – 6. April 1992 Augustin/USA). Mitbegründer der Polymerwissenschaften. [www.uni-kiel.de/anorg/laly/group/klausSchiver/mark.pdf, 03/29/2006].

Berg, Wieland/Gerstengarbe Sybille/Kleinert, Andreas/Parthier, Benno (2000): Vorträge und Abhandlungen zur Wissenschaftsgeschichte 1999/2000. Acta Historica Leopoldina Nummer 36. Heidelberg: Johann Ambrosius Barth.

Beyler, Richard A. (2004): "Reine" Wissenschaft und personelle "Säuberungen": Die Kaiser-Wilhelm-/Max-Plank-Gesellschaft 1933 und 1945. Geschichte der

Kaiser-Wilhelm-Gesellschaft im Nationalsozialismus, 16. Berlin: Max-Planck-Institut für Wissenschaftsgeschichte.
Bickel, Adolf (1916): Ueber das physiologische Verhalten des Perglycerins und Perkaglycerins. Berliner Klinische Wochenschrift 19 (1916), 495–496.
Björk, Ragnar (2001): Inside the Nobel Committee on Medicine: Prize Competition Procedures 1901–1950 and the Fate of Carl Neuberg. Minerva 39 (2001), 193–408.
Boas, Ismar (1916): Ueber "Glycerinersatz" bei Obstipationszuständen. Deutsche Medizinische Wochenschrift 22 (1916), 661.
Boletim da Academia das Ciências de Lisboa (1932): Nova Série, Vol. 4. Imprensa da Universidade. Coimbra.
Barnett, James A. (2003): A history of research on yeast 5. The fermentation pathway. Yeast 20 (2003), 509–543.
Bricka, Carl Frederik/Engelstoft, Povl/Dahl, Svend (1933–1944): Dansk biografisk leksikon. Kopenhagen: Schultz.
Brocke, Bernhard vom (1990a): Die Kaiser-Wilhelm-Gesellschaft im Kaiserreich. Vorgeschichte, Gründung und Entwicklung bis zum Ausbruch des Ersten Weltkrieges. In: Vierhaus/Brocke (1990), 17–162.
Brocke, Bernhard vom (1990b): Die Kaiser-Wilhelm-Gesellschaft in der Weimarer Republik. Ausbau zu einer gesamtdeutschen Forschungsorganisation (1918–1933). In: Vierhaus/Brocke (1990), 197–355.
Brocke, Bernhard vom (ed.) (1991): Wissenschaftsgeschichte und Wissenschaftspolitik im Industriezeitalter: Das "System Althoff" in historischer Perspektive. Geschichte von Bildung und Wissenschaft: Reihe B, Sammelwerk 5. Edition Bildung und Wissenschaft. Hildesheim: Lax.
Brocke, Bernhard vom (1996): Die Kaiser-Wilhelm-/Max-Planck-Gesellschaft und ihre Institute zwischen Universität und Akademie: Strukturprobleme und Historiographie. In: Brocke/Laitko (1996), 1–32.
Brocke, Bernhard vom/Laitko, Hubert (1996): Die Kaiser-Wilhelm-/Max-Planck-Gesellschaft und ihre Institute. Studien zu ihrer Geschichte: Das Harnack Prinzip. Berlin: de Gruyter.
Bruch, Rüdiger vom/Schaarschmidt, Rebecca (ed.) (2005): Die Berliner Universität in der NS-Zeit. Band II: Fachbereiche und Fakultäten. Stuttgart: Steiner Verlag.
Bundesarchiv, Koblenz/Internationaler Suchdienst, Bad Arolsen (ed.) (1986): Gedenkbuch: Opfer der Verfolgung der Juden unter der nationalsozialistischen Gewaltherrschaft in Deutschland 1933–1944. Koblenz: Bundesarchiv.
Buchner, Eduard/Ehrlich, Paul/Noorden, Karl v./Salkowski, Ernst (ed.) (1906): Biochemische Zeitschrift. Bd. 1. Redigiert von C. Neuberg. Berlin: Springer.
Burchardt, Lothar (1990): Die Kaiser-Wilhelm-Gesellschaft und deren Rolle während des Ersten Weltkrieges (1914–1918). In: Vierhaus/Brocke, 163–196.
Bussche, Hendrik van den (ed.) (1989): Medizinische Wissenschaft im "Dritten Reich": Kontinuität, Anpassung und Opposition an der Hamburger Medizinischen Fakultät. Hamburger Beiträge zur Wissenschaftsgeschichte, 5. Berlin, Hamburg: Reimer.

Butenandt, Adolf (1956): Carl Neuberg (29. 7. 1877 – 30. 5. 1956). Mitteilungen der Max-Planck-Gesellschaft 5 (1956), 223–227.

Butenandt, Adolf (1981): Das Werk eines Lebens. Band 2. Wissenschaftspolitische Aufsätze, Ansprachen und Reden. Göttingen: Vandenhoeck und Ruprecht.

Chargaff, Erwin (1978): Heraclitean Fire. Sketches from a Life before Nature. Rockefeller University Press, New York.

Chargaff, Erwin (1980): Das Feuer des Heraklit. Skizzen aus dem Leben vor der Natur. 2. Auflage. Stuttgart: Klett-Cotta.

Connstein, Wilhelm/Lüdecke, Karl (1919): Glyceringewinnung aus Zucker: Naturwissenschaften 7 (1919), 403–405.

Conrads, Hinderk (2002): Carl Neuberg (1877–1956). Eine Biographie. Diss. med. Hannover.

Conrads, Hinderk/Lohff, Brigitte (2006): Carl Neuberg – Biochemie, Politik und Geschichte. Lebenswege und Werk eines fast verdrängten Forschers. Stuttgart: Franz Steiner (2006)

Crawford, Elisabeth/Heilbron, John L. (1990): Die Kaiser-Wilhelm-Institute für Grundlagenforschung und die Nobelinstitution. In: Vierhaus/Brocke (1990), 835–857.

Crawford, Elisabeth/Heilbron, John L./Ullrich, Rebecca (1987): The Nobel population 1901–1937. A census of the nominators and nominees for the prizes in physics and chemistry (Reihe: Berkeley Papers in the History of Science 11, Reihe: Uppsala Studies in the History of Science 4). Berkeley: Office for History of Science and Technology, University of California.

Cuthbert, Alan W. (2005): Marthe Vogt. 8. September 1903 – 9. September 2003. Biographical Memories of the Fellows of the Royal Society 51 (2005), 409–423.

Curare [Zeitschrift des Allgemeinen Studierendenausschusses der Medizinischen Hochschule Hannover] Sonderausgabe (1994): Wer war Konstanty Gutschow? Hannover.

Dannenberg, Heinz (1962): Max-Planck-Institut für Biochemie in München. In: Generalverwaltung der Max-Planck-Gesellschaft zur Förderung der Wissenschaften e.V. Göttingen (1962), 90–110.

Degener, Hermann A. L. (ed.) (1911): Wer ist's? Unsere Zeitgenossen. Berlin, Leipzig: H. A. Ludwig Degener.

Deichmann, Ute (1993): Die alten Bande für die neue Karriere gut genutzt. Jüdische Allgemeine 48/15, 15 (15. April 1993).

Deichmann, Ute (2002): Chemists and Biochemists during the National Socialist Era. Angewandte Chemie International Edition 41 (2002), 1310–1328.

Deichmann, Ute/Müller-Hill, Benno (1998): The fraud of Abderhalden's enzymes. Nature 393 (1998), 109–111.

Düwell, Kurt (1990): Die deutsch-amerikanischen Wissenschaftsbeziehungen im Spiegel der Kaiser-Wilhelm- und Max-Planck-Gesellschaft. In: Vierhaus/Brocke (1990), 747–777.

Durig, Arnold/ Kulmer, W./Rainer, R./Reichel, H./Caspari, Wilhelm (1910): Physiologische Ergebnisse der im Jahre 1906 durchgeführten Monte Rosa-Expedition. Denkschriften der Kaiserlichen Akademie der Wissenschaften. Mathematisch-naturwissenschaftliche Klasse, 86. Wien: Komm. Hölder.

Dunn, Peter M. (2000): Sir Joseph Barcroft of Cambridge (1872–1947) and prenatal research. Archives of Disease in Childhood – Fetal and Neonatal Edition 82 (2000), 75–76.

Durth, Werner/Gutschow, Niels (1988): Träume in Trümmern. Planung zum Wiederaufbau zerstörter Städte im Westen Deutschlands 1940–1950. Braunschweig, Wiesbaden: Vieweg.

Eckart, Wolfgang U./Gradmann, Christoph (ed.) (1996): Die Medizin und der Erste Weltkrieg. Pfaffenweiler: Centaurus.

Eckart, Wolfgang U. (ed.) (2006): Man, Medicine, and the State. The Human Body as an Object of Government Sponsored Medical Research in the 20th Century. Beiträge zur Geschichte der Deutschen Forschungsgemeinschaft 2. Stuttgart: Franz Steiner Verlag.

Ehrlich, Felix (1937): In honorem Carl Neuberg. Festschrift zum 60. Geburtstage Carl Neubergs, Erster Teil. Enzymologia Bd. 3 (1937), XI–XII.

Engel, Michael (1982): Carl Neuberg (1877–1956). Bibliotheks-Informationen. Universitätsbibliothek der Freien Universität Berlin 3 (1982), 11–16.

Engel, Michael (1984): Geschichte Dahlems. Berlin: Verlag Arno Spitz.

Engel, Michael (1990): Dahlem als Wissenschaftszentrum. In: Vierhaus/Brocke, 552–578.

Engel, Michael (1996): Enzymologie und Gärungschemie: Alfred Wohls und Carl Neubergs Reaktionsschema der alkoholischen Gärung. Der "Chemische Gesichtspunkt" als Kennzeichen der Berliner Schule. Mitteilungen der Gesellschaft Deutscher Chemiker, Fachgruppe Geschichte der Chemie 12 (1996), 3–29.

Engel, Michael (1999): Neuberg, Carl. Neue Deutsche Biographie. Band 19. Berlin: Duncker & Humblot, 102–103.

Engel, Michael (2005): Sabalitschka, Theodor. Neue Deutsche Biographie. Band 22 Berlin: Duncker & Humblot, 319–320.

Engel, Michael/Conrads, Hinderk (1998): Paul Mayer (1872–1949). Ein vergessenes Mitglied des Kaiser-Wilhelm-Instituts für Biochemie und sein Vater Jacques Mayer – eine Spurensuche. In: Henning (1998), 71–92.

Engelhardt, Dietrich von (ed.) (2002): Biographische Enzyklopädie deutschsprachiger Mediziner. München: Saur.

Eulner, Hans-Heinz (1970): Die Entwicklung der medizinischen Spezialfächer an den Universitäten des deutschen Sprachgebietes. Enke: Stuttgart.

Fäßler, Peter E. (1997): Hans Spemann 1869–1941: Experimentelle Forschung im Spannungsfeld von Empirie und Theorie. Ein Beitrag zur Geschichte der Entwicklungsphysiologie zu Beginn des 20. Jahrhunderts. Schriften der Mathematisch-naturwissenschaftlichen Klasse der Heidelberger Akademie der Wissenschaften 1. Berlin: Springer.

Fischer, I. (ed.) (1962): Biographisches Lexikon der hervorragenden Ärzte der letzten fünfzig Jahre. 2. Aufl. München: Urban & Schwarzenberg.

Fischer, Wolfram/Hierholzer, Klaus/Hubenstorf Michael/Walther, Peter T./ Winau, Rolf (ed.) (1994): Exodus von Wissenschaften aus Berlin: Fragestellungen – Ergebnisse – Desiderate. Entwicklungen vor und nach 1933. Akademie der Wissenschaften zu Berlin. Arbeitsgruppe Exodus von Wissenschaftlern aus Berlin. Forschungsbericht Akademie der Wissenschaften zu Berlin, 7. Berlin, New York: de Gruyter.

Florkin, Marcel/Stotz, Elmer H./Neuberger, Albert/Deenen, Laurens L. M. van (ed.) (1975): Comprehensive biochemistry. Section 6, Vol. 30: A history of biochemistry Part III. Amsterdam: Elsevier.

Frewer, Andreas (2000): Medizin und Moral in Weimarer Republik und Nationalsozialismus. Die Zeitschrift "Ethik" unter Emil Abderhalden. Frankfurt a. M./ New York: Campus.

Gabaccia, Donna R. (2002): Immigration and American Diversity. A social and cultural history [Problems in American history, 7] Malden/Mass.: Blackwell.

Gausemeier, Bernd (2005): Natürliche Ordnungen und politische Allianzen: Biologische und chemische Forschung an Kaiser-Wilhelm-Instituten 1933–1945. Geschichte der Kaiser-Wilhelm-Gesellschaft im Nationalsozialismus 12. Göttingen: Wallstein.

Generalverwaltung der Max-Planck-Gesellschaft zur Förderung der Wissenschaften e. V. Göttingen (ed.) (1962): Jahrbuch der Max-Planck-Gesellschaft zur Förderung der Wissenschaften e. V. 1961, Teil II. Göttingen: Hubert & Co.

Gerstengarbe, Sybille/Seidler, Eduard (2002): "... den Erfordernissen der Zeit in vollem Ausmaß angepasst." In: Parthier/Engelhardt (2002), 227–262.

Glum, Friedrich (1964): Zwischen Wissenschaft, Wirtschaft und Politik. Erlebtes und Erdachtes in vier Reichen. Bonn: Bouvier.

Göbel, Wolfgang (1996): Max Bergmann und das Kaiser-Wilhelm-Institut für Lederforschung in Dresden. In: Brocke/Laitko (1996), 303–318.

Gradmann, Christoph (1996): 'Vornehmlich beängstigend' – Medizin, Gesundheit und chemische Kriegsführung im deutschen Heer 1914–1918. In: Eckart/ Gradmann (1996), 131–154.

Grau, Conrad (1996): Genie und Kärrner – zu den geistesgeschichtlichen Wurzeln des Harnack-Prinzips in der Berliner Akademietradition. In: Brocke/Laitko (1996), 139–144.

Grinstein, Louise S./Rose, Rose K./Rafailovich, Miriam H. (ed.) (1993): Women in Chemistry and Physics. A Bibliographic Sourcebook. Westport, Conn. London: Greenwood Press.

Habel, Walter (1951): Wer ist wer? Das deutsche Who's who. Lübeck, Schmidt-Römhild.

Haber, Charlotte (1970): Mein Leben mit Fritz Haber. Spiegelungen der Vergangenheit. Düsseldorf: Econ 1970.

Haber, Fritz (1923): Neue Arbeitsweisen. Wissenschaft und Wirtschaft nach dem Kriege. Die Naturwissenschaften 11 (1923), 753–756.

Hahn, Otto (1968): Mein Leben. 3. Aufl. München: Bruckmann.

Harnack, Adolf von (ed.) (1928): Handbuch der Kaiser-Wilhelm-Gesellschaft zur Förderung der Wissenschaften. Berlin: Hobbing.

Heiber, Helmut (1991): Universität unterm Hakenkreuz. Teil 1. Der Professor im Dritten Reich, Bilder aus der akademischen Provinz. München: Saur.

Hochberg, Edward (1993): Ines Hochmuth Mandl (1917–). In: Grinstein et al. (1993), 361–370.

Heinemann, Manfred (1990): Der Wiederaufbau der Kaiser-Wilhelm-Gesellschaft und die Neugründung der Max-Planck-Gesellschaft (1945–1949). In: Vierhaus/Brocke (1990), 407–470.

Henning, Eckart (1998): Dahlemer Archivgespräche 3. Berlin: Archiv zur Geschichte der Max-Planck-Gesellschaft.

Henning, Eckart/Kazemi, Marion (1990): Zeittafel zur Geschichte der Kaiser-Wilhelm-/Max-Planck-Gesellschaft und ihrer Institute (1901) 1911–1986. In Vierhaus/Brocke (1990), 913–951.

Henning, Eckart/Kazemi, Marion (2002): Dahlem – Domäne der Wissenschaft/ Dahlem – Domaine of Science. Archiv zur Geschichte der Max-Planck-Gesellschaft 16. Berlin: Archiv zur Geschichte der Max-Planck-Gesellschaft.

Herlitz, Georg/Kirschner, Bruno/Elbogen, Ismar (1987) Jüdisches Lexikon. Ein enzyklopädisches Handbuch des jüdischen Wissens in vier Bänden 2. Auflage. Frankfurt/Main: Jüdischer Verlag bei Athenäum.

Hess, Volker: Archibald Vivian Hill, Otto Fritz Meyerhof. In: Lexikonredaktion der F. A. Brockhaus (2001), 230–231.

Hirsch, Julius (1922): Ergebnisse der neueren Arbeiten über den Chemismus der alkoholischen Zuckerspaltung. Klinische Wochenschrift 1 (1922), 489–491.

Höffkes, Karl (1986): Hitlers politische Generale. Die Gauleiter des 3. Reiches. Tübingen: Grabert.

Höxtermann, E (2001): Otto Heinrich Warburg. In: Lexikonredaktion der F. A. Brockhaus (2001), 310–311.

Hoßfeld, Uwe/John, Jürgen/Lemuth, Oliver/Stutz, Rüdiger (2003): Kämpferische Wissenschaft: Studien zur Universität Jena im Nationalsozialismus. Köln, Weimar, Wien: Böhlau.

Jaenicke, Lothar (2002): Die Biochemie im Spiegel der Leopoldina. In: Parthier/Engelhardt (2002), 491–530.

Jahr, Christoph/Schaarschmidt, Rebecca (ed.) (2005): Die Berliner Universität in der NS-Zeit. Band 1: Strukturen und Personen. Stuttgart: Steiner Verlag.

Jahresberichte der Königlichen Landwirtschaftlichen Hochschule in Berlin (JKLHB) (1909–1915). Jahrgänge XVII–XXIII. Berlin.

Jakobi, Helga/Chroust, Peter/Hamann, Matthias (1989): Aeskulap & Hakenkreuz. Zur Geschichte der Medizinischen Fakultät in Gießen zwischen 1933 und 1945. Eine Dokumentation der Arbeitsgruppe "Medizin und Faschismus". Frankfurt a. M.: Mabuse.

Jöcher, Christian Gottlieb (1981): Allgemeines Gelehrten-Lexikon. Unveränderter Nachdruck der Ausgabe Leipzig 1750. Hildesheim, New York: Georg Olms Verlag.

Joseph, Max (1916): Ueber Glycerinersatz für dermatologische Zwecke. Berliner Klinische Wochenschrift 18 (1916), 481–482.

Jungbluth, Rüdiger (2004): Die Oetkers. Geschäfte und Geheimnisse der bekanntesten Wirtschaftsdynastie Deutschlands. Frankfurt am Main, New York: Campus.

Kaasch, Michael (2000): Sensation, Irrtum, Betrug? – Emil Abderhalden und die Geschichte des Abwehrferments. In: Berg et al. (2000), 145–200 [Diskussion 201–210].

Kaasch, Joachim/Kaasch, Michael (2003): Hallesche Naturwissenschaftler (Emil Abderhalden und Johannes Weigelt) in der Zeit des Nationalsozialismus: Eine Fallstudie mit Jenaer Beziehungen. In: Hoßfeld et al. (2003), 1027–1064.

Kaiser-Wilhelm-Gesellschaft (1926): Tätigkeitsbericht der Kaiser-Wilhelm-Gesellschaft zur Förderung der Wissenschaften (Oktober 1925 bis Oktober 1926). Naturwissenschaften 14 (1926), 1239–1254.

Kaiser-Wilhelm-Gesellschaft (1927): Tätigkeitsbericht der Kaiser-Wilhelm-Gesellschaft zur Förderung der Wissenschaften (Für das Halbjahr Oktober 1926 bis März 1927). Naturwissenschaften 15 (1927), 649–653.

Kaiser-Wilhelm-Gesellschaft (1928): Tätigkeitsbericht der Kaiser-Wilhelm-Gesellschaft zur Förderung der Wissenschaften (Für April 1927 bis März 1928). Naturwissenschaften 16 (1928), 427–432.

Kaiser-Wilhelm-Gesellschaft (1929): Tätigkeitsbericht der Kaiser-Wilhelm-Gesellschaft (Für April 1928 bis März 1929). Naturwissenschaften 17 (1929), 322–346.

Kaiser-Wilhelm-Gesellschaft (1930): Tätigkeitsbericht der Kaiser-Wilhelm-Gesellschaft zur Förderung der Wissenschaften (Für April 1929 bis März 1930. Naturwissenschaften 18 (1930), 482–487.

Kaiser-Wilhelm-Gesellschaft (1931): Tätigkeitsbericht der Kaiser-Wilhelm-Gesellschaft zur Förderung der Wissenschaften (Für April 1930 bis März 1931). Naturwissenschaften 19 (1931), 531–538.

Kaiser-Wilhelm-Gesellschaft (1932): Tätigkeitsbericht der Kaiser-Wilhelm-Gesellschaft zur Förderung der Wissenschaften (Für April 1931 bis März 1932). Naturwissenschaften 20 (1932), 428–436.

Kaiser-Wilhelm-Gesellschaft (1933): Tätigkeitsbericht der Kaiser-Wilhelm-Gesellschaft zur Förderung der Wissenschaften (Für April 1932 bis März 1933). Naturwissenschaften 21 (1933), 417–428.

Kaiser-Wilhelm-Gesellschaft (1934): Tätigkeitsbericht der Kaiser-Wilhelm-Gesellschaft zur Förderung der Wissenschaften (Für April 1933 bis März 1934). Naturwissenschaften 22 (1934), 339–350.

Kaiser-Wilhelm-Gesellschaft (1935a): Tätigkeitsbericht der Kaiser-Wilhelm-Gesellschaft zur Förderung der Wissenschaften (Für April 1934 bis März 1935). Naturwissenschaften 23 (1935), 411–425.

Kaiser-Wilhelm-Gesellschaft (1935b): Tätigkeitsbericht der Kaiser-Wilhelm-Gesellschaft zur Förderung der Wissenschaften für (April 1935 bis September 1935). Naturwissenschaften 24 (1935), 19–30.

Kaplan, Marion (ed.) (2003a): Geschichte des jüdischen Alltags in Deutschland. Vom 17. Jahrhundert bis 1945. München: C. H. Beck.

Kaplan, Marion (2003b): Konsolidierung bürgerlichen Lebens im kaiserlichen Deutschland 1817–1918. In: Kaplan (2003a), 225–344.

Karlson, Peter (1990): Adolf Butenandt. Biochemiker, Hormonforscher, Wissenschaftspolitiker. Stuttgart: Wissenschaftliche Verlagsgesellschaft.

Katterle, Heike/Peschel, Delia (1989): Carl Neuberg als Forscher und Hochschullehrer an der Charité und an der Landwirtschaftlichen Hochschule zu Berlin in den Jahren von 1898 bis 1913. Diplomarbeit. HU, Berlin.

Kay, Lily E. (1989): Molecules, cells, and life: an annotated bibliography of manuscript sources on physiology, biochemistry, and biophysics, 1900–1960, in the Library of the American Philosophical Society. Philadelphia: American Philosophical Society Library.

Kay, Lily E. (1993): The Molecular Vision of Life. Caltech, the Rockefeller Foundation and the Rise of the New Biology. New York, Oxford: Oxford University Press.

Kazemi, Marion (2002): Nobelpreisträger in der Kaiser-Wilhelm-/Max-Planck-Gesellschaft zur Förderung der Wissenschaften. Veröffentlichungen aus dem Archiv zur Geschichte der Max-Planck-Gesellschaft 15. Berlin: Archiv zur Geschichte der Max-Planck-Gesellschaft.

Keitel, Wolfgang (2003): Hans Reiter und das okulo-urethro-synoviale Syndrom. 1. Die Geschichte des Krankheitsbildes. Zeitschrift für Rheumatologie 62 (2003), 574–578.

Keitel, Wolfgang (2004a): Hans Reiter und das okulo-urethro-synoviale Syndrom. 3. Der unbekannte Hans Reiter. Wissenschaftler und NS-Propagandist. Zeitschrift für Rheumatologie 63 (2004), 244–249.

Keitel, Wolfgang (2004b): Hans Reiter und das okulo-urethro-synoviale Syndrom. 4. Nachspiel und Erbe. Zeitschrift für Rheumatologie 63 (2004), 341–345.

Kennedy, Eugen P. (2001): Hitler`s gift and the era of biosynthesis. The Journal of Biological Chemistry 276 (2001): 42619–42631.

Killy, Walter/Engelhardt, Dietrich von/Vierhaus, Rudolf (ed.) (1995–2003): Deutsche Biographische Enzyklopädie. München: Saur.

Kisch, Bruno (1947): Professor Neuberg's life and achievements up to date. Experimental Medicine and Surgery 5 (1947), 97–106.

Kisch, Bruno (1966): Wanderungen und Wandlungen. Die Geschichte eines Arztes im 20. Jahrhundert. Köln: Greven-Verlag.

Kleinkauf, Horst/Döhren, Hans v./Jaenicke, Lothar (ed.) (1988): The Roots of Modern Biochemistry. Berlin, New York: de Gruyter.

Krebs, Hans A. (1972): Otto Meyerhof's Ancestry. In: Weber (1972), 14–15.

Krebs, Hans A./Lipmann, Fritz A. (1988): Dahlem in the Late Nineteen Twenties. In: Kleinkauf et al. (1988), 111–124.

Laitko, Hubert (1996): Persönlichkeitszentrierte Forschungsorganisation als Leitgedanke der Kaiser-Wilhelm-Gesellschaft: Reichweite und Grenzen, Ideal und Wirklichkeit. In: Brocke/Laitko (1996), 583–632.

Lehrerkollegium Parey (ed.) (1906): Die Königliche Landwirtschaftliche Hochschule in Berlin. Festschrift zur Feier des 25jährigen Bestehens. Red. Wittmak, Ludwig. Berlin: Parey.

Lexikonredaktion der F. A. Brockhaus (2001): Nobelpreise. Chronik herausragender Leistungen. Mannheim, Leipzig: F. A. Brockhaus.

Lieben, Fritz (1970): Geschichte der Physiologischen Chemie. Hildesheim, New York: Georg Olms.

Lipmann, Fritz A. (1988): A long life in times of great upheaval. In: Kleinkauf et al. (1988), 9–36.

Lipmann, Fritz A./Nord, Friedrich F./Waelsch, Heinrich (1956): C. Neuberg, Biochemist [Nachruf]. Science 124 (1956), 1244–1245.

Lohff, Brigitte (2005): "[...] die Grundgedanken des Nationalsozialismus aufsaugen und verarbeiten." Die politisch-ideologische Funktion der Medizinischen Fakultät der Christian-Albrechts-Universität 1933–1945. Jahrbuch für Universitätsgeschichte 8 (2005), 211–234.

Macrakis, Kristie (1994): Exodus der Wissenschaftler aus der Kaiser-Wilhelm-Gesellschaft. In: Fischer et al. (1994), 267–283.

Maitra, Robin T. (2001): "... wer imstande und gewillt ist, dem Staate mit Höchstleistungen zu dienen!" Hans Reiter und der Wandel der Gesundheitskonzeption im Spiegel der Lehr- und Handbücher der Hygiene zwischen 1920 und 1960. Abhandlungen zur Geschichte der Medizin und Naturwissenschaften 88. Husum: Matthiesen.

Mandl, Ines (1978): Carl Neuberg – An Appreciation. Proceedings of the Virchow-Piquet Medical Society 32 (1978), 42–46.

Mangold, Otto (1953): Hans Spemann. Ein Meister der Entwicklungsphysiologie. Sein Leben und sein Werk. Große Naturforscher Band 11. Stuttgart: Wissenschaftliche Verlagsgesellschaft.

Mann, Golo/Kluke, Paul/Richter, Gert (ed.) (1964): Unser Jahrhundert im Bild. Gütersloh: Bertelsmann Lesering.

Martin, Werner (ed.) (1988): Verzeichnis der Nobelpreisträger 1901–1987. München: Saur.

Mayer, Christian (2004): Spurensuche im Deutschen Oxford. Max Planck Forschung 2 (2004), 52–57.

Medawar, Jean S./Pyke, David (2001): Hitler's gift: The true story of the scientists expelled by the Nazi regime. New York: Arcade Publisher.

Moser, Gabriele (2006): From deputy to 'Reichsbevollmächtigter and defendant at the Nuremberg medical trials: Dr. Kurt Blome and the cancer research in national socialist Germany. In: Eckart (2006), 199–220.

Munk, Fritz (1916): Ueber die Wirkung und Anwendung des "Nucleo-Hexyl" bei Fleckfieber. Münchener Medizinische Wochenschrift 34 (1916), 1239–1241.

Muscholl, Ernst (2003): Marthe Louise Vogt (1903–2003). Nachruf. [www.dgpt-online.de/material/Nachruf_Vogt.pdf, 01/22/2006]

Nachmansohn, David (1956): Carl Neuberg 1877–1956. Proceedings of the Rudolf Virchow Medical Society in the City of New York 25 (1956), 75–82.

Nachmansohn, David (1979): German-Jewish pioneers in science: 1900–1933. Highlights in atomic physics, chemistry and biochemistry. Berlin: Springer.
Nachmansohn, David/Schmid, Roswitha (1988): Die große Ära der Wissenschaft in Deutschland 1900 bis 1933. Jüdische und nichtjüdische Pioniere in der Atomphysik, Chemie und Biochemie. Stuttgart: Wissenschaftliche Verlagsgesellschaft.
Nagelschmidt, Franz (1916): Ein neues Glycerinersatzmittel (Perkaglycerin). Die Therapie der Gegenwart 5 (1916), 199–200.
Neuberg, Carl (ed.) (1906–1930): Biochemische Zeitschrift. Band 1–246. Berlin: Springer.
Neuberg, Carl/Jacoby, Martin (ed.) (1931–1933): Biochemische Zeitschrift. Band 247–264. Berlin: Springer.
Neuberg, Carl/Grassmann, Wolfgang (ed.) (1933–1935): Biochemische Zeitschrift. Band 265–279 Berlin: Springer.
Neuberg, Irene S. (1932): Untersuchungen in der 3-Kohlenstoffreihe und Gewinnung von Zuckern aus ihren Hydrazonen. Diss. phil. Berlin.
Neuberg, Oskar (1890): I. Eine Synthese des Dehydrothiotoluidins. II. Über Einwirkung von Säurechloriden auf Phenolaether. Diss. phil. Heidelberg.
Nord, Friedrich. F. (1958): Carl Neuberg, 1877–1956. Advances in Carbohydrate Chemistry 13 (1958), 1–7.
Nordwig, Arnold (1983): Vor fünfzig Jahren: der Fall Neuberg. Aus der Geschichte des Kaiser-Wilhelm-Instituts für Biochemie zur Zeit des Nationalsozialismus. MPG Spiegel 6 (1983), 49–53.
Notgemeinschaft der Deutschen Wissenschaft (ed.) (1928): Siebenter Bericht der Notgemeinschaft der deutschen Wissenschaft umfassend ihre Tätigkeit vom 1. April 1927 bis zum 31. März 1928. Wittenberg (Bez. Halle): Herrosé & Ziemsen.
Notgemeinschaft deutscher Wissenschaftler im Ausland (ed.) (1937): List of displaced German Scholars. Reproduction. San Bernadino, CA: Borgo Press.
Orth, J[ohannes] (1916): Glycerin-Ersatz. Berliner Klinische Wochenschrift 18 (1916), 482.
Parthier, Benno/Engelhardt, Dietrich von (ed.) (2002): 350 Jahre Leopoldina Anspruch und Wirklichkeit. Festschrift der Deutschen Akademie der Naturforscher Leopoldina 1652–2002. Halle: Druck-Zuck.
Planck, Max (ed.) (1936a): 25 Jahre Kaiser-Wilhelm-Gesellschaft zur Förderung der Wissenschaften. Band 1. Handbuch. Berlin: Springer.
Planck, Max (ed.) (1936b): 25 Jahre Kaiser-Wilhelm-Gesellschaft zur Förderung der Wissenschaften. Band 2. Die Naturwissenschaften. Berlin: Springer.
Pötsch, Winfried P. (ed.) (1988): Lexikon bedeutender Chemiker. Leipzig: VEB Bibliographisches Institut Leipzig.
Poggendorf, Johann C. (ed.) (1970): Biographisch-Literarisches Handwörterbuch zur Geschichte der exacten Wissenschaften: enthaltend Nachweisungen über Lebensverhältnisse und Leistungen von Mathematikern, Astronomen, Physikern, Chemikern, Mineralogen, Geologen usw. aller Völker und Zeiten. [Unchanged Reprint of the Edition. Leipzig 1863]. Amsterdam: Israël.

Portz, Thomas (2000): Großindustrie, Kriegszielbewegung und OHL, Siegfrieden und Kanzlersturz: Carl Duisberg und die deutsche Außenpolitik im Ersten Weltkrieg. Subsidia academica: Reihe A, Neuere und neueste Geschichte 1. Lauf a. d. Pegnitz: Europaforum-Verlag.

Posner, Carl (1916): Perkaglycerin in der Urologie. Berliner Klinische Wochenschrift 21 (1916), 577.

Proctor, Robert N. (1999): The Nazi war on cancer. Princeton, NJ: Princeton University Press.

Przyrembel, Alexandra (2004): Friedrich Glum und Ernst Telschow: Die Generalsekretäre der Kaiser-Wilhelm-Gesellschaft: Handlungsfelder und Handlungsoptionen der "Verwaltenden" von Wissen während des Nationalsozialismus. Geschichte der Kaiser-Wilhelm-Gesellschaft im Nationalsozialismus 20. Berlin: Max-Planck-Gesellschaft zur Förderung der Wissenschaften.

Pufendorf, Astrid von (2006): Die Plancks. Eine Familie zwischen Patriotismus und Widerstand. Berlin: Propyläen-Verlag.

Reichsgesetzblatt 34 (1933): Gesetz zur Wiederherstellung des Berufsbeamtentums, ausgegeben am 7. April 1933 Teil 1, hrsg. vom Reichsministerium des Innern. Berlin: Reichsdruckerei, 175–177.

Reichsgesetzblatt 68 (1933): Gesetz zur Änderung des Gesetzes zur Wiederherstellung des Berufsbeamtentums, ausgegeben am 24. Juni 1933 Teil 1, hrsg. vom Reichsministerium des Innern, Berlin: Reichsdruckerei, 389.

Remy, Steven P. (2002): The Heidelberg myth. The nazification and denazification of a German university. Cambridge, London: Harvard Univ. Press.

Richarz, Monika (ed.) (1982): Jüdisches Leben in Deutschland. Band 3: Selbstzeugnisse zur Sozialgeschichte: 1918–1945 [Veröffentlichung des Leo-Baeck-Instituts]. Stuttgart: Deutsche Verlags-Anstalt.

Ringer, Fritz (1983): Die Gelehrten. Der Niedergang der deutschen Mandarine 1890–1933. Stuttgart: Klett-Cotta.

Roelcke, Volker (2006): Funding the scientific foundations of race policies: Ernst Rüdin and the impact of career resources on psychiatric genetics, ca. 1910–1945. In: Eckart (2006), 73–87.

Rürup, Reinhard (ed.) (1993): Topographie des Terrors. Gestapo, SS und Reichssicherungshauptamt auf dem "Prinz-Albrecht-Gelände". Eine Dokumentation Berlin: Arenhövel.

Salkowski, Ernst L./Leube, Wilhelm (1882): Die Lehre vom Harn. Ein Handbuch für Ärzte und Studierende. Berlin: Hirschwald.

Salkowski, Ernst L. (1912): Practicum der physiologischen und pathologischen Chemie nebst einer Einführung zur anorganischen Analyse für Mediciner. Vierte vermehrte Auflage. Berlin: Hirschwald.

Sarkowski, Heinz (1992): Der Springer-Verlag. Stationen seiner Geschichte. Teil I: 1842–1945. Berlin: Springer.

Satzinger, Helga (2004a): Adolf Butenandt, Hormone und Geschlecht. Ingredienzen einer wissenschaftlichen Karriere. In: Schieder/Trunk (2004), 78–133.

Satzinger, Helga (2004b): Rasse, Gene und Geschlecht: Zur Konstituierung zentraler biologischer Begriffe bei Richard Goldschmidt und Fritz Lenz, 1916–

1936. Forschungsprogramm "Geschichte der Kaiser-Wilhelm-Gesellschaft im Nationalsozialismus" 15. Berlin: Max-Planck-Gesellschaft zur Förderung der Wissenschaften.

Sauer, Günter (2001): Hans Adolf Krebs. Fritz Albert Lipmann. In: Lexikonredaktion der F. A. Brockhaus (2001), 488–489.

Sauer, Günter/Rapoport, Samuel/Rost, Günther (1960): Zur Geschichte der Biochemie in Berlin. Zeitschrift für Geschichte der Naturwissenschaften, Technik und Medizin 1 (1960), 119–147.

Schaper, Wolfgang/Schaper, Jutta (1995): Bruno Kisch, Leben und Werk. Ein Versuch. Gedächtnisvorlesung anlässlich der 61. Jahrestagung der Deutschen Gesellschaft für Kardiologie, Herz- und Kreislaufforschung, April 1995. [www.dgk/organe/geschichte/kisch.aspk, 04/13/2006].

Schieder, Wolfgang (2004): Spitzenforschung und Politik. Adolf Butenandt in der Weimarer Republik und im Dritten Reich. In: Schieder/Trunk (2004), 23–77.

Schieder, Wolfgang/Trunk, Achim (ed.) (2004): Adolf Butenandt und die Kaiser-Wilhelm-Gesellschaft: Wissenschaft, Industrie und Politik im Dritten Reich. Geschichte der Kaiser-Wilhelm-Gesellschaft im Nationalsozialismus 7. Göttingen: Wallstein.

Schmaltz, Florian (2005): Kampfstoff-Forschung im Nationalsozialismus. Zur Kooperation von Kaiser-Wilhelm-Instituten, Militär und Industrie. Geschichte der Kaiser-Wilhelm-Gesellschaft im Nationalsozialismus 2. Göttingen: Wallstein.

Schmaltz, Florian (2006): Otto Bickenbach's human experiments with chemical warfare agents and the concentration camp Natweiler. In Eckart (2006), 139–156.

Schneiderman, Harry/Karpman [später Carmin] Itzhak J. (1972): Who is who in world Jewry? A Biographical Dictionary of Outstanding Jews. New York: Pitman Publishing Corporation.

Schottlaender, Rudolf (1988): Verfolgte Berliner Wissenschaft. Ein Gedenkwerk. Berlin: Hentrich.

Schroeder-Gudehus, Brigitte (1990): Wissenschaftsbeziehungen und auswärtige Kulturpolitik 1919–1933. In: Vierhaus/Brocke (1990), 877–883.

Schuder, Werner (ed.) (1987): Kürschners Deutscher Gelehrten Kalender: Biobibliographisches Verzeichnis deutschsprachiger Wissenschaftler der Gegenwart. Berlin, New York: de Gruyter.

Schüring, Michael (2003): Minervas verstoßene Kinder. Vertreibung, Entschädigung und die Vergangenheitspolitik der Max-Planck-Gesellschaft. Diss. phil. Berlin.

Schüring, Michael (2004): Carl Neubergs Verhältnis zu Adolf Butenandt. In: Schieder/Trunk (2004), 346–368.

Schulte, Uwe (2001a): Heinrich Otto Wieland. In: Lexikonredaktion der F. A. Brockhaus (2001), 270–271.

Schulte, Uwe (2001b): Hans von Euler-Chelpin, Arthur Harden. Lexikonredaktion der F. A. Brockhaus (2001), 288–289.

Schwabe, Klaus (1983): Deutsche Hochschullehrer 1815–1945. Deutsche Führungsschichten in der Neuzeit 17. Boppart am Rhein: Bold.

Singer, Isidore/Haneman, Frederick T. (2006): Posner, Carl. Jewish Encyclopedia [www.jewishencyclopedia.com/view.jsp?artid=470&letter, 02/22/2006].

Slater, Edward C. (1986): Biochimica et Biophisica Acta. The Story of a Biochemical Journal. Amsterdam, New York, Oxford: Elsevier.

Steudel, Johannes (1953): Zwerg auf der Schulter der Riesen. In: Artelt et al. (1953), 394–399.

Stiller, Hartmut (1995): Gaskrieg, Völkerrecht, Geheimrüstung. Zur Frage der Beschränkung von bakteriologischen und chemischen Waffen in der Zwischenkriegszeit. Wissenschaft und Frieden 12 (03/1995). Marburg: Verlag des Bundes demokratischer Wissenschaftlerinnen und Wissenschaftler. [www.uni-muenster.de/PeaCon/wuf/wf-95/9531001m.htm, 03/21/2006].

Stolzenberg, Dietrich (1996): Fritz Haber und das Kaiser-Wilhelm-Institut für physikalische und Elektrochemie. In: Brocke/Laitko (1996), 263–274.

Stolzenberg, Dietrich (1998): Fritz Haber: Chemiker, Nobelpreisträger, Deutscher, Jude. Eine Biographie. Weinheim: Willey-VCH.

Strauss, Herbert A./Roder, Werner. (1983): International Biographical Dictionary of Central European Emigres 1933–1945. München, New York, Paris: Saur.

Stürzbecher, Manfred (1983): Hans Reiter (1881–1969). Die Medizinische Welt 41 (1983), 3–4.

Szabó, Anikó (2000): Vertreibung, Rückkehr, Wiedergutmachung. Göttinger Hochschullehrer im Schatten des Nationalsozialismus. Veröffentlichungen des Arbeitskreises Geschichte des Landes Niedersachsen (nach 1945) 15. Göttingen: Wallstein.

Trunk, Achim (2004): Rassenforschung und Biochemie. Ein Projekt – und die Frage nach dem Beitrag Butenandts. In: Schieder/Trunk (2004), 247–285.

Vierhaus, Rudolf (1996): Bemerkungen zum sogenannten Harnack-Prinzip. Mythos und Realität. In Brocke/Laitko (1996), 129–138.

Vierhaus, Rudolf/Brocke, Bernhard vom (ed.) (1990): Forschung im Spannungsfeld von Politik und Gesellschaft. Geschichte und Struktur der Kaiser-Wilhelm-/Max-Planck-Gesellschaft. Stuttgart: Deutsche Verlags-Anstalt.

Vogt, Annette (1996): Lise Meitner und ihre Kolleginnen – Naturwissenschaftlerinnen in den Instituten der Kaiser-Wilhelm-Gesellschaft zwischen 1912 und 1945. Preprint 46. Berlin: Max-Planck-Institut für Wissenschaftsgeschichte.

Vogt, Annette (1997): Vom Hintereingang zum Hauptportal – Wissenschaftlerinnen in der Kaiser-Wilhelm-Gesellschaft. Preprint 67. Berlin: Max-Planck-Institut für Wissenschaftsgeschichte.

Vogt, Annette (1999): Fellow of the Royal Society. Die Wissenschaftlerin Marthe Vogt. Berlinische Monatsschrift 11 (1999), 44–49. [www.luise-berlin.de/Bms/BMSTXT99/9911prog.htm#seite44, 03/11/2006].

Völckel, Anja (2003): Die Anfänge der physiologischen Chemie: Ernst Felix Immanuel Hoppe-Seyler (1825–1895). Diss. rer. nat. TU Berlin.

Volkov, Shulamit (1997): Die Juden als wissenschaftliche "Mandarine" im Kaiserreich und in der Weimarer Republik. Archiv für Sozialgeschichte 37 (1997).

Volz, Robert Wilhelm/Tönnies, Ferdinand (1931): Reichshandbuch der deutschen Gesellschaft: Das Handbuch der Persönlichkeiten in Wort und Bild. Berlin: Deutscher Wirtschaftsverlag.

Voswinckel, Peter (ed.) (2002): Fischer, Isidor: Biographisches Lexikon der hervorragenden Ärzte der letzten fünfzig Jahre: Zugleich Fortsetzung des Biographischen Lexikons hervorragender Ärzte aller Zeiten und Völker. Band 3: Nachträge und Ergänzungen. Hildesheim, Zürich, New York: Georg Olms.

Walk, Joseph (ed.) (1988): Kurzbiographien zur Geschichte der Juden 1918–1945 [Leo Baeck Institute, Jerusalem]. München: Saur.

Wechselmann, Wilhelm: Perglycerin und Perkaglycerin als Ersatzmittel des Glycerins in der Dermatologie. Deutsche Medizinische Wochenschrift 17 (1916), 517.

Weber, Hans H. (1972): Molecular Bioenergetics and Macromolecular Biochemistry [Meyerhof Symposium, Heidelberg, July 5–8, 1970]. Berlin, Heidelberg, New York: Springer, 14–15.

Weiß, Hermann (ed.) (1998): Biographisches Lexikon zum Dritten Reich. Frankfurt a. M.: Fischer.

Wendel, Günter (1975): Die Kaiser-Wilhelm-Gesellschaft 1911–1914. Zur Anatomie einer imperialistischen Forschungsgesellschaft. Studien zur Geschichte der Akademie der Wissenschaften der DDR 4. Berlin (Ost): Akademie-Verlag.

Werner, Petra (1995): Wege zur Biochemie: Otto Meyerhofs Wechsel von der Philosophie zur experimentellen Naturwissenschaft. Berlin: Brandenburgische Akademie der Wissenschaften.

Werner, Petra (1996): Otto Warburg, Jacques Loeb und die Entstehung der Institutsidee des Kaiser-Wilhelm-Instituts für Zellphysiologie. In: Brocke/Laitko (1996), 319–330.

Who was who among English and European authors? (1978): 1931–1949 [Gale composite biographical dictionary series]. Detroit, Mich.: Gale.

Who was who in America? (1966): Volume III (1951–1960). New Providence: Marquis.

Wie is dat? (1938): Naamlijst van bekende personen op elk gebied in het Koninkrijk der Nederlanden met biografische aanteekeningen, opgave hunner voornaamste werken, adressen, enz., enz. s'Gravenhage: Nijhoff.

Willstätter, Richard (1949): Aus meinem Leben. Von Arbeit, Muße und Freunden. Weinheim/Bergstraße: Verlag Chemie.

Zähringer, Norbert (2004): Vor 70 Jahren. Das deutsche Tabakforschungsinstitut gibt den Anbau von nikotinfreiem Tabak bekannt. Deutschland Radio Berlin. Kalenderblatt vom 28. September 2004. [www.dradio.de/drl/sendung/kalender/306932, 03/06/2006].

Zuntz, Nathan (1906): Das Tierphysiologische Institut. In: Lehrerkollegium Parey (1906), 155–170.

13. LIST OF ILLUSTRATIONS

14. INDEX OF PERSONS

ILLUSTRATIONS

PROFESSOR CARL NEUBERG

Drawn from life for the ASEC Bulletin by RICHARD COLIN

MENS AGITAT MOLEM

Fig. 1: Drawing by Richard Colin on the occasion of Carl Neuberg's 70th birthday in 1947

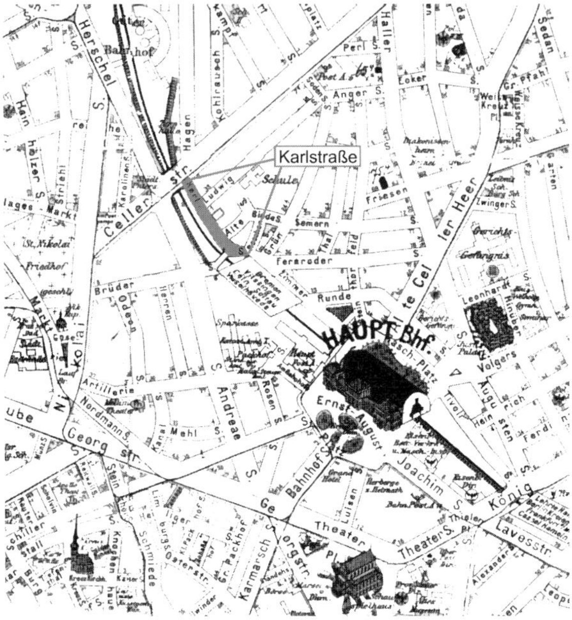

Fig. 2: Extract from a town plan of Hanover (c.1890)

Fig. 3: Remembering Hanover. Postcard from Neuberg to Bruno Kisch (1947, in exile)

Zur Kenntnis des Acroleïns und Glycerin-aldehyds.

Ueber saure Ester der Borsäure.

Ueber die Reinigung der Osazone und zur Bestimmung ihrer optischen Drehungsrichtung.

INAUGURAL-DISSERTATION

ZUR

ERLANGUNG DER DOCTORWÜRDE

VON DER PHILOSOPHISCHEN FACULTÄT

DER

FRIEDRICH-WILHELMS-UNIVERSITÄT ZU BERLIN

GENEHMIGT

UND

NEBST DEN BEIGEFÜGTEN THESEN

ÖFFENTLICH ZU VERTEIDIGEN

AM 21. JULI 1900

VON

Carl Neuberg

aus Hannover.

OPPONENTEN:

Hr. Dr. med. F. Blumenthal, Privatdozent.
- Dr. iur. A. Nussbaum, Referendar.
- Dr. phil. F. Ehrlich, Assistent.

Berlin.

Druck von E. Ebering.

Mittelstrasse 29.

Fig. 4: Title page of Neuberg's doctoral thesis of July 21, 1900

Fig. 5: View of the Institute of Animal Physiology, Royal Agricultural College, Berlin (1910)

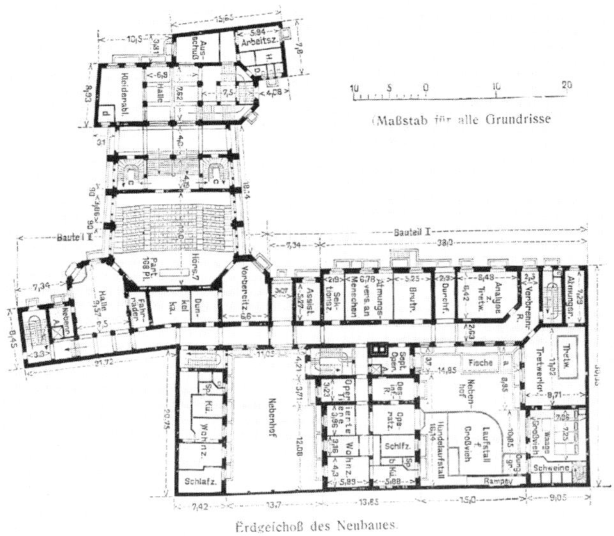

Fig. 6: Ground plan of the Institute of Animal Physiology (1910)

Fig. 7: The Kaiser Wilhelm Institute for Experimental Therapy on the day of its inauguration

Kaiser-Wilhelm-Institut für experimentelle Therapie in Dahlem, Thielallee

Eintrittskarte

für die am 28. Oktober 1913, vormittags 9¾ Uhr,
im Hauptgebäude des Instituts stattfindende
Einweihungsfeier.

Anzug: Überrock, Mitglieder der Kaiser-Wilhelm-Gesellschaft tragen das Abzeichen

Fig. 8: Admission ticket to the inauguration ceremony

Fig. 9: After the inauguration of the KWI for Experimental Therapy, Dahlem, Berlin: Carl Neuberg, Kaiser Wilhelm II, August von Wassermann, Adolf von Harnack (from left to right)

Fig. 10: German cavalry soldiers in the First World War with their horses and the gas masks developed by Neuberg

Fig. 11: Carl Neuberg in his laboratory (c.1915)

Fig. 12: Employees and staff of the Kaiser Wilhelm Institute for Biochemistry, photo taken probably in the 1930s. Carl Neuberg in a black suit, Maria Kobel on his left

Fig. 13: Hans von Euler-Chelpin

Fig. 14: Paul Mayer shortly before his death in London

Fig. 15: Adolf Butenandt in his laboratory, c.1935

Fig. 16: Adolf Butenandt at the time when he took over Neuberg's position, 1936

Fig. 17: The Neuberg family grave in the Jewish Cemetery, Weissensee, Berlin Alongside the gravestone of Neuberg's parents, the last resting place of Neuberg's wife Hela is to be seen in the foreground

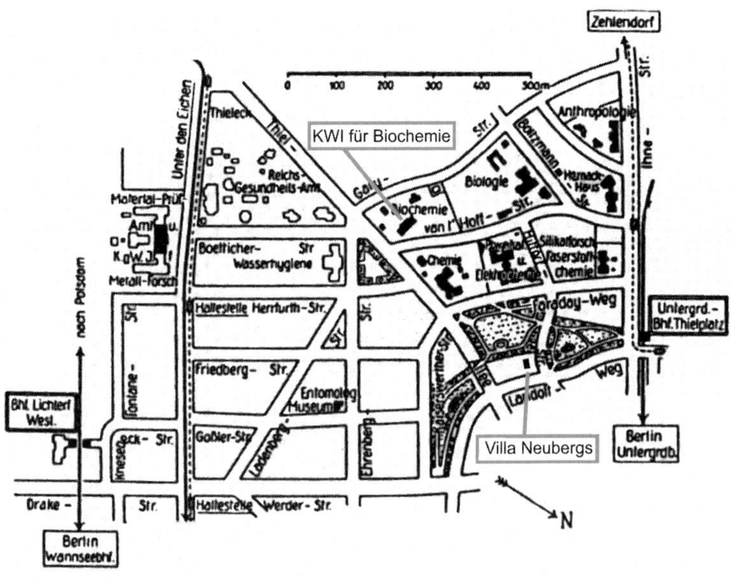

Fig. 18: Locational plan of Dahlem, Berlin, in the 1920s

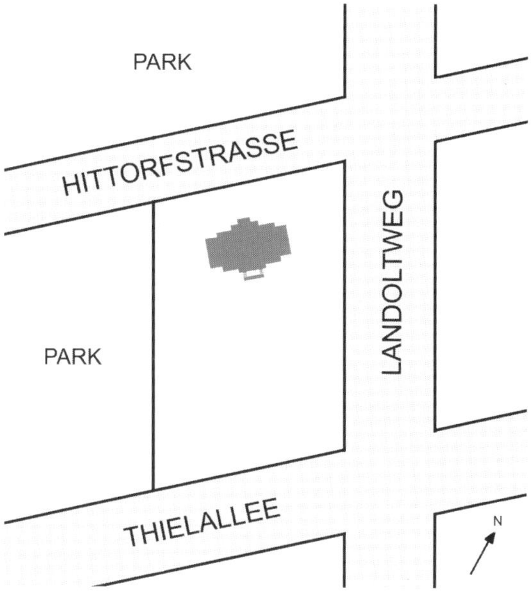

Fig. 19: Position of the site of Neuberg's house, see also Fig. 18

Fig. 20: Neuberg's house, Hittorfstrasse 18

Fig. 21: Carl Neuberg (2nd from left) with colleagues from the Department of Chemistry of New York University

Fig. 22: Carl Neuberg Medal. American Society of European Chemists and Pharmacists (1947)

Fig. 23: Ines Mandl in the 1940s

Fig. 24: Ines Mandl (1997)

Fig. 25: Otto Warburg on "Nixe"

Fig. 26: Carl Neuberg in New York, shortly before his death (1956)

Biochemische Zeitschrift.

Herausgegeben von

E. Buchner-Berlin, **P. Ehrlich**-Frankfurt a. M., **C. von Noorden**-Wien, **E. Salkowski**-Berlin, **N. Zuntz**-Berlin

unter Mitwirkung von

L. Asher-Bern, **J. Bang**-Lund, **P. Bergell**-Berlin, **G. Bertrand**-Paris, **A. Bickel**-Berlin, **F. Blumenthal**-Berlin, **Chr. Bohr**-Kopenhagen, **F. Ehrlich**-Berlin, **G. Embden**-Frankfurt a. M., **E. Freund**-Wien, **G. Galeotti**-Neapel, **H. J. Hamburger**-Groningen, **A. Heffter**-Marburg, **M. Jacoby**-Heidelberg, **R. Kobert**-Rostock, **M. Kumagawa**-Tokio, **L. Langstein**-Berlin, **L. von Liebermann**-Budapest, **J. Loeb**-Berkeley, **A. Loewy**-Berlin, **J. A. Mandel**-New-York, **L. Marchlewski**-Krakau, **P. Mayer**-Karlsbad, **L. Michaelis**-Berlin, **J. Morgenroth**-Berlin, **W. Nernst**-Berlin, **R. Pfeiffer**-Königsberg, **Ch. Porcher**-Lyon, **F. Roehmann**-Breslau, **S. Salaskin**-St. Petersburg, **N. Sieber**-St. Petersburg, **M. Siegfried**-Leipzig, **Zd. H. Skraup**-Graz, **S. P. L. Sörensen**-Kopenhagen, **E. H. Starling**-London, **H. v. Tappeiner**-München, **H. Thoms**-Berlin, **J. Traube**-Charlottenburg, **A. J. J. Vandevelde**-Gent, **A. Wohl**-Danzig, **J. Wohlgemuth**-Berlin.

Redigiert von

C. Neuberg-Berlin.

Erster Band.

Mit vier Tafeln.

Berlin.

Verlag von Julius Springer.

1906.

Fig. 27: Title page of the first volume of the *Biochemische Zeitschrift*

Fig. 28: January 1, 1997: Carl-Neuberg-Strasse in Hanover, as the postal address of Hanover Medical School